Student's Solutions Manual

Mathematical Excursions

FOURTH EDITION

Richard N. Aufmann
Palomar College

Joanne S. Lockwood
Nashua Community College

Richard D. Nation
Palomar College

Daniel K. Clegg
Palomar College

Prepared by

Christi Verity

 Cengage

Australia • Brazil • Canada • Mexico • Singapore • United Kingdom • United States

ISBN: 978-1-305-96561-4

Cengage
200 Pier 4 Boulevard
Boston, MA 02210
USA

Cengage is a leading provider of customized learning solutions with employees residing in nearly 40 different countries and sales in more than 125 countries around the world. Find your local representative at: **www.cengage.com.**

To learn more about Cengage platforms and services, register or access your online learning solution, or purchase materials for your course, visit **www.cengage.com.**

Printed at CLDPC, USA, 06-22

Contents

Chapter 1: Problem Solving

EXCURSION EXERCISES, SECTION 1.1

1.

3.

5.

EXERCISE SET 1.1

1. 28. Add 4 to obtain the next number.

3. 45. Add 2 more than the integer added to the previous integer.

5. 64. The numbers are the squares of consecutive integers. $8^2 = 64$.

7. $\frac{15}{17}$. Add 2 to the numerator and denominator.

9. −13. Use the pattern of adding 5, then subtracting 10 to obtain the next pair of numbers.

11. Correct.

13. Correct.

15. Incorrect. The resulting number will be 3 times the original number.

17. a. $8 - 0 = 8$ cm
    b. $32 - 8 = 24$ cm
    c. $72 - 32 = 40$ cm
    d. $128 - 72 = 56$ cm
    e. $200 - 128 = 72$ cm

19. a. 8 cm = 1 unit
    Therefore, $8 \cdot n = 1 \cdot n$
    24 cm = $8 \cdot 3$ cm
    $1 \cdot 3 = 3$ units
    b. 40 cm = $8 \cdot 5$
    $1 \cdot 5 = 5$ units
    c. 56 cm = $8 \cdot 7$ cm
    $1 \cdot 7 = 7$ units
    d. 72 cm = $8 \cdot 9$ cm
    $1 \cdot 9 = 9$ units

21. It appears that doubling the ball's time, quadruples the ball's distance. In the inclined plane time distance table, the ball's time of 2 seconds has a distance that is quadrupled the ball's distance of 1 second. The ball's time of 4 seconds has a distance that is quadrupled the ball's distance of 2 seconds.

23. 288 cm. The ball rolls 72 cm in 3 seconds. So in doubling 3 seconds to 6 seconds, we quadruple 72 get 288.

25. This argument reaches a conclusion based on a specific example, so it is an example of inductive reasoning.

27. The conclusion is a specific case of a general assumption, so this argument is an example of deductive reasoning.

29. The conclusion is a specific case of a general assumption, so this argument is an example of deductive reasoning.
$$1^3 + 5^3 + 3^3 = 1 + 125 + 27$$
$$= 153$$

31. This argument reaches a conclusion based on a specific example, so it is an example of inductive reasoning.

33. Any number less than or equal to − 1 or between 0 and 1 will provide a counterexample.

35. Any number less than −1 or between 0 and 1 will provide a counterexample.

1

37. Any negative number will provide a counterexample.

39. Consider any two odd numbers. Their sum is even, but their product is odd.

41.

16	3	2	13
5	10	11	8
9	6	7	12
4	15	14	1

43. Using deductive reasoning:

n	pick a number
$6n$	multiply by 6
$6n+8$	add 8
$\dfrac{6n+8}{2}=3n+4$	divide by 2
$3n+4-2n=n+4$	subtract twice the original number
$n+4-4=n$	subtract 4

45.

	Util	Auto	Tech	Oil
A	Xa	Xa	✓	Xc
T	Xa	Xa	Xc	✓
M	✓	Xb	Xb	Xb
J	Xb	✓	Xb	Xb

47.

	Coin	Stamp	Comic	Baseball
A	Xc	✓	Xc	Xd
C	Xb	Xd	Xa	✓
P	✓	Xc	Xc	Xb
S	Xb	Xc	✓	Xc

49. home, bookstore, supermarket, credit union, home; or home, credit union, supermarket, bookstore, home.

51. N. These are the first letters of the counting numbers: **O**ne, **T**wo, **T**hree, etc. N is the first letter of the next number, which is **N**ine.

53. a. 1010 is a multiple of 101, (10 × 101 = 1010) but 11 × 1010 =11,110. The digits of this product are not all the same.

   b. For $n = 11$, $n^2 - n +11 = 11^2 -11 +11 = 121$, which is not a prime number.

1. The sixth triangular number is 21.

The sixth square number is 36.

The sixth pentagonal number is 51.

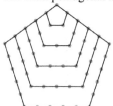

3. The fourth hexagonal number is 28.

EXERCISE SET 1.2

1. 1   1   17   31   39   71   97
       6   10   14   18   22   26
         4   4   4   4   4
    26 + 71 = 97

3. −1   4   21   56   115   204   329
       5   17   35   59   89   125
         12   18   24   30   36
           6   6   6   6
    125 + 204 = 329

5. 9   4   3   12   37   84   159
      −5   −1   9   25   47   75
        4   10   16   22   28
          6   6   6   6
    75 + 84 =159

7. Substitute in the appropriate values for n.

   For $n = 1$, $a_1 = \dfrac{1(2(1)+1)}{2} = \dfrac{3}{2}$

   For $n = 2$, $a_2 = \dfrac{2(2(2)+1)}{2} = 5$

   For $n = 3$, $a_3 = \dfrac{3(2(3)+1)}{2} = \dfrac{21}{2}$

   For $n = 4$, $a_4 = \dfrac{4(2(4)+1)}{2} = 18$

   For $n = 5$, $a_5 = \dfrac{5(2(5)+1)}{2} = \dfrac{55}{2}$

9. Substitute in the appropriate values for n in $a_n = 5n^2 - 3n$ to obtain 2, 14, 36, 68, 110.

11. Notice that each figure is square with side length n plus an "extra row" of length $n - 1$. Thus the nth figure will have $a_n = n^2 + (n - 1)$ tiles.

13. Each figure is composed of a horizontal group of n tiles, a horizontal group of $n - 1$ tiles, and a single "extra" tile. Thus the nth figure will have $a_n = n + n - 1 + 1 = 2n$ tiles.

15. a. There are 56 cannonballs in the sixth pyramid and 84 cannonballs in the seventh pyramid.

    b. The eighth pyramid has eight levels of cannonballs. The total number of cannonballs in the eighth pyramid is equal to the sum of the first 8 triangular numbers: $1 + 3 + 6 + 10 + 15 + 21 + 28 + 36 = 120$.

17. a. Five cuts produce six pieces and six cuts produce seven pieces.

    b. The number of pieces is one more than the number of cuts, so $a_n = n + 1$.

19. a. Substituting in $n = 5$:
$$P_5 = \frac{5^3 + 5(5) + 6}{6} = \frac{156}{6} = 26$$

    b. Substituting several values:
$$P_6 = \frac{6^3 + 5(6) + 6}{6} = 42 < 60$$
$$P_7 = \frac{7^3 + 5(7) + 6}{6} = 64 > 60$$
Thus the fewest number of straight cuts is 7.

21. Substituting:
$a_3 = 2 \cdot a_2 - a_1 = 10 - 3 = 7$
$a_4 = 2 \cdot a_3 - a_2 = 14 - 5 = 9$
$a_5 = 2 \cdot a_4 - a_3 = 18 - 7 = 11$

23. Substituting:
$F_{20} = 6765$
$F_{30} = 832,040$
$F_{40} = 102,334,155$

25. The drawing shows the nth square number. The question mark should be replaced by n^2.

27. a. For $n = 1$, we get $1 + 2(1) + 2 = 5 = F_5$
For $n = 2$, we get $1 + 2(2) + 3 = 8 = F_6$
For $n = 3$, we get $2 + 2(3) + 5 = 13 = F_7$.
Thus, $F_n + 2F_{n+1} + F_{n+2} = F_{n+4}$.

    b. For $n = 1$, we get $1 + 1 + 3 = 5 = F_5$
For $n = 2$, we get $1 + 2 + 5 = 8 = F_6$
For $n = 3$, we get $2 + 3 + 8 = 13 = F_7$.
Thus, $F_n + F_{n+1} + F_{n+3} = F_{n+4}$.

29. a.

row	total
0	1
1	2
2	4
3	8
4	16
5	32

Each row total is twice the number in the previous row. These numbers are powers of 2. It appears that the sum for the nth row is 2^n. The sum of the numbers in row 9 is $2^9 = 512$.

    b. They appear in the third diagonals.

31. a. 1 move

    b. 3 moves

    c. 7 moves (Start with the discs on post 1. Let A, B, C be the discs with A smaller than B and B smaller than C. Move A to post 2, B to post 3, A to post 3, C to post 2, A to post 1, B to post 2, and A to post 2.) This is 7 moves.

    d. 15 moves

    e. 31 moves

    f. $2^n - 1$ moves

    g. $n = 64$, so there are $2^{64} - 1 = 1.849 \times 10^{19}$ moves required. Since each move takes 1 second, it will take 1.849×10^{19} seconds to move the tower. Divide by 3600 to obtain the number of hours, then by 24 to obtain the days, then by 365 to obtain the number of years, about 5.85×10^{11}.

EXCURSION EXERCISES, SECTION 1.3

1. There is one route to point B, that of all left turns. Add the two numbers above point C to obtain $1 + 3 = 4$ routes. Add the two numbers above point D to obtain $3 + 3 = 6$ routes. Add the two numbers above point E to obtain $3 + 1 = 4$ routes. As with point B, there is only one route to point F, that of all right turns.

3. The figure is symmetrical about a vertical line from A to K. Since J is the same distance to the left of the line AK as L is to the right of AK, the same number of routes lead from A to J as lead from A to L.

5. By adding adjacent pairs, the number of routes from A to P: 1 route; A to Q: 9 routes; A to R: 36 routes; A to S: 84 routes; A to T: 126 routes; A to U: 126 routes.

EXERCISE SET 1.3

1. Let g be the number of first grade girls, and let b be the number of first grade boys. Then $b + g = 364$ and $g = b + 26$. Solving gives $g = 195$, so there are 195 girls.

3. There are 36 1×1 squares, 25 2×2 squares, 16 3×3 squares, 9 4×4 squares, 4 5×5 squares and 1 6×6 square in the figure, making a total of 91 squares.

5. Solving:
    $x =$ cost of the shirt
    $x - 30 =$ cost of the tie
    $$(x - 30) + x = 50$$
    $$2x - 30 = 50$$
    $$2x = 80$$
    $$x = 40$$
    The shirt costs \$40.

7. There are 14 different routes to get to Fourth Avenue and Gateway Boulevard and 4 different routes to get to Second Avenue and Crest Boulevard. Adding gives that there are 18 different routes altogether.

9. Try solving a simpler problem to find a pattern. If the test had only 2 questions, there would be 4 ways. If the test had 3 questions, there would be 8 ways. Further experimentation shows that for an n question test, there are 2^n ways to answer. Letting $n = 12$, there are $2^{12} = 4096$ ways

11. 8 people shake hands with 7 other people. Multiply 8 and 7 and divide by 2 to eliminate repetitions to obtain 28 handshakes.

13. Let p be the number of pigs and let d be the number of ducks. Then $p + d = 35$ and $4p + 2d = 98$. Solving gives $d = 21$ and $p = 14$, so there are 21 ducks and 14 pigs.

15.

Dimes	Nickels	Pennies
0	0	25
0	1	20
0	2	15
0	3	10
0	4	5
0	5	0
1	0	15
1	1	10
1	2	5
1	3	0
2	0	5
2	1	0

There are 12 ways.

17. The units digits of powers of 4 form the sequence 4, 6, 4, 6,…. Even powers end in 6. Therefore, the units digit of 4^{300} is 6.

19. The units digits of powers of 3 form the sequence 3, 9, 7, 1, 3, 9, 7, 1, …. Divide 412 by 4 to obtain the remainder 0, which corresponds to 1. Therefore the units digit of 3^{412} is 1.

21. a. Add the numbers in pairs: 1 and 400, 2 and 399, 3 and 398, and so on. There are 200 pair sums equal to 401.
    $200 \times 401 = 80{,}200$.

    b. Add the numbers in pairs: 1 and 550, 2 and 549, 3 and 548 and so on. There are 275 pair sums equal to 551.
    $275 \times 551 = 151{,}525$.

    c. Add the numbers in pairs: 2 and 84, 4 and 82, and so on, leaving off the 86. There are 21 sums of 86 plus one additional 86.
    $21 \times 86 + 86 = 1892$.

23. a. 121, 484, and 676 are the only three-digit perfect square palindromes.

    b. 1331 is the only four-digit perfect cube palindrome.

25. Draw a simpler picture:

Start   Finish

Note that the first page of the first volume is the second dot on the line, and the last page of the third volume is the seventh dot on the line. This is because when books sit on a shelf, the first pages are on the right side of the book and their last pages are on the left side.
$$\frac{1}{8} + \frac{1}{8} + 1 + \frac{1}{8} + \frac{1}{8} = 1\frac{1}{2} \text{ inches.}$$

27. a. 1.4 billion admissions
    b. 2014
    c. 2009

29. a. PG-13
    b. \$2.2 billion

31. Since there is one blue tile in each column, there are n blue tiles on the diagonal that starts in the upper left hand corner. Similarly, there are n blue tiles in the diagonal that starts in the upper right hand corner. The two diagonals have one tile in common, so the actual total number of blue tiles is $2n - 1$. Since $2n - 1 = 101$, we can solve to find $n = 51$. The total number of tiles is n^2. Substituting the value for n yields 2601.

33. Let b be the number of boys in the family and g be the number of girls. The first two statements imply that the speaker is a girl. Thus, $g - 1 = b + 2$. Solving for b, $b = g - 3$. To answer the last question, we must omit the youngest brother, so $b - 1 = g - 4$. There are four more sisters than brothers.

35. The bacteria population doubles every day, so on the 11th day there are half as many bacteria as on the 12th day.

37. Let x be the score that Dana needs on the fourth exam:
$$\frac{82 + 91 + 76 + x}{4} = 85$$
$$249 + x = 340$$
$$x = 91$$

39. a. Place four coins on the left balance pan and the other 4 coins on the right balance pan. The pan that is the higher contains the fake coin. Take the four coins from the higher pan and use the balance scale to compare the weight of two of these coins to the weight of the other two coins. The pan that is the higher contains the fake coin. Take the two coins from the higher pan and use the balance scale to compare the weights. The pan that is the higher contains the fake coin. This procedure enables you to determine the fake coin in 3 weighings.

    b. Place 3 of the coins on one of the balance pans and 3 coins on the other balance pan. If the pans balance, then the fake coin is one of the two remaining coins. You can put each one of these coins on a balance pan and the higher pan contains the fake coin. If the 3 coins on the left do not balance with the 3 coins on the right, then the higher pan contains the fake coin. Pick any 2 of these 3 coins and use the balance scale to compare their weights. If these 2 coins do not balance, then the higher pan contains the fake coin. If these two coins balance, then the 3rd coin (the one that you did not place on the balance pan) is the fake. In either case this procedure enables you to determine the fake coin in 2 weighings.

41. The correct answer is a., 1600. Sally likes perfect squares.

43. The correct answer is d., 64. The numbers are all perfect cubes. The missing number is the cube of 4.

45. a. Write an equation. When the people who were born in 1980 are x years old, it will be the year $1980 + x$. We are looking for the year that satisfies $1980 + x = x^2$. Solving gives $x = 45$ and $x = -44$. The solution must be a natural number, so $x = 45$. Therefore when the people born in 1980 are 45 years old, the year will be $45^2 = 2025$.

    b. 2070, because people born in 2070 will be 46 in $2116 = 46^2$.

47. It takes 9 1-digit numbers for pages 1-9, 180 digits for pages 10-99, and 423 digits for pages 100-240. The total is 612.

49. Answers will vary.

CHAPTER 1 REVIEW EXERCISES

1. This argument reaches a conclusion based on a case of a general assumption, so it is an example of deductive reasoning.

2. This argument reaches a conclusion based on specific examples, so it is an example of inductive reasoning.

3. This argument reaches a conclusion based on a specific example, so it is an example of inductive reasoning.

4. This argument reaches a conclusion based on a case of a general assumption, so it is an example of deductive reasoning.

5. Any number from 0 to 1 provides a counterexample. For example, $x = \frac{1}{2}$ provides a counterexample because $\left(\frac{1}{2}\right)^4 = \frac{1}{16}$ and $\frac{1}{16}$ is not greater than $\frac{1}{2}$.

6. $n = 4$ provides a counterexample because $\frac{(4)^3 + 5(4) + 6}{6} = \frac{90}{6} = 15,$ which is not even.

7. $x = 1$ provides a counterexample because $(1 + 4)^2 = (5)^2 = 25$ and $1^2 + 4^2 = 1 + 16 = 17$

8. $a = 1$ and $b = 1$ provides a counterexample because $(1 + 1)^3 = 2^3 = 8$, but $1^3 + 1^3 = 1 + 1 = 2$.

9. a. $\quad$–2 $\quad$ 2 $\quad$ 12 $\quad$ 28 $\quad$ 50 $\quad$ 78 $\quad$ **112**
$\qquad$ 4 $\quad$ 10 $\quad$ 16 $\quad$ 22 $\quad$ 28 $\quad$ 34
$\qquad\quad$ 6 $\quad$ 6 $\quad$ 6 $\quad$ 6 $\quad$ 6
$\qquad$ Add 34 and 78 to obtain 112.

  b. $\quad$–4 $\quad$ –1 $\quad$ 14 $\quad$ 47 $\quad$ 104 $\quad$ 191 $\quad$ 314 $\quad$ **47**
$\qquad$ 3 $\quad$ 15 $\quad$ 33 $\quad$ 57 $\quad$ 87 $\quad$ 123 $\quad$ 165
$\qquad\quad$ 12 $\quad$ 18 $\quad$ 24 $\quad$ 30 $\quad$ 36 $\quad$ 42
$\qquad\qquad$ 6 $\quad$ 6 $\quad$ 6 $\quad$ 6 $\quad$ 6
$\qquad$ Add 165 and 314 to obtain 479.

10. a. $\quad$5 $\quad$ 6 $\quad$ 3 $\quad$ –4 $\quad$ –15 $\quad$ –30 $\quad$ –49 $\quad$ **–72**
$\qquad$ 1 $\quad$ –3 $\quad$ –7 $\quad$ –11 $\quad$ –15 $\quad$ –19 $\quad$ –23
$\qquad$ –4 $\quad$ –4 $\quad$ –4 $\quad$ –4 $\quad$ –4 $\quad$ –4
$\qquad$ Add –23 and –49 to obtain –72.

  b. $\quad$2 $\quad$ 0 $\quad$ –18 $\quad$ –64 $\quad$ –150 $\quad$ –288 $\quad$ –490 $\quad$ **–768**
$\qquad$ –2 $\quad$ –18 $\quad$ –46 $\quad$ –86 $\quad$ –138 $\quad$ –202 $\quad$ –278
$\qquad$ –16 $\quad$ –28 $\quad$ –40 $\quad$ –52 $\quad$ –64 $\quad$ –76
$\qquad$ –12 $\quad$ –12 $\quad$ –12 $\quad$ –12 $\quad$ –12
$\qquad$ Add –278 and –490 to obtain –768.

11. Substituting:
$$a_1 = 4(1)^2 - 1 - 2 = 4 - 3 = 1$$
$$a_2 = 4(2)^2 - 2 - 2 = 16 - 4 = 12$$
$$a_3 = 4(3)^2 - 3 - 2 = 36 - 5 = 31$$
$$a_4 = 4(4)^2 - 4 - 2 = 64 - 6 = 58$$
$$a_5 = 4(5)^2 - 5 - 2 = 100 - 7 = 93$$
$$a_{20} = 4(20)^2 - 20 - 2$$
$$= 4(400) - 20 - 2$$
$$= 1600 - 22 = 1578$$

12. $\quad F_7 = F_6 + F_5 = 8 + 5 = 13$
$\quad F_8 = F_7 + F_6 = 13 + 8 = 21$
$\quad F_9 = F_8 + F_7 = 21 + 13 = 34$
$\quad F_{10} = F_9 + F_8 = 34 + 21 = 55$
$\quad F_{11} = F_{10} + F_9 = 55 + 34 = 89$
$\quad F_{12} = F_{11} + F_{10} = 89 + 55 = 144$

13. Each figure has a horizontal section with $n + 1$ tiles, a horizontal section with n tiles, and a vertical section with $n - 1$ tiles.
$a_n = n + 1 + n + n - 1 = 3n$

14. Each figure is a square with side length $n + 2$ and n tiles removed.
$$a_n = (n + 2)^2 - n = n^2 + 4n + 4 - n$$
$$= n^2 + 3n + 4$$

15. Each figure is a square with side length $n + 1$ with an attached diagonal with $n + 1$ tiles.
$a_n = (n + 1)^2 + (n + 1) = n^2 + 3n + 2$

16. Each figure made up of four sides of length n with a diagonal piece in the middle with length $n - 1$.
$a_n = 4n + (n - 1) = 5n - 1$

17. Let x be the width. Then $5x$ is the length. Since one length already exists, only 3 sides of fencing are needed. The total perimeter is $5x + x + x = 2240$. Solving, we find $x = 320$. The dimensions are 320 feet by 1600 feet.

18. Solve a simpler problem. If the test has 1 question, there are 3 ways to answer. If the test has 2 questions, there are 9 ways to answer. If the test has 3 questions, there are 27 ways to answer. It appears that for a test with n questions, there are 3^n ways to answer. In this case, $n = 15$, so there are $3^{15} = 14{,}348{,}907$ ways to answer the test.

19. If the 11th and 35th are opposite each other, there must be 23 more skyboxes between them (going in each direction). The total is $23 + 1 + 23 + 1 = 48$ skyboxes.

20. On the first trip the rancher takes the rabbit across the river. The rancher returns alone. The rancher takes the dog across the river and returns with the rabbit. The rancher next takes the carrots across the river and returns alone. On the final trip the rancher takes the rabbit across the river.

21. $1400 - $1200 = $200 profit.
$1900 - $1800 = $100 profit.
Total profit = $200 + $100 = $300

22. Multiply 15 and 14 and divide by 2 to eliminate repetitions. 105 handshakes will take place.

23. Answers will vary. Possible answers include: make a list, draw a diagram, make a table, work backwards, solve a simpler similar problem, look for a pattern, write an equation, perform an experiment, guess and check, and use indirect reasoning.

24. Answers will vary. Possible answers include: ensure that the solution is consistent with the facts of the problem, interpret the solution in the context of the problem, and ask yourself whether there are generalizations of the solution that could apply to other problems.

25.

	CS	Chem	Bus	Bio
M	Xa	Xd	Xd	✓
C	Xd	Xb	✓	Xb
R	✓	Xb	Xd	Xb
E	Xd	✓	Xd	Xc

26.

	Bank	Super	Service	Drug
D	Xd	Xd	Xd	✓
P	Xb	✓	Xd	Xb
T	✓	Xd	Xa	Xd
G	Xb	Xc	✓	Xb

27. a. Yes. Answers will vary.

    b. No. The countries of India, Bangladesh, and Myanmar all share borders with each of the other two countries. Thus, at least three colors are needed to color the map.

28. a. The following figure shows a route that starts from North Bay and passes over each bridge once and only once.

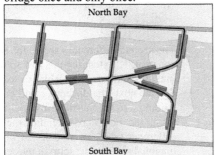

    b. No.

29. Draw the three possible pictures (one with *x* diagonal from 2, one diagonal from 5 and one diagonal from 10) to find the three possible values for *x*: 1 square inch, 4 square inches, 25 square inches.

30. a. Adding smaller line segments to each end of the shortest line doubles the total number of line segments. Thus the nth figure has 2^n line segments. For $n = 10$, $a_{10} = 1024$.

    b. $a_{30} = 2^{30} = 1,073,741,824$

31. A represents 1, B represents 9, D represents 0

32. Making a table:

quarters	nickels
4	0
3	5
2	10
1	15
0	20

There are 5 ways.

33. Use a list. WWWLL, WWLWL, WWLLW, WLWLW, WLLWW, WLWWL, LWWWL, LWLWW, LWWWL, LLWWW.
There are 10 different orders.

34. The units digit of powers of 7 are 7, 9, 3, 1, 7, 9, 3, 1, ... Divide 56 by 4 to obtain the remainder of 0, which corresponds to 1. Therefore the units digit of 7^{56} is 1.

35. The units digit of powers of 23 are 3, 9, 7, 1, 3, 9, 7, 1, ... Divide 85 by 4 to obtain the remainder of 1, which corresponds to 3. Therefore the units digit of 23^{85} is 3.

36. Pick a number *n*:

n

$4n$ multiply by 4

$4n+12$ add 12

$\dfrac{4n+12}{2} = 2n+6$ divide by 2

$2n+6-6 = 2n$ subtract 6

37. Each nickel is worth 5 cents. Thus 2004 nickels are worth $2004 \times 5 = 10,020$ cents, or $100.20.

38. a. $3.64 per gallon in 2012

    b. 2010 to 2011

39. a. $3.8 million

    b. $\dfrac{\$4.5 \text{ million}}{118.5 \text{ million viewers}} \approx \$0.04$, or 4 cents per viewer

40. a. (18.6 billion)(0.638) ≈ 11.9 billion searches

    b. $\dfrac{12.7\%}{1.8\%} \approx 7.1$

41. Every multiple of 5 ends in a 5 or a zero. Every palindromic number begins with the same digit it ends with. We cannot begin a number with 0, so the number must end in 5. Thus it must begin with 5. The smallest such number is 5005.

42. Checking all of the two-digit natural numbers shows that there are no narcissistic numbers.

    a. 10 intersections.

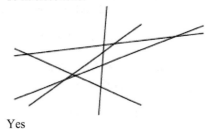

    b. Yes

44. Construct a difference table as shown below.

    2   6   12   20   **30**
      4   6   8   **10**
        2   2   2

    The second differences are all the same constant, 2. Extending this row so that it includes additional 2 enables us to predict that the next first difference will be 10. Adding 10 to the fourth term 20 yields 30. Using the method of extending the difference table, we predict that 30 is the next term in the sequence.

45. a.   22. 9^{22} has 21 digits.

    b. $9^9 = 387{,}420{,}489$, so $9^{\left(9^9\right)}$ is the product of 387,420,489 nines. At one multiplication per second this would take about 12.3 years. It is probably not a worthwhile project.

CHAPTER 1 TEST

1.   This argument reaches a conclusion based on a specific example, so it is an example of inductive reasoning.

2.   This conclusion is based on a specific case of a general assumption, so this argument is an example of deductive reasoning.

3.   –1   0   9   32   75   144   245   **384**
       1   9   23   43   69   101   **139**
         8   14   20   26   32   **38**
           6   6   6   6   **6**

    Add 139 and 245 to obtain 384.

4.   1, 1, 2, 3, 5, 8, 13, 21, 34, 55

5.   a.   Each figure contains a horizontal group of n tiles, a horizontal group of $n + 1$ tiles, and a vertical group of $n - 1$ tiles. $a_n = n + n + 1 + 2n - 1 = 4n$.

    b.   Each figure contains a horizontal group of $n + 1$ tiles and 2 horizontal groups of n tiles.
    $a_n = n + 1 + n + n = 3n + 1$.

6. $a_1 = 0$, $a_2 = 1$, $a_3 = -3$, $a_4 = 6$, $a_5 = -10$,

    $$a_{105} = (-1)^{105}\left(\frac{105(104)}{2}\right)$$
    $$= -1(105 \cdot 52) = -5460$$

7. $a_3 = 2a_{3-1} + a_{3-2}$
    $\quad = 2a_2 + a_1$
    $\quad = 2(7) + 3$
    $\quad = 17$
    $a_4 = 2a_{4-1} + a_{4-2}$
    $\quad = 2a_3 + a_2$
    $\quad = 2(17) + 7$
    $\quad = 41$
    $a_5 = 2a_{5-1} + a_{5-2}$
    $\quad = 2a_4 + a_3$
    $\quad = 2(41) + 17$
    $\quad = 99$

8.   a.   Construct a difference table as shown below.

    0   2   5   9   **14**
      2   3   4   **5**
        1   1   **1**

    The second differences are all the same constant, 1. Extending this row so that it includes additional 1 enables us to predict that the next first difference will be 5. Adding 5 to the fourth term 9 yields 14. Using the method of extending the difference table, we predict that 14 diagonals is the next term in the sequence.

    b.   Construct a difference table as shown below.

    0   2   5   9   14   **20**
      2   3   4   5   **6**
        1   1   1   **1**

    The second differences are all the same constant, 1. Extending this row so that it includes additional 1 enables us to predict that the next first difference will be 6. Adding 6 to the fifth term 14 yields 20. Using the method of extending the difference table, we predict that 20 diagonals is the next term in the sequence.

9.   Understand the problem. Devise a plan. Carry out the plan. Review the solution.

10.   Making a table:

Half-dollars	Quarters	Dimes
2	0	0
1	0	5
1	2	0
0	4	0
0	2	5
0	0	10

There are 6 ways.

11. Make a list.

    LLWWW LWLWWW LWWLWW
    LWWWLW LWWWWL WLWWWL
    WLWWLW WLWLWW WLLWWW
    WWLLWW WWWLWL WWWWLL
    WWLWWL WWWLLW WWLWLW

    There are 15 ways.

12. The units digits form a sequence with 4 terms that repeat, so divide the powers by 4 and look at the remainders. A remainder of 1 corresponds to a 3.

13. Work backwards. Subtract $150 from $326. This is $176. Let x be the amount of money Shelly had before renting the room. Then

    $x - \frac{x}{3} = \$176$, so $x = \$264$. Adding on $22 and $50 gives $336. Since this is half of her savings (the other half was spent on the plane ticket), double to get $672.

14. 126 routes. Add successive pairs of vertices. The last pair give 70 routes + 56 routes or 126 routes.

15. Multiply 9 times 8 and divide by 2 to eliminate repetitions. There will be a total of 36 league games.

16.

	5	7	13	15
Rey	Xa	Xc	✓	Xd
Ram	✓	Xc	Xc	Xc
Shak	Xc	Xc	Xd	✓
Sash	Xc	✓	Xc	Xb

17. $x = 4$ gives $\frac{(4-4)(4+3)}{(4-4)} = \frac{0}{0}$, which makes the left side of the equation meaningless since division by zero is undefined, but in any case not equal to $4 + 3 = 7$.

18. $\frac{1}{2}$ provides a counterexample because

    $\frac{1}{2} > \left(\frac{1}{2}\right)^2$.

19. $\frac{500(500+1)}{2} = 125,250$

20. a. 2009

    b. $717 - 700 = 17$
        17,000 motor vehicle thefts

    c. 2009 to 2010

Chapter 2: Sets

EXCURSION EXERCISES, SECTION 2.1

1. a. Erica. Since B, C, D, and F were assigned membership value 0, Erica is certain they don't belong to the set good grade.

    b. Larry. A, B, C, and D were assigned membership value 1 so he is certain they belong to the set good grade.

    c. Answers will vary.

3. a. 0 since (2, 0) is on the graph.

    b. 0.5 since (3.5, 0.5) is on the graph.

    c. 0 since (7, 0) is on the graph.

    d. (3.5, 0.5) and (4.5, 0.5) since 0.5 is the membership value.

5. Answers will vary.

EXERCISE SET 2.1

1. {penny, nickel, dime, quarter}

3. {Mercury, Mars}

5. {George W. Bush, Barack Obama}

7. The negative integers greater than –6 are –5, –4, –3, –2, –1. Using the roster method, write the set as {–5, –4, –3, –2, –1}.

9. Adding 4 to each side of the equation produces $x = 7$. {7} is the solution set.

11. Solving:
    $$x + 4 = 1$$
    $$x = -3$$
    But –3 is not a counting number so the solution set is empty, Ø.

In exercises 13 – 20, only one possible answer is given. Your answers may vary from the given answers.

13. {a, e, i, o, u}

15. the set of days of the week that begin with the letter T

17. the set consisting of the two planets in our solar system that are closest to the sun

19. the set of single digit natural numbers

21. the set of natural numbers less than or equal to 7

23. the set of odd natural numbers less than 10

25. Because b is an element of the given set, the statement is true.

27. False; although b ∈ {a, b, c}, {b} ∉ {a, b, c}.

29. False; {0} contains 1 element but Ø contains no elements.

31. False; "good" is subjective.

33. True; the set of natural numbers is equal to the set of whole numbers greater than 0

35. True; both sets contain the same elements

37. $\{x \mid x \in N \text{ and } x < 13\}$

39. $\{x \mid x \text{ is a multiple of 5 and } 4 < x < 16\}$

41. $\{x \mid x \text{ is the name of a month that has 31 days}\}$

43. $\{x \mid x \text{ is the name of a U.S. state that begins with the letter A}\}$

45. $\{x \mid x \text{ is a season that starts with the letter s}\}$

47. {February, April, June, September, November}

49. a. {2013, 2014, 2015}

    b. {2008, 2009}

    c. {2010, 2011, 2012}

51. a. {May, June, July, August}

    b. {March, April, September}

    c. {January, November}

53. 11 since set A has 11 elements

55. The cardinality of the empty set is 0.

57. 4 since 4 states border Minnesota.

59. 16. There are 16 baseball teams in the league.

61. 121

63. Neither. The sets are not equal, nor do they have the same number of elements.

65. Both.

67. Equivalent. The sets are not equal but each has 3 elements.

69. Equivalent. Each set has 2 elements.

10

71. Not well-defined since the word "good" is not precise.

73. Not well-defined since "tall" is not precise.

75. Well-defined.

77. Well-defined.

79. Not well-defined; "small" is not precise.

81. Not well-defined; "best" is not precise.

83. Identify the natural numbers less than 5, which are 1, 2, 3, and 4.
Replace x with those numbers and simplify.
When $x = 1$,
$$3(1)^2 - 1 = 3(1) - 1 = 3 - 1 = 2.$$
When $x = 2$,
$$3(2)^2 - 1 = 3(4) - 1 = 12 - 1 = 11.$$
When $x = 3$,
$$3(3)^2 - 1 = 3(9) - 1 = 27 - 1 = 26.$$
When $x = 4$,
$$3(4)^2 - 1 = 3(16) - 1 = 48 - 1 = 47.$$
Therefore, $D = \{2, 11, 26, 47\}$.

85. Identify the natural numbers greater than or equal to 2, which are 2, 3, 4, 5, 6, 7, ...
Replace x with those numbers and simplify.
When $x = 2$,
$$(-1)^2 (2)^3 = (1)(8) = 8.$$
When $x = 3$,
$$(-1)^3 (3)^3 = (-1)(27) = -27.$$
When $x = 4$,
$$(-1)^4 (4)^3 = (1)(64) = 64.$$
When $x = 5$,
$$(-1)^5 (5)^3 = (-1)(125) = -125.$$
When $x = 6$,
$$(-1)^6 (6)^3 = (1)(216) = 216.$$
When $x = 7$,
$$(-1)^7 (7)^3 = (-1)(343) = -343.$$
Therefore,
$$F = \{8, -27, 64, -125, 216, -343, \ldots\}.$$

87. $A = B$. Replacing n with whole numbers, starting with 0, $A = \{1, 3, 5, \ldots\}$. Replacing n with natural numbers, starting with one, $B = \{1, 3, 5, \ldots\}$.

89. $A \neq B$.
$$A = \{2(1) - 1,\ 2(2) - 1,\ 2(3) - 1,\ \ldots\}$$
$$= \{1,\ 3,\ 5,\ \ldots\}$$
$$B = \left\{ \frac{1(1+1)}{2}, \frac{2(2+1)}{2}, \frac{3(3+1)}{2},\ \ldots \right\}$$
$$= \{1,\ 3,\ 6,\ \ldots\}$$

EXCURSION EXERCISES, SECTION 2.2

1. Yes, because the membership value of each element of J is less than or equal to its membership value in set K.
$(0.3 \le 0.4, 0.6 \le 0.6, 0.5 \le 0.8, 0.1 \le 1)$.

3. $G' = \{(A, 1 - 1), (B, 1 - 0.7), (C, 1 - 0.4),$
$(D, 1 - 0.1), (F, 1 - 0)\}$
$= \{(A, 0), (B, 0.3), (C, 0.6), (D, 0.9), (F, 1)\}$

5. The dashed line is COLD. The solid line is COLD'.

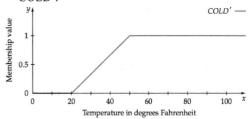

EXERCISE SET 2.2

1. The complement of $\{2, 4, 6, 7\}$ contains elements in U but not in the set: $\{0, 1, 3, 5, 8\}$.

3. $\varnothing' = U = \{0, 1, 2, 3, 4, 5, 6, 7, 8\}$

5. $\{x | x \in W \text{ and } x < 4\} = \{0, 1, 2, 3\}$
$\{x | x \in W \text{ and } x < 4\}' = \{4, 5, 6, 7, 8\}$

7. The set of odd natural numbers less than 8
$= \{1, 3, 5, 7\}$.
$\{1, 3, 5, 7\}' = \{0, 2, 4, 6, 8\}$

9. $\{x | x \in I \text{ and } -3 \le x \le 7\}$
$= \{-3, -2, -1, 0, 1, 2, 3, 4, 5, 6, 7\}$
Then $\{-2, 0, 2, 4, 6, 7\}' = \{-3, -1, 1, 3, 5\}$

11. $\{x | x \in I \text{ and } -3 \le x \le 7\}$
$= \{-3, -2, -1, 0, 1, 2, 3, 4, 5, 6, 7\}$
$\{x | x \in I \text{ and } -2 \le x < 3\} = \{-2, -1, 0, 1, 2\}$
Then
$\{x | x \in I \text{ and } -2 \le x < 3\}' = \{-3, 3, 4, 5, 6, 7\}$

13. $\{a, b, c, d\} \subseteq \{a, b, c, d, e, f, g\}$ since all elements of the first set are contained in the second set.

15. $\{big, small, little\} \not\subseteq \{large, petite, short\}$

17. $I \subseteq Q$

19. $\subseteq$ since the empty set is a subset of every set.

21. $\subseteq$ since every element of the first set is an element of the second set.

23. True; every element of F is an element of D.

25. True; $F \neq D$.

27. True; s is an element of E and $G \neq E$.

29. $G' = \{p, q, r, t\}$. Since q is not an element of D, the statement is false.

31. True; the empty set is a subset of every set.

33. True; $D' = \{q\}$ and $D' \neq E$.

35. False; D does not contain sets.

37. False; D has 4 elements, $2^4 = 16$ subsets and $2^4 - 1 = 15$ proper subsets.

39. False; $F' = \{q, r, s\}$ so F' has $2^3 = 8$ subsets.

41. $2^{16} = 65536$ subsets.
65536 seconds
$= 65536 \div 60$ seconds per minute
≈ 1092 minutes $\div 60$ minutes per hour
$= 18$ hours (to the nearest hour).

43. $\emptyset, \{\alpha\}, \{\beta\}, \{\alpha, \beta\}$

45. $\emptyset, \{I\}, \{II\}, \{III\}, \{I, II\}, \{I, III\}, \{II, III\},$
$\{I, II, III\}$

47. The number of subsets is 2^n where n is the number of elements in the set. $2^2 = 4$.

49. List the elements in the set:
$\{8, 10, 12, 14, 16, 18, 20\}$.
$2^7 = 128$.

51. $2^{11} = 2,048$

53. There are no negative whole numbers. $2^0 = 1$.

55. a. This is equivalent to finding the number of proper subsets for a set with 4 elements.
$2^4 - 1 = 16 - 1 = 15$.

b. The sets that contain the 1976 dime or the 1992 dime produce duplicate amounts of money.
There are 8 sets that contain one dime producing 4 sets with the same value.
$15 - 4 = 11$. The set containing the nickel and two dimes has the same value as the set containing the quarter. $11 - 1 = 10$ different sums.

c. Two different sets of coins can have the same value.

57. a. $2^8 = 256$ different types

b. Solve $2^x > 2000$ by guessing and checking.
$2^{10} = 1,024$
$2^{11} = 2,048$
So at least 11 condiments.

59. a. $2^{10} = 1,024$ types of omelets

b. Solve $2^x > 4,000$ by guessing and checking.
$2^{11} = 2,048$
$2^{12} = 4,096$
At least 12 ingredients must be available.

61. a. $\{2\}$ is the set containing 2, not the element 2. $\{1, 2, 3\}$ has only three elements, namely 1, 2, and 3. Because $\{2\}$ is not equal to 1, 2, or 3, $\{2\} \notin \{1, 2, 3\}$.

b. 1 is not a set, so it cannot be a subset.

c. The given set has the elements 1 and $\{1\}$. Because $1 \neq \{1\}$, there are exactly two elements in $\{1, \{1\}\}$.

63. a. $\{A, B, C\}, \{A, B, D\}, \{A, B, E\},$
$\{A, C, D\}, \{A, C, E\}, \{A, D, E\},$
$\{B, C, D\}, \{B, C, E\}, \{B, D, E\},$
$\{C, D, E\}, \{A, B, C, D\}, \{A, B, C, E\},$
$\{A, B, D, E\}, \{A, C, D, E\}, \{B, C, D, E\},$
$\{A, B, C, D, E\}$

b. $\{A\}, \{B\}, \{C\}, \{D\}, \{E\}, \{A,B\}, \{A, C\},$
$\{A, D\}, \{A, E\}, \{B, C\}, \{B, D\}, \{B, E\},$
$\{C, D\}, \{C, E\}, \{D, E\}$

EXCURSION EXERCISES SECTION 2.3

1. To form the union, use the maximum membership value.
$M \cup J$
$= \{(A, 1), (B, 0.8), (C, 0.6), (D, 0.5), (F, 0)\}$

3. $J' = \{(A, 1-1), (B, 1-0.8), (C, 1-0.6),$
$(D, 1-0.1), (F, 1-0)\}$
$= \{(A, 0), (B, 0.2), (C, 0.4), (D, 0.9),$
$(F, 1)\}$
$E \cup J'$
$= \{(A, 1), (B, 0.2), (C, 0.4), (D, 0.9), (F, 1)\}$

5. $M' = \{(A, 1-1), (B, 1-0.75), (C, 1-0.5),$
      $(D, 1-0.5), (F, 1-0)\}$
      $= \{(A, 0), (B, 0.25), (C, 0.5), (D, 0.5),$
      $(F, 1)\}$

   Since $M' \cup L'$ is in parentheses, form the union first.
      $L' = \{(A, 1-1), (B, 1-1), (C, 1-1),$
         $(D, 1-1), (F, 1-0)\}$
      $= \{(A, 0), (B, 0), (C, 0),$
         $(D, 0), (F, 1)\}$
      so $M' \cup L' = M'$. Now, $J \cap M' = \{(A, 0),$
      $(B, 0.25), (C, 0.5), (D, 0.1), (F, 0)\}$.

7. $A \cap B$
      $= \{(a, 0.3), (b, 0.4), (c, 0.9), (d, 0.2), (e, 0.45)\}.$
      $(A \cap B)' = \{(a, 1-0.3), (b, 1-0.4),$
         $(c, 1-0.9), (d, 1-0.2),$
         $(e, 1-0.45)\}$
      $= \{(a, 0.7), (b, 0.6), (c, 0.1),$
         $(d, 0.8), (e, 0.55)\}$
   $A' = \{(a, 0.7), (b, 0.2), (c, 0), (d, 0.8), (e, 0.25)\}$
   $B' = \{(a, 0.5), (b, 0.6), (c, 0.1), (d, 0.3), (e, 0.55)\}$
   $A' \cup B'$
      $= \{(a, 0.7), (b, 0.6), (c, 0.1), (d, 0.8), (e, 0.55)\}$
   Since $(A \cap B)' = A' \cup B'$ De Morgan's Law holds.

EXERCISE SET 2.3

1. $A \cup B = \{2, 4, 6\} \cup \{1, 2, 5, 8\}$
      $= \{1, 2, 4, 5, 6, 8\}$

3. $A \cap B' = \{2, 4, 6\} \cap \{3, 4, 6, 7\} = \{4, 6\}$

5. $(A \cup B)' = (\{1, 2, 4, 5, 6, 8\})' = \{3, 7\}$

7. $B \cup C = \{1, 2, 5, 8\} \cup \{1, 3, 7\}$
      $= \{1, 2, 3, 5, 7, 8\}.$
   $A \cup (B \cup C) = \{2, 4, 6\} \cup \{1, 2, 3, 5, 7, 8\}$
      $= \{1, 2, 3, 4, 5, 6, 7, 8\} = U$

9. $B \cap C = \{1, 2, 5, 8\} \cap \{1, 3, 7\} = \{1\}.$
   $A \cap (B \cap C) = \{2, 4, 6\} \cap \{1\} = \emptyset$

11. $B \cap (B \cup C) = \{1, 2, 5, 8\} \cap \{1, 2, 3, 5, 7, 8\}$
      $= \{1, 2, 5, 8\} = B$

13. $B \cup B' = \{1, 2, 5, 8\} \cup \{3, 4, 6, 7\}$
      $= \{1, 2, 3, 4, 5, 6, 7, 8\} = U$

15. $A \cup C' = \{2, 4, 6\} \cup \{2, 4, 5, 6, 8\}$
      $= \{2, 4, 5, 6, 8\}.$
   $B \cup A' = \{1, 2, 5, 8\} \cup \{1, 3, 5, 7, 8\}$
      $= \{1, 2, 3, 5, 7, 8\}$
   $(A \cup C') \cap (B \cup A')$
      $= \{2, 4, 5, 6, 8\} \cap \{1, 2, 3, 5, 7, 8\}$
      $= \{2, 5, 8\}$

17. $C \cup B' = \{1, 3, 7\} \cup \{3, 4, 6, 7\}$
      $= \{1, 3, 4, 6, 7\}.$
   $(C \cup B') \cup \emptyset = \{1, 3, 4, 6, 7\} \cup \{\}$
      $= \{1, 3, 4, 6, 7\} = (C \cup B')$

19. $A \cup B = \{1, 2, 4, 5, 6, 8\}$
   $B \cap C' = \{1, 2, 5, 8\} \cap \{2, 4, 5, 6, 8\}$
      $= \{2, 5, 8\}$
   $(A \cup B) \cap (B \cap C')$
      $= \{1, 2, 4, 5, 6, 8\} \cap \{2, 5, 8\}$
      $= \{2, 5, 8\}$

In Exercises 21-28, one possible answer is given. Your answers may vary from the given answers.

21. The set of all elements that are not in L or are in T.

23. The set of all elements that are in A, or are in C, but not in B.

25. The set of all elements that are in T, and are also in J or not in K.

27. The set of all elements that are in both W and V, or are in both W and Z.

29.
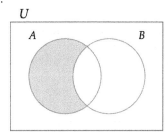

31.
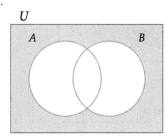

33.
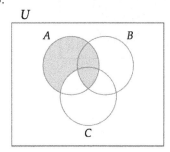

35.

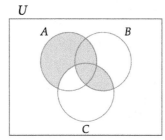

37.

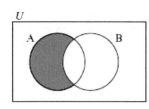

$A \cap B'$

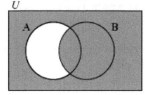

$A' \cup B$

Because the sets $A \cap B'$ and $A' \cup B$ are represented by different regions, $(A \cup B') \neq (A' \cup B)$.

39.

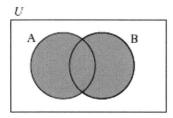

$A \cup (A' \cap B)$

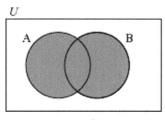

$A \cup B$
$A \cup (A' \cap B) = A \cup B$

41.

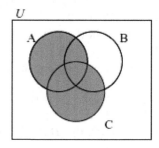

$(A \cup C) \cap B'$

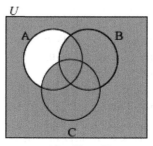

$A' \cup (B \cup C)$

$(A \cup C) \cap B' \neq A' \cup (B \cup C)$

43.

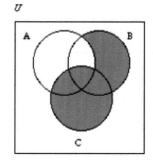

$(A' \cap B) \cup C$

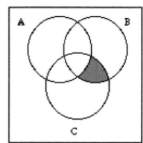

$(A' \cap C) \cap (A' \cap B)$
$(A' \cap B) \cup C \neq (A' \cap C) \cap (A' \cap B)$

45.

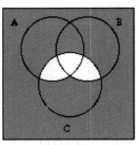

$$((A \cup B) \cap C)'$$

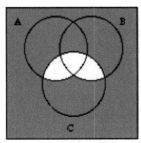

$$(A' \cap B') \cup C'$$
$$((A \cup B) \cap C)' = (A' \cap B') \cup C'$$

47. $R \cap G \cap B'$
$B' = \{R, Y, G\}$
$R = \{R, M, W, Y\}$
$G = \{G, Y, W, C\}$
$R \cap G \cap B' = Y$ (yellow)

49. $R' \cap G \cap B$
$R' = \{B, C, G\}$
$G = \{G, Y, W, C\}$
$B = \{B, M, W, C\}$
$R \cap G \cap B = C$ (cyan)

51. $C' = \{Y, R, M\}$
$M = \{M, B, K, R\}$
$Y = \{Y, G, K, R\}$
$C' \cup M \cap Y = R$ (red)

In Exercises 53–62, one possible answer is given.
Your answers may vary from the given answers.
53. $A \cap B'$

55. $(A \cup B)'$

57. $B \cup C$

59. $C \cap (A \cup B)'$

61. $(A \cup B)' \cup (A \cap B \cap C)$

63. a.

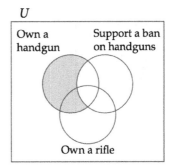

b.

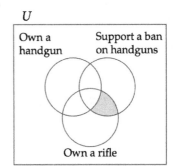

c.

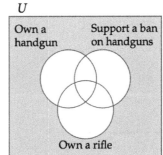

65.

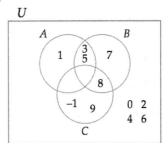

67.

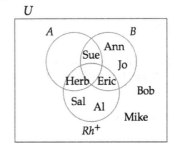

69.

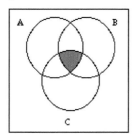

71. $B - A = \{2, 3, 8, 9\} - \{2, 4, 6, 8\}$
      $= \{3, 9\}$

73. $A - B' = \{2, 4, 6, 8\} - \{1, 4, 5, 6, 7\}$
      $= \{2, 8\}$

75. $A' - B' = \{1, 3, 5, 7, 9\} - \{1, 4, 5, 6, 7\}$
      $= \{3, 9\}$

77.   Responses will vary.

79.

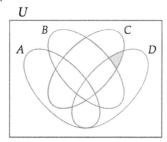

EXCURSION EXERCISES 2.4

1. a.   {Ryan, Susan}, {Ryan, Trevor},
         {Susan, Trevor}, {Ryan, Susan, Trevor}

    b.   Ø, {Ryan}, {Susan}, {Trevor}

3. In $\{M, N\}$, if either voter leaves the coalition,
    the coalition becomes a losing coalition.
    The same is true for $\{M, P\}$. However, in
    $\{M, N, P\}$, if N or P leaves, the coalition still
    wins. The minimal winning coalition is $\{M, N\}$
    and $\{M, P\}$.

EXERCISE SET 2.4

1. $B \cup C = \{$Math, Physics, Chemistry,
    Psychology, Drama, French, History$\}$
    so $n(B \cup C) = 7$

3. $n(B) + n(C) = 5 + 3 = 8$

5. $(A \cup B) \cup C = \{$English, History, Psychology,
    Drama, Math, Physics, Chemistry, French$\}$ so
    $n[(A \cup B) \cup C] = 8$

7. $n(A) + n(B) + n(C) = 4 + 5 + 3 = 12$

9. $B \cap C = \{$Chemistry$\}$
    $A \cup (B \cap C) = \{$English, History, Psychology,
    Drama, Chemistry$\}$
    so $n[A \cup (B \cap C)] = 5$

11. $n(A \cup B) = n(A) + n(B) - n(A \cap B)$
    $= 4 + 5 - 2 = 7$

13. Using the formula:
    $n(J \cup K) = n(J) + n(K) - n(J \cap K)$
    $310 = 245 + 178 - n(J \cap K)$
    $310 = 423 - n(J \cap K)$
    $-113 = -n(J \cap K)$
    $n(J \cap K) = 113$

15. Using the formula:
    $n(A \cup B) = n(A) + n(B) - n(A \cap B)$
    $2250 = 1500 + n(B) - 310$
    $2250 = 1190 + n(B)$
    $1060 = n(B)$

17.

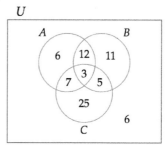

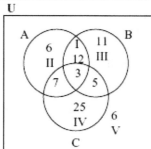

To find n(I): Since $n(A \cap B) = 15$ and 3 has
been accounted for, $15 - 3 = 12$ for n(I).
To find n(II): $n(A) = 28$. $7 + 3 + 12 = 22$ is
accounted for so $28 - 22 = 6$ for n(II).
To find n(III): $n(B) = 31$. $5 + 3 + 12 = 20$.
$31 - 20 = 11$.
To find n(IV): $n(C) = 40$. $40 - (7 + 3 + 5) = 25$.
To find n(V): $n(U) = 75$ so
n(V) $= 75 - (6 + 12 + 11 + 7 + 3 + 5 + 25) = 6$.

19. a. $S = \{\text{investors in stocks}\}$ and let
$B = \{\text{investors in bonds}\}$. Since 75 had not
invested in either stocks or bonds,
$n(S \cup B) = 600 - 75 = 525$.
$$n(S \cup B) = n(S) + n(B) - n(S \cap B)$$
$$525 = 380 + 325 - n(S \cap B)$$
$$525 = 705 - n(S \cap B)$$
$$-180 = -n(S \cap B)$$
$$n(S \cap B) = 180$$

$n(S \cap B) = 180$ represents the number of
investors in both stocks and bonds.

b. $n(S \text{ only}) = 380 - 180 = 200$

21. Draw a Venn diagram to represent the data:
Since 44% responded favorably to both forms,
place 44% in the intersection of the two sets.
A total of 72% responded favorably to an
analgesic, so 72% – 44% = 28% who responded
favorably to only the analgesic. Similarly,
59% – 44% = 15% who responded favorably
only to the muscle relaxant.

a. 15% responded favorably to the muscle
relaxant but not the analgesic.

b. Since the universe must contain 100%,
subtract the known values from 100%:
100% – (28% + 44% + 15%) =13%. This
represents the percent of athletes who were
treated who did not respond favorably to
either form of treatment.

23. Draw a Venn diagram to represent the data. Fill
in the diagram starting from the innermost
region.

  i: 85
  ii: 150 – 85 = 65
  iii: 135 – 85 = 50
  iv: 110 – 85 = 25
  v: 390 – (65 + 85 + 50) = 190
  vi: 290 – (25 + 85 + 50) = 130
  vii: 305 – (65 + 85 + 25) = 130
  viii: 770 – (130+25+130+65+85+50 +190)
      = 95

a. exactly one of these forms of advertising is
represented by v, vi, and vii:
190 + 130 + 130 = 450

b. exactly two of these forms are represented
by ii, iii, and iv: 65 + 50 + 25 = 140

c. PC World and neither of the other two
forms is represented by vii: 130.

25. Draw a Venn diagram to represent the data. Fill
in the diagram starting from the innermost
region.

  i: 52
  ii: 10 – 52 = 88
  iii: 437 – (52 + 88 + 202) = 95
  iv: 74 – 52 = 22
  v: 271 – (22 + 88 + 52) = 109
  vi: 202
  vii: 497 – (22 + 52 + 95) = 328
  viii: 1000 – (109+88+52+22+202+95+328)
      = 104

a. this group is represented by v: 109

b. this group is represented by vii: 328

c. this group is represented by viii: 104

27. a. 101

b. 124 + 82 + 65 + 51 + 48 = 370

c. 124+82+133 + 41 = 380

d. 124+82+101 + 66 = 373

e. 124 +101 = 225

f. 124 + 82 + 65 + 101 + 66 + 51 + 41 = 530

29. Given $n(A) = 47$ and $n(B) = 25$.

a. If A and B are disjoint sets,
$n(A \cup B) = n(A) + n(B) = 47 + 25 = 72$.

b. If $B \subset A$, then $A \cup B = A$ and $n(A) = 47$.

c. If $B \subset A$, then $A \cap B = B$ and $n(B) = 25$.

d. If A and B are disjoint sets, then $A \cap B = \varnothing$
and $n(A \cap B) = 0$.

31. Complete the Venn diagram. Since there are
450 users of Webcrawler,
450 – (45 + 41 + 30 + 50 + 60 + 80 + 45) = 99
gives the total who use only Webcrawler.
Similarly, 585 – (55 + 50 + 60 + 41 + 34 + 100
+ 45) = 200 gives the total who use only Bing.
To find Yahoo only users:
620 – (55 + 50 + 100 + 60 +80 +41 +30) = 204.
To find Ask only users:
560 – (50 +80 +50 +60 + 34 + 100 + 45) = 141.

a. only Google: 200

b. To find the number who use exactly three
search engines, add the numbers given for
people who use only 3 search engines:
100 + 41 + 50 + 80 = 271

c. Total number in the regions: 204 + 55+ 200
+ 141 + 50 + 50 + 34 + 45 + 80 + 60 + 100
+ 99 + 30 + 41 + 45 = 1234.
Since 1250 people were surveyed, this
means 1250 – 1234 = 16 people do not use
any of the search engines.

EXCURSION EXERCISES 2.5

1. Let $C = \{3, 4, 5, 6, ..., n+2, ...\}$. C and N have the same cardinality because the elements of C can be paired with the elements of N using the general correspondence $(n+2) \leftrightarrow n$.

    Let $D = \{1, 2\}$. Because $C \cup D = N$, we can establish the following equations.
    $$n(C) + n(D) = n(N)$$
    $$\aleph_0 + 2 = \aleph_0$$

3. Let $C = \{1, 2, 3, 4, 5, 6\}$. Then
    $$n(N) - n(C) = n(N \cap C')$$
    $$\aleph_0 - 6 = n(\{7,\ 8,\ 9,\ 10,\ ...\})$$
    $$\aleph_0 - 6 = \aleph_0$$

EXERCISE SET 2.5

1. a. Comparing:
    $$V = \{a, e, i\}$$
    $$\updownarrow \updownarrow \updownarrow$$
    $$M = \{3, 6, 9\}$$

    b. The possible one-to-one correspondences (listed as ordered pairs) are: $\{(a, 6), (i, 3), (e, 9)\}$, $\{(a, 9), (i, 3), (e, 6)\}$, $\{(a, 3), (i, 9), (e, 6)\}$, $\{(a, 6), (i, 9), (e, 3)\}$, $\{(a, 9), (i, 6), (e, 3)\}$ plus the pairing shown in part a. produce 6 one-to-one correspondences.

3. Write the sets so that one is aligned below the other. One possible pairing is shown below.
    $$D = \{1, 3, 5, ..., 2n-1, ...\}$$
    $$\updownarrow \updownarrow \updownarrow \quad \updownarrow$$
    $$M = \{3, 6, 9, ... 3n, ...\}$$
    Pair $(2n-1)$ of D with $(3n)$ of M to establish a one-to-one correspondence.

5. The general correspondence $(n) \leftrightarrow (7n-5)$ establishes a one-to-one correspondence between the elements of N and the elements of the given set. Thus the cardinality is $\aleph_0$.

7. c

9. c. Any set of the form $\{x \mid a \le x \le b\}$ where a and b are real numbers and $a \ne b$ has cardinality c.

11. Sets with equal cardinality are equivalent. The cardinality of N = cardinality of $I = \aleph_0$ therefore the sets are equivalent.

13. The sets are equivalent since the set of rational numbers and the set of integers have cardinality $\aleph_0$.

15. Let $S = \{10, 20, 30, ..., 10n, ...\}$. Then S is a proper subset of A. A rule for a one-to-one correspondence between A and S is $(5n) \leftrightarrow (10n)$. Because A can be placed in a one-to-one correspondence with a proper subset of itself, A is an infinite set.

17. Let $R = \left\{ \frac{3}{4}, \frac{5}{6}, \frac{7}{8}, ..., \frac{2n+1}{2n+2}, ... \right\}$. Then R is a proper subset of C. A rule for a one-to-one correspondence between C and R is $\left(\frac{2n-1}{2n} \right) \leftrightarrow \left(\frac{2n+1}{2n+2} \right)$. Because C can be placed in a one-to-one correspondence with a proper subset of itself, C is an infinite set.

In Exercises 19-26, let $N = \{1, 2, 3, 4, .., n, ..\}$. Then a one-to-one correspondence between the given sets and the set of natural numbers N is given by the following general correspondences.

19. $(n + 49) \leftrightarrow (n)$

21. $\left(\dfrac{1}{3^{n-1}} \right) \leftrightarrow (n)$

23. $(10^n) \leftrightarrow (n)$

25. $(n^3) \leftrightarrow (n)$

27. a. For any natural number n, the two natural numbers preceding $3n$ are not multiples of 3. Pair these two numbers, $3n-2$ and $3n-1$, with the multiples of 3 given by $6n-3$ and $6n$, respectively. Using the two general correspondences $(6n-3) \leftrightarrow (3n-2)$ and $(6n) \leftrightarrow (3n-1)$, we can establish a one-to-one correspondence between the multiples of 3 (set M) and the set K of all natural numbers that are not multiples of 3.

    b. First find n. $6n-606$, $n = 101$. Then $(6n) \leftrightarrow (3n-1)$ means $(6(101)) \leftrightarrow (3(101)-1) = 302$.

    c. Solve.
    $$3n - 1 = 899$$
    $$3n = 900$$
    $$n = 300$$
    Then $(6n) \leftrightarrow (3n-1)$ means $(6(300)) \leftrightarrow (3(300)-1)$ and $(1800) \leftrightarrow (899)$.

29. The set of real numbers x such that $0 < x < 1$ is equivalent to the set of all real numbers.

CHAPTER 2 REVIEW EXERCISES

1. {January, June, July}

2. {Alaska, Hawaii}

3. {0, 1, 2, 3, 4, 5, 6, 7}

4. {−8, 8}. Since the set of integers includes positives and negatives, −8 and 8 satisfy $x^2 = 64$.

5. Solving:
    $$x + 3 \le 7$$
    $$x \le 4$$
    {1, 2, 3, 4}.

6. {1, 2, 3, 4, 5, 6}. Counting numbers begin at 1.

7. $\{x \mid x \in I \text{ and } x > -6\}$

8. $\{x \mid x$ is the name of a month with exactly 30 days$\}$

9. $\{x \mid x$ is the name of a U.S. state that begins with the letter $K\}$

10. $\{x^3 \mid x = 1, 2, 3, 4, 5\}$

11. The sets are equivalent, since each set has exactly four elements and are not equal.

12. The sets are both equal and equivalent.

13. False. The set contains numbers, not sets.
    $\{3\} \notin \{1, 2, 3, 4\}$.

14. True. The set of integers includes positive and negative integers.

15. True. The symbol ~ means equivalent and sets with the same number of elements are equivalent.

16. False. The word small is not precise.

17. $A \cap B = \{2, 6, 10\} \cap \{6, 10, 16, 18\} = \{6, 10\}$

18. $A \cup B = \{2, 6, 10\} \cup \{6, 10, 16, 18\}$
    $= \{2, 6, 10, 16, 18\}$

19. $A' \cap C = \{8, 12, 14, 16, 18\} \cap \{14, 16\}$
    $= \{14, 16\} = C$

20. $B \cup C' = \{6, 10, 16, 18\} \cup \{2, 6, 8, 10, 12, 18\}$
    $= \{2, 6, 8, 10, 12, 16, 18\}$

21. $B \cap C = \{6, 10, 16, 18\} \cap \{14, 16\} = \{16\}$
    $A \cup \{16\} = \{2, 6, 10\} \cup \{16\} = \{2, 6, 10, 16\}$

22. $A \cup C = \{2, 6, 10\} \cup \{14, 16\}$
    $= \{2, 6, 10, 14, 16\}$
    $(A \cup C)' = \{8, 12, 18\}$
    $\{8, 12, 18\} \cap \{2, 8, 12, 14\} = \{8, 12\}$

23. $A \cap B' = \{2, 6, 10\} \cap \{2, 8, 12, 14\} = \{2\}$
    $(\{2\})' = \{6, 8, 10, 12, 14, 16, 18\}$

24. $A \cup B \cup C = \{2, 6, 10, 14, 16, 18\}$
    $(A \cup B \cup C)' = \{8, 12\}$

25. No, the first set is not a subset of the second set because 0 is not in the set of natural numbers.

26. No, the first set is not a subset of the second set because 9.5 is not in the set of integers.

27. Proper subset. All natural numbers are whole numbers, but 0 is not a natural number so $N \subset W$.

28. Proper subset. All integers are real numbers, but $\frac{1}{2}$ is a real number that is not an integer so $I \subset R$.

29. Counting numbers and natural numbers represent the same set of numbers. The set of counting numbers is not a proper subset of the set of natural numbers.

30. The set of real numbers is not a proper subset of the set of rational numbers.

31. Ø, {I}, {II}, {I, II}

32. Ø, {s} {u}, {n}, {s, u}, {s, n}, {u, n}, { s, u, n}

33. Ø, {penny}, {nickel}, {dime}, {quarter}, {penny, nickel}, {penny, dime}, {penny, quarter}, {nickel, dime}, {nickel, quarter}, {dime, quarter}, {penny, nickel, dime}, {penny, nickel, quarter}, {penny, dime, quarter}, {nickel, dime, quarter}, {penny, nickel, dime, quarter}

34. Ø, {A}, {B}, {C}, {D}, {E}, {A, B}, {A, C}, {A, D}, {A, E}, {B, C}, {B, D}, {B, E}, {C, D}, {C, E}, {D, E}, {A, B, C}, {A, B, D}, {A, B, E}, {A, C, D}, {A, C, E}, {A, D, E}, {B, C, D}, {B, C, E}, {B, D, E}, {C, D, E}, {A, B, C, D}, {A, B, C, E}, {A, B, D, E}, {A, C, D, E}, {B, C, D, E}, {A, B, C, D, E}

35. The number of subsets of a set with n elements is 2^n. The set of four musketeers has 4 elements and $2^4 = 16$ subsets.

36. $n = 26$.
    $2^{26} = 67,108,864$ subsets

37. The number of letters is 15.
$2^{15} = 32,768$ subsets

38. $n = 7$. $2^7 = 128$ subsets

39. True, by De Morgan's Law: $(A \cup B')' = A' \cap B$

40. True, by De Morgan's Law: $(A' \cap B')' = A \cup B$

41.

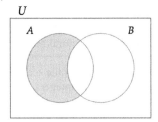

42.

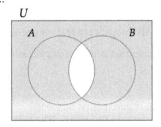

43.

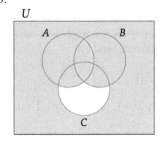

44.

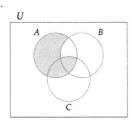

45.

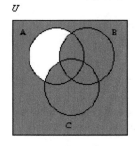

$$A' \cup (B \cup C)$$

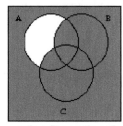

$$(A' \cup B) \cup (A' \cup C)$$

$$A' \cup B \cup C = (A' \cup B) \cup (B' \cup C)$$

46.

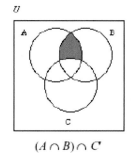

$$(A \cap B) \cap C$$

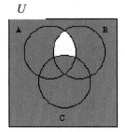

$$(A' \cup B') \cup C$$
$$(A \cap B) \cap C \neq (A' \cup B') \cup C$$

47.

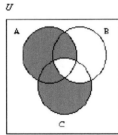

$$A \cap (B' \cap C)$$

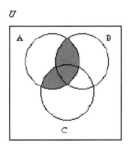

$$(A \cup B') \cap (A \cup C)$$
$$A \cap (B' \cap C) \neq (A \cup B') \cap (A \cup C)$$

48.

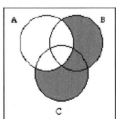

$$A \cap (B \cup C)$$

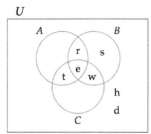

$$A' \cap (B \cup C)$$

$$A \cap (B \cup C) \neq A' \cap (B \cup C)$$

49. $(A \cup B)' \cap C$ or $C \cap (A' \cap B')$

50. $(A \cap B) \cup (B \cap C')$

51.

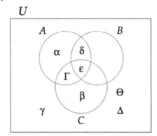

52.

(Venn diagram)

53. Use a Venn diagram to represent the survey results. Total the numbers from each region to find the number of members surveyed:
$111 + 97 + 48 + 135 = 391$

54. Use a Venn diagram to represent the survey results: After placing 96 in the intersection of all three types, use the information on customers who like two types of coffee:
116 espresso and cappuccino – 96 like all three = 20
136 espresso and chocolate-flavored – 96 = 40
127 cappuccino and chocolate-flavored – 96=31
221 espresso – (20 + 96 + 40) = 65 who like only espresso.
182 cappuccino – (20 + 96 + 31) = 35 like only cappuccino.
209 chocolate-flavored coffee – (40 + 96 + 31) = 42 who like only chocolate- flavored coffee.

a. 42 customers

b. 31 customers

c. 20 customers

d. $65 + 35 + 42 = 142$ customers

55. Let O = (athletes playing offense} and D = {athletes playing defense}.
$$n(O \cup D) = n(O) + n(D) - n(O \cap D)$$
$$43 = 27 + 22 - n(O \cap D)$$
$$43 = 49 - n(O \cap D)$$
$$6 = n(O \cap D)$$

Therefore, 6 athletes play both offense and defense.

56. Let B = {students registered in biology} and let P = {students registered in psychology}.
$$n(B \cup P) = n(B) + n(P) - n(B \cap P)$$
$$= 625 + 433 - 184$$
$$= 874$$
Therefore, 874 students are registered in biology or psychology.

57. One possible one-to-one correspondence between {1, 3, 6, 10} and {1, 2, 3, 4} is given by
$$\{1, 3, 6, 10\}$$
$$\updownarrow \updownarrow \updownarrow \quad \updownarrow$$
$$\{1, 2, 3, 4\}$$

58. $\{x \mid x > 10 \text{ and } x \in N\} = \{11, 12, 13, 14,..., n + 10, ...\}$
Thus a one-to-one correspondence between the sets is given by
$$\{11, 12, 13, 14,...n + 10,...\}$$
$$\updownarrow \updownarrow \updownarrow \updownarrow \quad \updownarrow$$
$$\{2, 4, 6, 8, ..., 2n,...\}$$

59. One possible one-to-one correspondence between the set is given by
$$\{3, \quad 6, \quad 9, \quad ... \quad 3n,...\}$$
$$\updownarrow \quad \updownarrow \quad \updownarrow \qquad \updownarrow$$
$$\{10, 100, 1000, \ ..., 10^n,...\}$$

60. In the following figure, the line from E that passes through $\overline{AB}$ and $\overline{CD}$ illustrates a method of establishing a one-to-one correspondence between the sets $\{x \mid 0 \leq x \leq 1\}$ and $\{x \mid 0 \leq x \leq 4\}$.

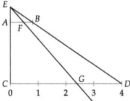

61. A proper subset of A is $S = \{10, 14, 18,...,4n + 6, ...\}$. A one-to-one correspondence between A and S is given by
$$A = \{6, 10, 14, 18,...,4n + 2,...\}$$
$$\updownarrow \updownarrow \updownarrow \updownarrow \qquad \updownarrow$$
$$S = \{10, 14, 18, 22,...,4n + 6,...\}$$
Because A can be placed in a one-to-one correspondence with a proper subset of itself, A is an infinite set.

62. A proper subset of B is $T = \left\{\dfrac{1}{2}, \dfrac{1}{4}, \dfrac{1}{8}, \dfrac{1}{16},...,\dfrac{1}{2^n},...\right\}$. A one-to-one correspondence between B and T is given by:

$$B = \left\{1, \ \frac{1}{2}, \ \frac{1}{4}, \ \frac{1}{8},...,\frac{1}{2^{n+1}},...\right\}$$
$$\updownarrow \quad \updownarrow \quad \updownarrow \quad \updownarrow \qquad \updownarrow$$
$$T = \left\{\frac{1}{2}, \ \frac{1}{4}, \ \frac{1}{8}, \ \frac{1}{16},...,\frac{1}{2^n},...\right\}$$
Because B can be placed in a one-to-one correspondence with a proper subset of itself, B is an infinite set.

63. 5

64. 10

65. 2

66. 5

67. $\aleph_0$

68. $\aleph_0$

69. c

70. c

71. $\aleph_0$

72. c

CHAPTER 2 TEST

1. $(A \cap B)' = (\{3, 5, 7, 8\} \cap \{2, 3, 8, 9, 10\})'$
$$= \{3, 8\}' = \{1, 2, 4, 5, 6, 7, 9, 10\}$$

2. $A' \cap B = \{1, 2, 4, 6, 9, 10\} \cap \{2, 3, 8, 9, 10\}$
$$= \{2, 9, 10\}$$

3. $A' \cup (B \cap C')$
$$= \{1, 2, 4, 6, 9, 10\} \cup (\{2, 3, 8, 9 \ 10\}$$
$$\cap \{2, 3, 5, 6, 9, 10\})$$
$$= \{1, 2, 4, 6, 9, 10\} \cup \{2, 3, 9, 10\}$$
$$= \{1, 2, 3, 4, 6, 9, 10\}$$

4. $A \cap (B' \cup C)$
$$= \{3, 5, 7, 8\} \cap (\{1, 4, 5, 6, 7\} \cup \{1, 4, 7, 8\})$$
$$= \{3, 5, 7, 8\} \cap \{1, 4, 5, 6, 7, 8\}$$
$$= \{5, 7, 8\}$$

5. $\{x \mid x \in W \text{ and } x < 7\}$

6. $\{x \mid x \in I \text{ and } -3 \leq x \leq 2\}$

7. a. The set of whole numbers less than 4
$$= \{0, 1, 2, 3\}.$$
$$n(\{0, 1, 2, 3\}) = 4$$

b. $\aleph_0$

8. a. Neither, the sets do not have the same number of elements and are not equal.

   b. Equivalent, the sets have the same number of elements but are not equal.

9. a. Equivalent. The set of natural numbers and the set of integers have cardinality $\aleph_0$. The sets are not equal since integers such as -3 and 0 are not natural numbers.

   b. Equivalent. Both sets have cardinality $\aleph_0$. The sets are not equal because $0 \in W$ but 0 is not a positive integer.

10. Ø, {a}, {b}, {c}, {d}, {a, b}, {a, c}, {a, d}, {b, c}, {b, d}, {c, d}, {a, b, c,}, {a, b, d}, {a, c, d}, {b, c, d}, {a, b, c, d}

11. The number of subsets of a set of n elements is 2^n.
    $2^{21} = 2{,}097{,}152$ subsets

12.

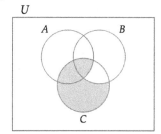

13.

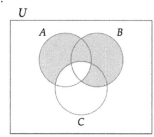

14. $(A \cup B)' = A \cap B'$ by De Morgan's Laws

15. a. $2^9 = 512$ different versions of this sedan

    b. Use the method of guess and check to find the smallest natural numbers n for which $2^n > 2500$.

       $2^n > 2500$

       $2^{10} = 1024$

       $2^{11} = 2048$

       $2^{12} = 4096$

       The company must provide a minimum of <u>12</u> upgrade options if it wishes to offer at least 2500 versions of this sedan.

16. Let F = {students receiving financial aid} and let B = {students who are business majors}
    $$n(F \cup B) = n(F) + n(B) - n(F \cap B)$$
    $$= 841 + 525 - 202$$
    $$= 1164$$
    Therefore, 1164 students are receiving financial aid or are business majors.

17. a. {2007, 2008, 2014}

    b. {2009, 2010, 2013, 2014}

    c. Ø

18. Draw a Venn diagram to represent the data. Fill in the diagram starting from the innermost region.

    i: 105
    ii: $412 - 105 = 307$
    iii: $280 - (80 + 105 + 64) = 232$
    iv: $185 - 105 = 80$
    v: $724 - (105 + 80 + 307) = 190$
    vi: $545 - (31 + 105 + 307) = 102$
    vii: 64
    viii: $1000 - (105 + 307 + 31 + 80 + 232 + 102 + 64) = 79$

    a. this group is represented by v: 232 households

    b. this group is represented by vi: 102 households

    c. this group is represented by i, ii, iii, iv, v, and vi:
       $105 + 307 + 31 + 80 + 232 + 102$
       $= 857$ households

    d. this group is represented by viii: 79 households

19. A possible correspondence:
    $\{5, 10, 15, 20, 25, \ldots, 5n, \ldots\}$
    $\updownarrow \; \updownarrow \; \updownarrow \; \updownarrow \quad \updownarrow$
    $\{0, 1, 2, 3, 4, \ldots, n-1, \ldots\}$
    $(5n) \leftrightarrow (n-1)$

20. A possible correspondence:
    $\{3, 6, 9, 12, \ldots, 3n, \ldots\}$
    $\updownarrow \; \updownarrow \; \updownarrow \; \updownarrow \quad \updownarrow$
    $\{6, 12, 18, 24, \ldots, n-1, \ldots\}$
    $(3n) \leftrightarrow (6n)$

Chapter 3: Logic

EXCURSION EXERCISES, SECTION 3.1

1. $S \wedge (\sim R \vee \sim Q \vee P)$

3. $(\sim P \vee Q) \wedge (\sim R \vee S)$

5. $[\sim R \vee (Q \wedge S)] \vee (R \wedge \sim S)$

7. 1 is not a closed network since the series circuit containing S is not closed.
2 is a closed network, since $\sim S$ is closed, $\sim Q$ is closed, and R is closed.
3 is not a closed network. Since both $\sim P$ and Q are open, the parallel circuit containing them is open.
4 is not a closed network. Since Q is open, the series circuit containing it is open.
5 is a closed network. Since R is closed and $\sim S$ is closed, the lower series circuit containing them is closed.
6 is a closed network. The upper parallel circuit is closed because P is closed or because $\sim Q$ is closed.

9.

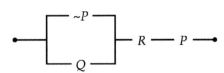

11.

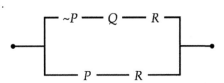

13.

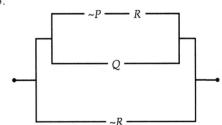

15. a. On, since P and $\sim Q$ are both closed.

    b. Off

    c. On, since $\sim P$ and Q are both closed.

    d. On, since $\sim P$ and $\sim Q$ are both closed.

1. The sentence is an opinion. It can be true for one person and false for another person. Since it can be both true and false, it is not a statement.

3. You may not know the area code for Storm Lake, Iowa; however, you do know that it is either 512 or it is not 512. The sentence is either true or it is false, and it is not both true and false, so it is a statement.

5. The sentence "Have a fun trip," is a command, so it is not a statement.

7. The sentence is either true or false, and it is not both true and false, so it is a statement.

9. The equation is either true or false, and it is not both true and false, so it is a statement.

11. One component is "The principal will attend the class on Tuesday." The other component is "The principal will attend the class on Wednesday."

13. One component is "A triangle is an acute triangle." The other component is "A triangle has three acute angles."

15. The Giants did not lose the game.

17. The game went into overtime.

19. $w \rightarrow t$; conditional

21. $l \leftrightarrow a$; biconditional

23. $d \rightarrow f$; conditional

25. $m \rightarrow c$; disjunction

27. The tour goes to Italy and the tour does not go to Spain.

29. If we go to Venice, then we will not go to Florence.

31. We will go to Florence if and only if we do not go to Venice.

33. Taylor Swift is a singer or an actress, and is not a songwriter.

35. If Taylor Swift is a singer, then she is not a songwriter and not an actress.

37. Taylor Swift is an actress and a singer if and only if she is not a songwriter.

39. $(p \vee q) \wedge \sim r$

41. $(q \wedge r) \rightarrow \sim p$

43. $s \rightarrow (q \wedge \sim p)$

45. $3 > 1$ is true, so the statement is true.

47. $(-1)^{50} = 1$ is true and $(-1)^{99} = -1$ is true, so the statement is true.

49. $-5 > -11$, so the statement is true.

51. "2 is an even number" is a true statement, so the statement is true.

53. All cats have claws.

55. Some classic movies were not first produced in black and white.

57. Some of the numbers were even numbers.

59. Some cars do not run on gasoline.

61. $p \rightarrow q$, where p represents "you can count your money" and q represents "you don't have a billion dollars."

63. $p \rightarrow q$, where p represents "people concentrated on the really important things in life" and q represents "there'd be a shortage of fishing poles."

65. $p \leftrightarrow q$, where p represents "an angle is a right angle" and q represents "its measure is 90°."

67. $p \rightarrow q$, where p represents "two sides of a triangle are equal in length" and q represents "the angles opposite those sides are congruent."

69. $p \rightarrow q$, where p represents "it is a square" and q represents "it is a rectangle."

71. 6 cups. in any teapot the tea level cannot rise above its spout opening, because any extra tea will flow out the spout. Because both spout openings are at the same height, the maximum number of cups that can hold is the same.

EXCURSION EXERCISES, SECTION 3.2

1. The network is closed if P and Q are both closed. Otherwise the network is open.

P	Q	P	$\wedge$	$(\sim P \vee Q)$
1	1	1	1	1
1	0	1	0	0
0	1	0	0	1
0	0	0	0	1

3. The network is closed if Q is closed or if both P and R are open.

P	Q	R	$(\sim P \vee Q)$	$\wedge$	$(\sim R \vee Q)$
1	1	1	1	1	1
1	1	0	1	1	1
1	0	1	0	0	0
1	0	0	0	0	1
0	1	1	1	1	1
0	1	0	1	1	1
0	0	1	1	0	0
0	0	0	1	1	1

5. The network is closed except when P is open and R is closed.

P	Q	R	$[\sim R$	$\vee$	$(Q \wedge P)]$	$\vee$	$(R \wedge P)$
1	1	1	0	1	1	1	1
1	1	0	1	1	1	1	0
1	0	1	0	0	0	1	1
1	0	0	1	1	0	1	0
0	1	1	0	0	0	0	0
0	1	0	1	1	0	1	0
0	0	1	0	0	0	0	0
0	0	0	1	1	0	1	0

7.

a.   The closure table is shown below:

P	Q	R	{P	∧	[(Q ∧ ~R)	∨	(~Q ∧ R)]}	∨	[(~P ∨ ~Q)	∧	R]
1	1	1	1	0	0	0	0	0	0	0	1
1	1	0	1	1	1	1	0	1	0	0	0
1	0	1	1	1	0	1	1	1	0	0	1
1	0	0	1	0	0	0	0	0	0	0	0
0	1	1	0	0	0	0	0	0	0	0	1
0	1	0	0	0	1	1	0	0	0	0	0
0	0	1	0	0	0	1	1	1	1	1	1
0	0	0	0	0	0	0	0	0	1	0	0

b.   The closure table is the same as the one for 7a. This means that both of the circuits will be closed under the exact same conditions and both circuits will be open under the exact same conditions. Thus, the circuit in 7b. could be used in place of the one in 7a.

P	Q	R	[(P ∨ Q)	∧	~R]	∨	(~Q ∧ R)]}
1	1	1	1	0	0	0	0
1	1	0	1	1	1	1	0
1	0	1	0	0	0	1	1
1	0	0	0	0	1	0	0
0	1	1	0	0	0	0	0
0	1	0	0	0	1	0	0
0	0	1	0	0	0	1	1
0	0	0	0	0	1	0	0

EXERCISE SET 3.2

1.   True.

p	q	r	p	∨	(~q	∨	r)
F	T	T	T	T	F	T	T
			1	5	2	4	3

3.   False.

p	q	r	(p	∧	q)	∨	(~p	∧	~q)
F	T	T	F	F	T	F	T	F	F
			1	6	2	7	3	5	4

5.   False.

p	q	r	[~	(p	∧	~q)	∨	r]	∧	(p	∧	~r)
F	T	T	T	F	F	F	T	T	F	F	F	F
			8	1	7	2	9	3	10	4	6	5

7.   False.

p	q	r	[(p	∧	~q)	∨	~r]	∧	(q	∧	r)
F	T	T	F	F	F	F	F	F	T	T	T
			1	7	2	8	3	9	4	6	5

9.   False.

p	q	r	[(p	∧	q)	∧	r]	∨	[p	∨	(q	∧	~r)]
F	T	T	F	F	T	F	T	F	F	F	T	F	F
			1	7	2	8	3	11	4	10	5	9	6

11. a.   If p is false, then p ∧ (q ∨ r) must be a false statement.

b.   For a conjunctive statement to be true, it is necessary that all components of the statement be true. Because it is given that one of the components (p) is false, p ∧ (q ∨ r) must be a false statement.

13. The truth table is given below:

p	q	~p	V	q
T	T	F	T	T
T	F	F	F	F
F	T	T	T	T
F	F	T	T	F
		1	3	2

15. The truth table is given below:

p	q	p	∧	~q
T	T	T	F	F
T	F	T	T	T
F	T	F	F	F
F	F	F	F	T
		1	3	2

17. The truth table is given below:

p	q	(p	∧	~q)	V	[~	(p	∧	q)]
T	T	T	F	F	F	F	T	T	T
T	F	T	T	T	T	T	T	F	F
F	T	F	F	F	T	T	F	F	T
F	F	F	F	T	T	T	F	F	F
		1	7	2	8	6	3	5	4

19. The truth table is given below:

p	q	r	~	(p	V	q)	∧	(~r	V	q)
T	T	T	F	T	T	T	F	F	T	T
T	T	F	F	T	T	T	F	T	T	T
T	F	T	F	T	T	F	F	F	F	F
T	F	F	F	T	T	F	F	T	T	F
F	T	T	F	F	T	T	F	F	T	T
F	T	F	F	F	T	T	F	T	T	T
F	F	T	T	F	F	F	F	F	F	F
F	F	F	T	F	F	F	T	T	T	F
			7	1	6	2	8	3	5	4

21. The truth table is given below:

p	q	r	(p	∧	~r)	V	[~q	V	(p	∧	r)]
T	T	T	T	F	F	T	F	T	T	T	T
T	T	F	T	T	T	T	F	F	T	F	F
T	F	T	T	F	F	T	T	T	T	T	T
T	F	F	T	T	T	T	T	T	T	F	F
F	T	T	F	F	F	F	F	F	F	F	T
F	T	F	F	F	T	F	F	F	F	F	F
F	F	T	F	F	F	T	T	T	F	F	T
F	F	F	F	F	T	T	T	T	F	F	F
			1	8	2	9	3	7	4	6	5

23. The truth table is given below:

p	q	r	[(p	∧	q)	V	(r	∧	~p)]	∧	(r	V	~q)
T	T	T	T	T	T	T	T	F	F	T	T	T	F
T	T	F	T	T	T	T	F	F	F	F	F	F	F
T	F	T	T	F	F	F	T	F	F	F	T	T	T
T	F	F	T	F	F	F	F	F	F	F	F	T	T
F	T	T	F	F	T	T	T	T	T	T	T	T	F
F	T	F	F	F	T	F	F	F	T	F	F	F	F
F	F	T	F	F	F	T	T	T	T	T	T	T	T
F	F	F	F	F	F	F	F	F	T	F	F	T	T
			1	7	2	9	3	8	4	11	5	10	6

25. The truth table is given below:

p	q	r	q	V	[~r	V	(p	∧	r)]
T	T	T	T	T	F	T	T	T	T
T	T	F	T	T	T	T	T	F	F
T	F	T	F	T	F	T	T	T	T
T	F	F	F	T	T	T	T	F	F
F	T	T	T	T	F	F	F	F	T
F	T	F	T	T	T	T	F	F	F
F	F	T	F	F	F	F	F	F	T
F	F	F	F	T	T	T	F	F	F
			1	7	2	6	3	5	4

27. The truth table is given below:

p	q	r	(~q	∧	r)	V	[p	∧	(q	∧	~r)]
T	T	T	F	F	T	F	T	F	T	F	F
T	T	F	F	F	F	T	T	T	T	T	T
T	F	T	T	T	T	T	T	F	F	F	F
T	F	F	T	F	F	F	T	F	F	F	T
F	T	T	F	F	T	F	F	F	T	F	F
F	T	F	F	F	F	F	F	F	T	T	T
F	F	T	T	T	T	T	F	F	F	F	F
F	F	F	T	F	F	F	F	F	F	F	T
			1	6	2	9	3	8	4	7	5

29. The truth tables are given below:

p	r	p	V	(p	∧	r)
T	T	T	T	T	T	T
T	F	T	T	T	F	F
F	T	F	F	F	F	T
F	F	F	F	F	F	F
		1	5	2	4	3

p
T
T
F
F

31. The truth tables are given below:

p	q	r	p	∧	(q	V	r)
T	T	T	T	T	T	T	T
T	T	F	T	T	T	T	F
T	F	T	T	T	F	T	T
T	F	F	T	F	F	F	F
F	T	T	F	F	T	T	T
F	T	F	F	F	T	T	F
F	F	T	F	F	F	T	T
F	F	F	F	F	F	F	F
			1	5	2	4	3

(p	∧	q)	V	(p	∧	r)
T	T	T	T	T	T	T
T	T	T	T	T	F	F
T	F	F	T	T	T	T
T	F	F	F	T	F	F
F	F	T	F	F	F	T
F	F	T	F	F	F	F
F	F	F	F	F	F	T
F	F	F	F	F	F	F
1	5	2	7	3	6	4

33. The truth tables are given below:

p	q	p	V	(q	∧	~p)
T	T	T	T	T	F	F
T	F	T	T	F	F	F
F	T	F	T	T	T	T
F	F	F	F	F	F	T
		1	5	2	4	3

p	V	q
T	T	T
T	T	F
F	T	T
F	F	F
1	3	2

35. The truth tables are given below:

p	q	r	[(p	∧	q)	∧	r]	∨	[p	∧	(q	∧	~r)]
T	T	T	T	T	T	T	T	T	T	F	T	F	F
T	T	F	T	T	T	F	F	T	T	T	T	T	T
T	F	T	T	F	F	F	T	F	T	F	F	F	F
T	F	F	T	F	F	F	F	F	T	F	F	F	T
F	T	T	F	F	T	F	T	F	F	F	T	F	F
F	T	F	F	F	T	F	F	F	F	F	T	T	T
F	F	T	F	F	F	F	T	F	F	F	F	F	F
F	F	F	F	F	F	F	F	F	F	F	F	F	T
			1	7	2	8	3	11	4	10	5	9	6

(p	∧	q)
T	T	T
T	T	T
T	F	F
T	F	F
F	F	T
F	F	T
F	F	F
F	F	F
1	3	2

37. Let *p* represent the statement "It rained." Let *q* represent the statement "It snowed." In symbolic form, the original sentence is ~(p ∨ q). One of De Morgan's laws states that this is equivalent to ~p ∧ ~q. Thus an equivalent sentence is "It did not rain and it did not snow."

39. Let *p* represent the statement "She visited France." Let *q* represent the statement "She visited Italy." In symbolic form, the original sentence is ~p ∧ ~q. One of De Morgan's laws states that this is equivalent to ~(p ∨ q). Thus an equivalent sentence is "She did not visit either France or Italy."

41. Let *p* represent the statement "She received a promotion." Let *q* represent the statement "She received a raise." In symbolic form, the original sentence is ~(p ∨ q). One of De Morgan's laws states that this is equivalent to ~p ∧ ~q. Thus an equivalent sentence is "She did not receive a promotion and she did not receive a raise."

43. The statement is always true, so it is a tautology.

p	p	∨	~p
T	T	T	F
F	F	T	T
	1	3	2

45. The statement is always true, so it is a tautology.

p	q	(p	∨	q)	∨	(~p	∨	q)
T	T	T	T	T	T	F	T	T
T	F	T	T	F	T	F	F	F
F	T	F	T	T	T	T	T	T
F	F	F	F	F	T	T	T	F
		1	5	2	7	3	6	4

47. The statement is always true, so it is a tautology.

p	q	r	(~p	∨	q)	∨	(~q	∨	r)
T	T	T	F	T	T	T	F	T	T
T	T	F	F	T	T	T	F	F	F
T	F	T	F	F	F	T	T	T	T
T	F	F	F	F	F	T	T	T	F
F	T	T	T	T	T	T	F	T	T
F	T	F	T	T	T	T	F	F	F
F	F	T	T	T	F	T	T	T	T
F	F	F	T	T	F	T	T	T	F
			1	5	2	7	3	6	4

49. The statement is always false, so it is a self-contradiction.

r	~r	∧	r
T	F	F	T
F	T	F	F
	1	3	2

51. The statement is always false, so it is a self-contradiction.

p	q	p	∧	(~p	∧	q)
T	T	T	F	F	F	T
T	F	T	F	F	F	F
F	T	F	F	T	T	T
F	F	F	F	T	F	F
		1	5	2	4	3

53. The statement is not always false, so it is not a self-contradiction.

p	q	[p	∧	(~p	∨	q)]	∨	q
T	T	T	T	F	T	T	T	T
T	F	T	F	F	F	F	F	F
F	T	F	F	T	T	T	T	T
F	F	F	F	T	T	F	F	F
		1	6	2	5	3	7	4

55. The symbol ≤ means "less than *or* equal to."

57. There are 5 simple statements, *p*, *q*, *r*, *s*, and *t*, so $2^5 = 32$ rows are needed.

59. The truth table is given below:

p	q	r	s	[(p	∧	~q)	∨	(q	∧	~r)]	∧	(r	∧	~s)
T	T	T	T	T	F	F	F	T	F	F	F	T	T	F
T	T	T	F	T	F	F	F	T	F	F	F	T	T	T
T	T	F	T	T	F	F	T	T	T	T	F	F	F	F
T	T	F	F	T	F	F	T	T	T	T	T	F	T	T
T	F	T	T	T	T	T	T	F	F	F	T	T	T	F
T	F	T	F	T	T	T	T	F	F	F	T	T	T	T
T	F	F	T	T	T	T	T	F	F	T	F	F	F	F
T	F	F	F	T	T	T	T	F	F	T	T	F	T	T
F	T	T	T	F	F	F	F	T	F	F	F	T	T	F
F	T	T	F	F	F	F	F	T	F	F	F	T	T	T
F	T	F	T	F	F	F	T	T	T	T	F	F	F	F
F	T	F	F	F	F	F	T	T	T	T	T	F	T	T
F	F	T	T	F	F	T	F	F	F	F	F	T	T	F
F	F	T	F	F	F	T	F	F	F	F	F	T	T	T
F	F	F	T	F	F	T	F	F	F	T	F	F	F	F
F	F	F	F	F	F	T	F	F	F	T	F	F	T	T
				1	7	2	9	3	8	4	11	5	10	6

61. Circle the 1 and the three 9s. Then invert the paper so that the digits are upside down and hand it back to your friend.

EXCURSION EXERCISES, SECTION 3.3

1. a. 1011

   b. 0001

   c. 10101011

EXERCISE SET 3.3

1. *Antecedent:* I had the money
   Consequent: I would buy the painting

3. *Antecedent:* They had a guard dog
   Consequent: No one would trespass on their property

5. *Antecedent:* I change my major
   Consequent: I must reapply for admission

7. Because the consequent is true when the antecedent is true, this is a true statement.

9. Because the antecedent is false, this is a true statement.

11. Because the antecedent is false, this is a true statement.

13. Consider the case $x = -6$: $|-6| = 6$ is true, but $-6 = 6$ is false. Because the consequent is false when the antecedent is true, this is a false statement.

15. The truth table is given below:

p	q	(p	∧	~q)	→	[~	(p	∧	q)
T	T	T	F	F	T	F	T	T	T
T	F	T	T	T	T	T	T	F	F
F	T	F	F	F	T	T	F	F	T
F	F	F	F	T	T	T	F	F	F
		1	5	2	8	7	3	6	4

17. The truth table is given below:

p	q	[(p	→	q)	∧	p]	→	q
T	T	T	T	T	T	T	T	T
T	F	T	F	F	F	T	T	F
F	T	F	T	T	F	F	T	T
F	F	F	T	F	F	F	T	F
		1	5	2	6	3	7	4

19. The truth table is given below:

p	q	r	[r	∧	(~p	∨	q)]	→	(r	∨	~q)
T	T	T	T	T	F	T	T	T	T	T	F
T	T	F	F	F	F	T	T	T	F	F	F
T	F	T	T	F	F	F	F	T	T	T	T
T	F	F	F	F	F	F	F	T	F	T	T
F	T	T	T	T	T	T	T	T	T	T	F
F	T	F	F	F	T	T	T	T	F	F	F
F	F	T	T	T	T	T	F	T	T	T	T
F	F	F	F	F	T	T	F	T	F	T	T
			1	7	2	6	3	9	4	8	5

21. The truth table is given below:

p	q	r	[(p	→	q)	∨	(r	∧	~p)]	→	(r	∨	~q)
T	T	T	T	T	T	T	T	F	F	T	T	T	F
T	T	F	T	T	T	T	F	F	F	F	F	F	F
T	F	T	T	F	F	F	T	F	F	T	T	T	T
T	F	F	T	F	F	F	F	F	F	T	F	T	T
F	T	T	F	T	T	T	T	T	T	T	T	T	F
F	T	F	F	T	T	T	F	F	T	F	F	F	F
F	F	T	F	T	F	T	T	T	T	T	T	T	T
F	F	F	F	T	F	T	F	F	T	T	F	T	T
			1	7	2	9	3	8	4	11	5	10	6

23. The truth table is given below:

p	q	r	[~	(p	→	~r)	∧	~q]	→	r
T	T	T	T	T	F	F	F	F	T	T
T	T	F	F	T	T	T	F	F	T	F
T	F	T	T	T	F	F	T	T	T	T
T	F	F	F	T	T	T	F	T	T	F
F	T	T	F	F	T	F	F	F	T	T
F	T	F	F	F	T	T	F	F	T	F
F	F	T	F	F	T	F	F	T	T	T
F	F	F	F	F	T	T	F	T	T	F
			6	1	5	2	7	3	8	4

25. She cannot sing or she would be perfect for the part.

27. Either x is not an irrational number or x is not a terminating decimal.

29. The fog must lift or our flight will be canceled.

31. They offered me the contract and I didn't accept.

33. Pigs have wings and they still can't fly.

35. She traveled to Italy and she didn't visit her relatives.

37. If $x = -3$, the first component is true and the second component is false. Thus this is a false statement.

39. Both components are true when $x \neq 0$ and false when $x = 0$. Both components have the same truth value for any value of x, so this is a true statement.

41. The number $\frac{2}{3}$ is a rational number. Its decimal form is $0.\overline{6}$, which does not terminate. The first component is true while the second component is false. Thus this is a false statement.

43. Both components are always false, so this statement is true.

45. Both components are true when triangle ABC is equilateral and false when triangle ABC is not equilateral. Thus this statement is true.

47. $v \leftrightarrow p$

49. $p \rightarrow v$

51. $t \rightarrow \sim v$

53. $(\sim t \vee p) \rightarrow v$

55. The statements are not equivalent.

p	r	p	→	~r
T	T	T	F	F
T	F	T	T	T
F	T	F	T	F
F	F	F	T	T
		1	3	2

r	∨	~p
T	T	F
F	F	F
T	T	T
F	T	T
1	3	2

57. The statements are not equivalent.

p	r	~p	→	(p	∨	r)
T	T	F	T	T	T	T
T	F	F	T	T	T	F
F	T	T	T	F	T	T
F	F	T	F	F	F	F
		1	5	2	4	3

r
T
F
T
F
1

59. The statements are equivalent.

p	q	r	p	→	(q	∨	r)
T	T	T	T	T	T	T	T
T	T	F	T	T	T	T	F
T	F	T	T	T	F	T	T
T	F	F	T	F	F	F	F
F	T	T	F	T	T	T	T
F	T	F	F	T	T	T	F
F	F	T	F	T	F	T	T
F	F	F	F	T	F	F	F
			1	5	2	4	3

(p	→	q)	∨	(p	→	r)
T	T	T	T	T	T	T
T	T	T	T	T	F	F
T	F	F	T	T	T	T
T	F	F	F	T	F	F
F	T	T	T	F	T	T
F	T	T	T	F	T	F
F	T	F	T	F	T	T
F	T	F	T	F	T	F
1	5	2	7	3	6	4

61. If a number is a rational number, then it is a real number.

63. If an animal is a sauropod, then it is herbivorous.

65. Turn two of the valves on. After one minute turn off one of these valves. When you get up the field the sprinklers will be running on the region that is controlled by the valve that is still in the on position. The region that is wet but not receiving any water is controlled by the valve you turned off. The region that is completely dry is the region that is controlled by the valve that you left in the off position.

EXCURSION EXERCISES, SECTION 3.4

1. a. The truth table is given below:

p	q	p			(q|q)
T	T	T	T	F
T	F	T	F	T
F	T	F	T	F
F	F	F	T	T
		1	3	2

b. $p|(q|q) \equiv p \rightarrow q$

3. a. 1000

b. AND gate

EXERCISE SET 3.4

1. If we take the aerobics class, then we will be in good shape for the ski trip.

3. If the number is an odd prime number, then it is greater than 2.

5. If he has the talent to play a keyboard, then he can join the band.

7. If I was able to prepare for the test, then I had the textbook.

9. If you ran the Boston marathon, then you are in excellent shape.

11. **a.** If I quit this job, then I am rich.

b. If I were not rich, then I would not quit this job.

c. If I would not quit this job, then I would not be rich.

13. **a.** If we are not able to attend the party, then she did not return soon.

b. If she returns soon, then we will be able to attend the party.

c. If we are able to attend the party, then she returned soon.

15. **a.** If a figure is a quadrilateral, then it is a parallelogram.

b. If a figure is not a parallelogram, then it is not a quadrilateral.

c. If a figure is not a quadrilateral, then it is not a parallelogram.

17. **a.** If I am able to get current information about astronomy, then I have access to the Internet.

b. If I do not have access to the Internet, then I will not be able to get current information about astronomy.

c. If I am not able to get current information about astronomy, then I don't have access to the Internet.

19. **a.** If we don't have enough money for dinner, then we took a taxi.

b. If we did not take a taxi, then we will have enough money for dinner.

c. If we have enough money for dinner, then we did not take a taxi.

21. a. If she can extend her vacation for at least two days, then she will visit Kauai.

    b. If she does not visit Kauai, then she could not extend her vacation for at least two days.

    c. If she cannot extend her vacation for at least two days, then she will not visit Kauai.

23. a. If two lines are parallel, then the two lines are perpendicular to a given line.

    b. If two lines are not perpendicular to a given line, then the two lines are not parallel.

    c. If two lines are not parallel, then the two lines are not both perpendicular to a given line.

25. The second statement is the converse of the first. The statements are not equivalent.

27. The second statement is the contrapositive of the first. The statements are equivalent.

29. The second statement is the inverse of the first. The statements are not equivalent.

31. If $x = 7$, then $3x - 7 \neq 11$. The original statement is true.

33. If $|a| = 3$, then $a = 3$. The original statement is false.

35. If $a + b = 25$, then $\sqrt{a+b} = 5$. The original statement is true.

37. The contrapositive of $p \to q$ is $\sim q \to \sim p$. The inverse of $\sim q \to \sim p$ is $\sim(\sim q) \to \sim(\sim p)$ or $q \to p$. The converse of $q \to p$ is $p \to q$.

For exercise 39, a sample answer is given. Your answer may vary from the given answer.
39. a. "If a figure is a triangle, then it is a 3-sided polygon." The converse, "If a figure is a 3-sided polygon, then it is a triangle" is true.

    b. "If it is an apple, then it is a fruit." The converse, "If it is a fruit, then it is an apple" is false.

41. If you can dream it, then you can do it.

43. If I were a dancer, then I would not be a singer.

45. A conditional statement and its contrapositive are equivalent. They always have the same truth value.

47. Suppose the Hatter is the only person telling the truth. Then the statements of the Dodo and the March Hare must be false. Therefore, negating the Dodo's statement gives "The Hatter does not lie," which is true. Negating the March Hare's statement gives "Either the Dodo or the Hatter tells lies," which is also true. The Hatter's statement is also true. Since all three statements are true, the Hatter is telling the truth.

EXCURSION EXERCISES, SECTION 3.5

1. Answers will vary.

3. Answers will vary.

5. The step in which each side is divided by $(a - b)$ is not valid because $a - b = 0$, and division by zero is an undefined operation.

EXERCISE SET 3.5

1. In symbolic form:
$r \to c$
$\dfrac{r}{\therefore c}$

3. In symbolic form:
$g \to s$
$\dfrac{\sim g}{\therefore \sim s}$

5. In symbolic form:
$s \to i$
$\dfrac{s}{\therefore i}$

7. In symbolic form:
$\sim p \to \sim a$
$\dfrac{a}{\therefore p}$

9.

p	q	First premise $p \vee \sim q$	Second premise $\sim q$	Conclusion p
T	T	T	F	T
T	F	T	T	T
F	T	F	F	F
F	F	T	T	F

The premises are true in rows 2 and 4. Because the conclusion in row 4 is false and the premises are both true, we know the argument is invalid.

11.

p	q	First premise $p \to {\sim}q$	Second premise ${\sim}q$	Conclusion p
T	T	F	F	T
T	F	T	T	T
F	T	T	F	F
F	F	T	T	F

The premises are true in rows 2 and 4. Because the conclusion in row 4 is false and the premises are both true, we know the argument is invalid.

13.

p	q	First premise ${\sim}p \to {\sim}q$	Second premise ${\sim}p$	Conclusion ${\sim}q$
T	T	T	F	F
T	F	T	F	T
F	T	F	T	F
F	F	T	T	T

The premises are true in row 4. Because the conclusion is true and the premises are both true, the argument is valid.

15.

p	q	First premise $(p \to q) \wedge ({\sim}p \to q)$	Second premise q	Conclusion p
T	T	T	T	T
T	F	F	F	T
F	T	T	T	F
F	F	F	F	F

The premises are true in rows 1 and 3. Because the conclusion in row 3 is false and the premises are both true, we know the argument is invalid.

17.

p	q	First premise $(p \wedge {\sim}q) \vee (p \to q)$	Second premise $q \vee p$	Conclusion ${\sim}p \wedge q$
T	T	T	T	F
T	F	T	T	F
F	T	T	T	T
F	F	T	F	F

The premises are true in rows 1, 2, and 3. Because the conclusions in rows 1 and 2 are false and the premises are true, we know the argument is invalid.

19.

p	q	r	First premise $(p \wedge {\sim}q) \vee (p \vee r)$	Second premise r	Conclusion $p \vee q$
T	T	T	T	T	T
T	T	F	T	F	T
T	F	T	T	T	T
T	F	F	T	F	T
F	T	T	T	T	T
F	T	F	F	F	T
F	F	T	T	T	F
F	F	F	F	F	F

The premises are true in rows 1, 3, 5, and 7. Because the conclusion in row 7 is false and the premises are both true, we know the argument is invalid.

21.

p	q	r	First premise $(p \leftrightarrow q)$	Second premise $p \to r$	Conclusion ${\sim}r \to {\sim}p$
T	T	T	T	T	T
T	T	F	T	F	F
T	F	T	F	T	T
T	F	F	F	F	F
F	T	T	F	T	T
F	T	F	F	T	T
F	F	T	T	T	T
F	F	F	T	T	T

The premises are true in rows 1, 7 and 8. Because the conclusions are true and the premises are true, the argument is valid.

23.

p	q	r	First premise $(p \wedge {\sim}q)$	Second premise $p \leftrightarrow r$	Conclusion $q \vee r$
T	T	T	F	T	T
T	T	F	F	F	T
T	F	T	T	T	T
T	F	F	T	F	F
F	T	T	F	F	T
F	T	F	F	T	T
F	F	T	F	F	T
F	F	F	F	T	F

The premises are true in row 3. Because the conclusion is true and the premises are true, the argument is valid.

25. In symbolic form:

$$h \to r$$
$$\underline{\sim h}$$
$$\therefore \sim r$$

h	r	First premise $h \to r$	Second premise $\sim h$	Conclusion $\sim r$
T	T	T	F	F
T	F	F	F	T
F	T	T	T	F
F	F	T	T	T

The premises are true in rows 3 and 4. Because the conclusion in row 3 is false and the premises are both true, we know the argument is invalid.

27. In symbolic form:

$$\sim b \to d$$
$$\underline{b \vee d}$$
$$\therefore b$$

b	d	First premise $\sim b \to d$	Second premise $b \vee d$	Conclusion b
T	T	T	T	T
T	F	T	T	T
F	T	T	T	F
F	F	F	F	F

The premises are true in rows 1, 2, and 3. Because the conclusion in row 3 is false and the premises are both true, we know the argument is invalid.

29. In symbolic form:

$$c \to t$$
$$\underline{t}$$
$$\therefore c$$

c	t	First premise $c \to t$	Second premise t	Conclusion c
T	T	T	T	T
T	F	F	F	T
F	T	T	T	F
F	F	T	F	F

The premises are true in rows 1 and 3. Because the conclusion in row 3 is false and the premises are both true, we know the argument is invalid.

31. Label the statements

   f: You take Art 151 in the fall.
   s: You will be eligible to take Art 152 in the spring.

In symbolic form:

$$f \to s$$
$$\underline{\sim s}$$
$$\therefore \sim f$$

The argument is valid using modus tollens.

33. Label the statements

   n: I had a nickel for every logic problem I have solved.
   r: I would be rich.

In symbolic form:

$$n \to r$$
$$\underline{\sim n}$$
$$\therefore \sim r$$

The argument is invalid using the fallacy of the inverse.

35. Label the statements

   s: We serve salmon.
   v: Vicky will join us for lunch.
   m: Marilyn will join us for lunch.

In symbolic form:

$$s \to v$$
$$\underline{v \to \sim m}$$
$$\therefore s \to \sim m$$

The argument is valid using the law of syllogism.

37. Label the statements

   c: My cat is left alone in the apartment.
   s: She claws the sofa.

In symbolic form:

$$c \to s$$
$$\underline{c}$$
$$\therefore s$$

The argument is valid using modus ponens.

39. Label the statements

   n: Rita buys a new car.
   c: Rita goes on a cruise.

In symbolic form:

$$n \to \sim c$$
$$\underline{c}$$
$$\therefore \sim n$$

The argument is valid using modus tollens.

41. Applying the law of syllogism to Premise 1 and Premise 2 produces

$\sim p \to r$	Premise 1
$\underline{r \to t}$	Premise 2
$\therefore \sim p \to t$	Law of Syllogism

The above conclusion $\sim p \to t$ can be written as $\sim t \to p$, using the contrapositive form. Combining the conclusion above with Premise 3 gives

$\sim t \to p$	Equivalent form of conclusion
$\underline{\sim t}$	Premise 3
$\therefore p$	Modus Ponens

This sequence of valid arguments has produced the desired conclusion. Thus the original argument is valid.

43. In symbolic form the argument is

$$s \rightarrow \sim r \quad \text{Premise 1}$$
$$\sim r \rightarrow c \quad \text{Premise 2}$$
$$\underline{\sim t \rightarrow \sim c} \quad \text{Premise 3}$$
$$\therefore s \rightarrow t \quad \text{Conclusion}$$

Applying the law of syllogism to Premise 1 and Premise 2 produces:

$$s \rightarrow \sim r \quad \text{Premise 1}$$
$$\underline{\sim r \rightarrow c} \quad \text{Premise 2}$$
$$\therefore s \rightarrow c \quad \text{Law of Syllogism}$$

Premise 3 can be written as $c \rightarrow t$ using the contrapositive form. Combining Premise 3 with the conclusion from above gives

$$s \rightarrow c \quad \text{Conclusion from above}$$
$$\underline{c \rightarrow t} \quad \text{Equivalent form of Premise 3}$$
$$\therefore s \rightarrow t \quad \text{Law of Syllogism}$$

This sequence of valid arguments has produced the desired conclusion. Thus the original argument is valid.

45. In symbolic form the argument is

$$\sim o \rightarrow \sim f \quad \text{Premise 1}$$
$$\underline{c \rightarrow \sim o} \quad \text{Premise 2}$$
$$\therefore f \rightarrow \sim c \quad \text{Conclusion}$$

Premise 1 can be written as $f \rightarrow o$ using the contrapositive form. Premise 2 can be written as $o \rightarrow \sim c$ using the contrapositive form.
Applying the equivalent forms and the law of syllogism to Premise 1 and Premise 2 produces

$$f \rightarrow o \quad \text{Equivalent form of Premise 1}$$
$$\underline{o \rightarrow \sim c} \quad \text{Equivalent form of Premise 2}$$
$$\therefore f \rightarrow \sim c \quad \text{Law of Syllogism}$$

This sequence of valid arguments has produced the desired conclusion. Thus the original argument is valid.

47. In symbolic form the argument is

$$\sim (p \wedge \sim q) \quad \text{Premise 1}$$
$$\underline{p} \quad \text{Premise 2}$$
$$\therefore ?$$

De Morgan's law gives $\sim p \vee q$ as an equivalent form of Premise 1. Combining this form with premise 2 creates a disjunctive syllogism:

$$\sim p \vee q \quad \text{Equivalent form of Premise 1}$$
$$\underline{p} \quad \text{Premise 2}$$
$$\therefore q \quad \text{Disjunctive Syllogism}$$

The conclusion is q.

49. Label the statements
    t: It is a theropod.
    h: It is a herbivorous.
    s: It is a sauropod.
    In symbolic form the argument is:

$$t \rightarrow \sim h \quad \text{Premise 1}$$
$$\sim h \rightarrow \sim s \quad \text{Premise 2}$$
$$\underline{s} \quad \text{Premise 3}$$
$$\therefore ?$$

Applying the law of syllogisms to Premise 1 and Premise 2 gives

$$t \rightarrow \sim h \quad \text{Premise 1}$$
$$\underline{\sim h \rightarrow \sim s} \quad \text{Premise 2}$$
$$\therefore t \rightarrow \sim s \quad \text{Law of Syllogism}$$

Write the conclusion from above in its equivalent contrapositive form, $s \rightarrow \sim t$. Use the equivalent form of the conclusion from above and Premise 3 to give

$$s \rightarrow \sim t \quad \text{Equivalent form of conclusion}$$
$$\underline{s} \quad \text{Premise 3}$$
$$\therefore \sim t \quad \text{Modus Ponens}$$

Therefore, it is not a theropod.

51. 12. Any number multiplied by 0 produces 0. Thus the only arithmetic you need to perform is the last two operations.

EXCURSION EXERCISES, SECTION 3.6

1. Assume T = 1, since the sum of two single digits plus a carry of at most 1 is 19 or less. Then assume O = 0, since only 0 + 0 = 0. Finally, assume S = 5, since no other single digit, when added to itself, sums to a number ending in 0. A check shows that the solutions work. So, T = 1, S = 5, and O = 0.

3. Assume O = 1, since the sum of two single digits plus a carry of at most 1 is 19 or less. In the third column from the left, the two 1's result in an S. Then S can be either 2 or 3. But since the right column also results in an S, and the two A's are identical digits, S must be even, so S cannot be 3. Assume S = 2. Since A cannot equal 1, since O = 1, assume A = 6.

$$\begin{array}{r} 1 \\ C\,1\,C\,6 \\ +C\,1\,L\,6 \\ \hline 1\,6\,2\,1\,2 \end{array}$$

Now assume C = 8, since the second column from the left results in a 6. Because the result of the fourth column does not carry, L = 0, since it cannot be 1 because O = 1. This leaves I = 9. A check shows that this solution works. So, C = 8, O = 1, A = 6, L = 0, S = 2, and I = 9.

EXERCISE SET 3.6

1. The argument is valid.

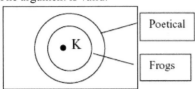

3. The argument is valid.

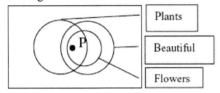

5. The argument is valid.

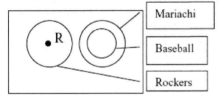

7. The argument is valid.

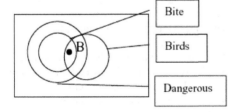

9. The argument is invalid.

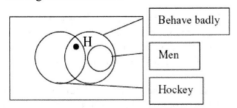

11. The argument is invalid.

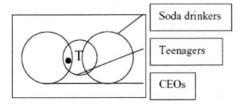

13. The argument is invalid.

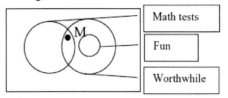

15. The argument is valid.

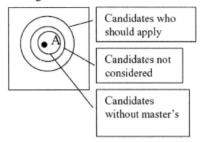

17. The argument is valid.

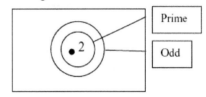

19. The argument is invalid.

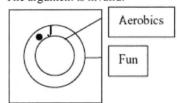

21. All Reuben sandwiches need mustard.

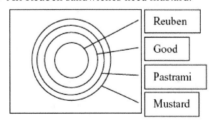

23. 1001 ends with a 5.

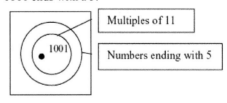

25. Some horses are grey.

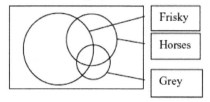

27. a. Invalid

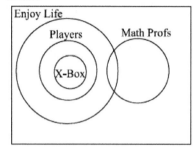

b. Invalid

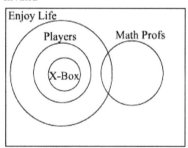

c. Invalid

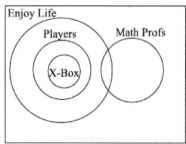

d. Invalid

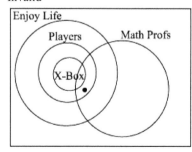

e. Valid

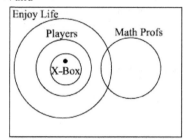

f. Valid

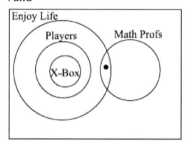

29. The only invalid statement is C. Of the seven different types of phones, each phone has either a touch screen keyboard or a push button keyboard and the majority must have a touch screen keyboard. Therefore, the push button keyboard phones cannot be more than 3.

CHAPTER 3 REVIEW EXERCISES

1. The sentence "How much is a ticket to London?" is a question, not a declarative sentence. Thus it is not a statement.

2. You may not know if 91 is a prime number or not; however, you do know that it is a whole number larger than 1, so it is either a prime number or it is not a prime number. The sentence is either true or it is false, and it is not both true and false, so it is a statement.

3. We do not know whether the given sentence is true or false, but we know that the sentence is either true or false and that it is not both true and false. Thus the sentence is a statement.

4. Since the square of any real number is either positive or zero, this is a true sentence. Thus the sentence is a statement.

5. The sentence "Lock the car" is a command. It is not a statement.

6. The sentence "Clark Kent is Superman" is either true or false, so it is a statement.

7. $m \wedge b$; conjunction

8. $d \rightarrow e$; conditional

9. $g \leftrightarrow$; biconditional

10. $t \rightarrow s$; conditional

11. No dogs bite

12. Some desserts at the Cove restaurant are not good.

13. Some winners do not receive a prize.

14. All cameras use film.

15. Some students finished the assignment.

16. Nobody enjoyed the story.

17. $5 > 2$ is true, so the statement is true.

18. $3 \neq 5$ is true and "7 is a prime number" is true, so the statement is true.

19. $4 < 7$ is true, so the statement is true.

20. $-3 < -1$ is true, so the statement is true.

21. False.

p	q	$(p$	$\wedge$	$q)$	$\vee$	$(\sim p$	$\vee$	$q)$
T	F	T	F	F	F	F	F	F
		1	5	2	7	3	6	4

22. False.

p	q	$(p$	$\rightarrow$	$\sim q)$	$\leftrightarrow$	$\sim$	$(p$	$\vee$	$q)$
T	F	T	T	T	F	F	T	T	F
		1	5	2	8	7	3	6	4

23. True.

p	q	r	$(p$	$\wedge$	$\sim q)$	$\wedge$	$(\sim r$	$\vee$	$q)$
T	F	F	T	T	T	T	T	T	F
			1	5	2	7	3	6	4

24. True.

p	q	r	$(r$	$\vee$	$\sim p)$	$\vee$	$[(p$	$\vee$	$\sim q)$	$\leftrightarrow$	$(q$	$\rightarrow$	$r)]$
T	F	F	F	F	F	T	T	T	T	T	T	F	T	F
			1	10	2	11	3	7	4	9	5	8	6

25. True.

p	q	r	$[p$	$\wedge$	$(r$	$\rightarrow$	$q)]$	$\rightarrow$	$(q$	$\vee$	$\sim r)$
T	F	F	T	T	F	T	F	T	F	T	T
			1	7	2	6	3	9	4	8	5

26. False.

p	q	r	$(\sim q$	$\wedge$	$\sim r)$	$\rightarrow$	$[(p$	$\leftrightarrow$	$\sim r)$	$\wedge$	$q]$
T	F	F	T	T	T	F	T	T	T	F	F
			1	8	2	9	3	6	4	7	5

27.

p	q	$(\sim p$	$\rightarrow$	$q)$	$\vee$	$(\sim q$	$\wedge$	$p)]$
T	T	F	T	T	T	F	F	T
T	F	F	T	F	T	T	T	T
F	T	T	T	T	T	F	F	F
F	F	T	F	F	F	T	F	F
		1	5	2	7	3	6	4

28.

p	q	$\sim p$	$\leftrightarrow$	$(q$	$\vee$	$p)$
T	T	F	F	T	T	T
T	F	F	F	F	T	T
F	T	T	T	T	T	F
F	F	T	F	F	F	F
		1	5	2	4	3

29.

p	q	~	(p	∨	~q)	∧	(q	→	p)]
T	T	F	T	T	F	F	T	T	T
T	F	F	T	T	T	F	F	T	T
F	T	T	F	F	F	F	T	F	F
F	F	F	F	T	T	F	F	T	F
		6	1	5	2	8	3	7	4

30.

p	q	(p	↔	q)	∨	(~q	∧	p)]
T	T	T	T	T	T	F	F	T
T	F	T	F	F	T	T	T	T
F	T	F	F	T	F	F	F	F
F	F	F	T	F	T	T	F	F
		1	5	2	7	3	6	4

31.

p	q	r	(r	↔	~q)	∨	(p	→	q)
T	T	T	T	F	F	T	T	T	T
T	T	F	F	T	F	T	T	T	T
T	F	T	T	T	T	T	T	F	F
T	F	F	F	F	T	F	T	F	F
F	T	T	T	F	F	T	F	T	T
F	T	F	F	T	F	T	F	T	T
F	F	T	T	T	T	T	F	T	F
F	F	F	F	F	T	T	F	T	F
			1	5	2	7	3	6	4

32.

p	q	r	(~r	∨	~q)	∧	(q	→	p)
T	T	T	F	F	F	F	T	T	T
T	T	F	T	T	F	T	T	T	T
T	F	T	F	T	T	T	F	T	T
T	F	F	T	T	T	T	F	T	T
F	T	T	F	F	F	F	T	F	F
F	T	F	T	T	F	F	T	F	F
F	F	T	F	T	T	T	F	T	F
F	F	F	T	T	T	T	F	T	F
			1	5	2	7	3	6	4

33.

p	q	r	[p	↔	(q	→	~r)]	∧	~q
T	T	T	T	F	T	F	F	F	F
T	T	F	T	T	T	T	T	F	F
T	F	T	T	T	F	T	F	T	T
T	F	F	T	T	F	T	T	T	T
F	T	T	F	T	T	F	F	F	F
F	T	F	F	F	T	T	T	F	F
F	F	T	F	F	F	T	F	F	T
F	F	F	F	F	F	T	T	F	T
			1	6	2	5	3	7	4

34.

p	q	r	~	(p	∧	q)	→	(~q	∨	~r)
T	T	T	F	T	T	T	T	F	F	F
T	T	F	F	T	T	T	T	F	T	T
T	F	T	T	T	F	F	T	T	T	F
T	F	F	T	T	F	F	T	T	T	T
F	T	T	T	F	F	T	F	F	F	F
F	T	F	T	F	F	T	T	F	T	T
F	F	T	T	F	F	F	T	T	T	F
F	F	F	T	F	F	F	T	T	T	T
			6	1	5	2	8	3	7	4

35. Let *p* represent the statement "Bob failed the English proficiency test." Let *q* represent the statement "He registered for a speech course." In symbolic form, the original sentence is ~(p ∧ q). One of De Morgan's laws states that this is equivalent to ~p ∨ ~q. Thus an equivalent sentence is "Bob passed the English proficiency test or he did not register for a speech course."

36. Let *p* represent the statement "Ellen went to work this morning." Let *q* represent the statement "She took her medication." In symbolic form, the original sentence is ~p ∧ ~q. One of De Morgan's laws states that this is equivalent to ~(p ∨ q). Thus an equivalent sentence is "It is not true that Ellen went to work this morning or she took her medication."

37. Let *p* represent the statement "Wendy will not go to the store this afternoon." Let *q* represent the statement "She will be able to prepare her fettuccine al pesto recipe." In symbolic form, the original sentence is ~p ∨ ~q. One of De Morgan's laws states that this is equivalent to ~(p ∧ q). Thus an equivalent sentence is "It is not the case that Wendy will not go to the store this afternoon and she will be able to prepare her fettuccine al pesto recipe."

38. Let *p* represent the statement "Gina did not enjoy the movie." Let *q* represent the statement "She did enjoy the party." In symbolic form, the original sentence is ~p ∧ ~q. One of De Morgan's laws states that this is equivalent to ~(p ∨ q). Thus an equivalent sentence is "It is not the case that Gina did not enjoy the movie or she enjoyed the party."

39.

p	q	~p	→	~q
T	T	F	T	F
T	F	F	T	T
F	T	T	F	F
F	F	T	T	T
		1	3	2

p	∨	~q
T	T	F
T	T	T
F	F	F
F	T	T
1	3	2

40.

p	q	~p	∨	q
T	T	F	T	T
T	F	F	F	F
F	T	T	T	T
F	F	T	T	F
		1	3	2

~	(p	∧	~q)
T	T	F	F
F	T	T	T
T	F	F	F
T	F	F	T
4	1	3	2

41.

p	q	p	∨	(q	∧	~p)
T	T	T	T	T	F	F
T	F	T	T	F	F	F
F	T	F	T	T	T	T
F	F	F	F	F	F	T
		1	7	2	6	3

p	q	p	∨	q
T	T	T	T	T
T	F	T	T	F
F	T	F	T	T
F	F	F	F	F
		4	8	5

42.

p	q	p	↔	q
T	T	T	T	T
T	F	T	F	F
F	T	F	F	T
F	F	F	T	F
		1	7	2

p	q	(p	∧	q)	∨	(~p	∧	~q)
T	T	T	T	T	T	F	F	F
T	F	T	F	F	F	F	F	T
F	T	F	F	T	F	T	F	F
F	F	F	F	F	T	T	T	T
		3	8	4	10	5	9	6

43. The statement is always false, so it is a self-contradiction.

p	q	p	∧	(q	∧	~p)
T	T	T	F	T	F	F
T	F	T	F	F	F	F
F	T	F	F	T	T	T
F	F	F	F	F	F	T
		1	5	2	4	3

44. The statement is always true, so it is a tautology.

p	q	(p	∧	q)	∨	(p	→	~q)
T	T	T	T	T	T	T	F	F
T	F	T	F	F	T	T	T	T
F	T	F	F	T	T	F	T	F
F	F	F	F	F	T	F	T	T
		1	5	2	7	3	6	4

45. The statement is always true, so it is a tautology.

p	q	[~	(p	→	q)]	↔	(p	∧	~q)
T	T	F	T	T	T	T	T	F	F
T	F	T	T	F	F	T	T	T	T
F	T	F	F	T	T	T	F	F	F
F	F	F	F	T	F	T	F	F	T
		6	1	5	2	8	3	7	4

46. The statement is always true, so it is a tautology.

p	q	p	∨	(p	→	q)
T	T	T	T	T	T	T
T	F	T	T	T	F	F
F	T	F	T	F	T	T
F	F	F	T	F	T	F
		1	5	2	4	3

47. *Antecedent:* He has talent
Consequent: He will succeed

48. *Antecedent:* I had a credential
Consequent: I could get the job

49. *Antecedent:* I join the fitness club
Consequent: I will follow the exercise program

50. *Antecedent:* I will attend
Consequent: It is free

51. She is not tall or she would be on the volleyball team.

52. He cannot stay awake or he would finish the report.

53. Rob is ill or he would start.

54. Sharon will not be promoted or she closes the deal.

55. I get my paycheck and I do not purchase a ticket.

56. The tomatoes will get big and you did not provide plenty of water.

57. You entered Cleggmore University and you did not have a high score on the SAT exam.

58. Ryan enrolled at a university and he did not enroll at Yale.

59. When $x = 3$ and $y = -3$, the second statement is true and the first statement is false. Thus this is a false statement.

60. Both components are true when $x > y$ and false when $x \leq y$. Both components have the same truth value for any values of x and y, so this is a true statement.

61. When $x = -1$, the antecedent is true and the consequent is false. So this is a false statement.

62. When $x = 2$ and $y = -2$, the antecedent is true and the consequent is false. So this is a false statement.

63. If a real number has a nonrepeating, nonterminating decimal form, then the real number is irrational.

64. If you are a politician, then you are well known.

65. If I can sell my condominium, then I can buy the house.

66. If a number is divisible by 9, then the number is divisible by 3.

67. a.   *Converse:* If $x > 3$, then $x + 4 > 7$.

    b.   *Inverse:* If $x + 4 \leq 7$, then $x \leq 3$.

    c.   *Contrapositive:* If $x \leq 3$, then $x + 4 \leq 7$.

68. a.   *Converse:* If the recipe can be prepared in less than 20 minutes, then the recipe is in this book.

    b.   *Inverse:* If the recipe is not in this book, then the recipe cannot be prepared in less than 20 minutes.

    c.   *Contrapositive:* If the recipe cannot be prepared in less than 20 minutes, then the recipe is not in this book.

69. a.   *Converse:* If $(a + b)$ is divisible by 3, then a and b are both divisible by 3.

    b.   *Inverse:* If a and b are not both divisible by 3, then $(a + b)$ is not divisible by 3.

    c.   *Contrapositive:* If $(a + b)$ is not divisible by 3, then a and b are not both divisible by 3.

70. a.   *Converse:* If they come, then you built it.

    b.   *Inverse:* If you do not build it, then they will not come.

    c.   *Contrapositive:* If they do not come, then you did not build it.

71. a. *Converse:* If it has exactly two parallel sides, then it is a trapezoid.

    b. *Inverse:* If it is not a trapezoid, then it does not have exactly two parallel sides.

    c. *Contrapositive:* If it does not have exactly two parallel sides, then it is not a trapezoid.

72. a. *Converse:* If they returned, then they liked it.

    b. *Inverse:* If they do not like it, then they will not return.

    c. *Contrapositive:* If they do not return, then they did not like it.

73. The contrapositive of $p \rightarrow q$ is $\sim q \rightarrow \sim p$. The inverse of $\sim q \rightarrow \sim p$ is $q \rightarrow p$, the converse of the original statement.

74. True. If yesterday was Sunday, then today is Monday.

75. The original statement is "If x is an odd prime number, then $x > 2$."

76. The negation of $p \wedge \sim q$ is $\sim p \vee q$, which is equivalent to $p \rightarrow q$. The original statement is "If the senator attends the meeting, then she will vote on the motion."

77. The original statement is "If their manager contacts me, then I will purchase some of their products."

78. The original statement is "If I can rollerblade, then Ginny can rollerblade."

79.

p	q	First premise $(p \wedge \sim q)$ $\wedge$ $(\sim p \rightarrow q)$	Second premise p	Conclusion $\sim q$
T	T	F	T	F
T	F	T	T	T
F	T	F	F	F
F	F	F	F	T

The premises are true in row 2. Because the conclusion is true and the premises are both true, the argument is valid.

80.

p	q	First premise $p \rightarrow \sim q$	Second premise q	Conclusion $\sim p$
T	T	F	T	F
T	F	T	F	F
F	T	T	T	T
F	F	T	F	T

The premises are true in row 3. Because the conclusion is true and the premises are both true, the argument is valid.

81.

p	q	r	First premise r	Second premise $p \rightarrow \sim r$	Third premise $\sim p \rightarrow q$	Conclusion $p \wedge q$
T	T	T	T	F	T	T
T	T	F	F	T	T	T
T	F	T	T	F	T	F
T	F	F	F	T	T	F
F	T	T	T	T	T	F
F	T	F	F	T	T	F
F	F	T	T	T	F	F
F	F	F	F	T	F	F

The premises are true in row 5. Because the conclusion is false and the premises are true, we know the argument is invalid.

82.

p	q	r	First premise $(p \vee \sim r) \rightarrow (q \wedge r)$	Second premise $r \wedge p$	Conclusion $p \vee q$
T	T	T	T	T	T
T	T	F	F	F	T
T	F	T	F	T	T
T	F	F	F	F	T
F	T	T	T	F	T
F	T	F	F	F	T
F	F	T	T	F	F
F	F	F	F	F	F

The premises are true in row 1. Because each conclusion is true and the premises are true, the argument is valid.

83. Label the statements.

        f: We will serve fish for lunch.

        c: We will serve chicken for lunch.

    In symbolic form the argument is:

    $f \vee c$

    $\dfrac{\sim f}{\therefore c}$

    The argument is valid using the disjunctive syllogism.

84. Label the statements.

        c: Mike is a CEO.

        d: He can afford to make a donation.

        s: He loves to ski.

    In symbolic form the argument is:

    $c \rightarrow d$

    $\dfrac{d \rightarrow s}{\therefore \sim s \rightarrow \sim c}$

    The argument is valid using the law of syllogism, since $\sim s \rightarrow \sim c$ is equivalent to $c \rightarrow s$.

85. Label the statements.

        w: We wish to win the lottery.

        b: We must buy a lottery ticket.

    In symbolic form the argument is:

    $w \rightarrow b$

    $\dfrac{\sim w}{\therefore \sim b}$

    The argument is invalid using the fallacy of the inverse.

86. Label the statements.

        m: Robert can charge it on his MasterCard.

        v: Robert can charge it on his Visa.

    In symbolic form the argument is:

    $m \vee v$

    $\dfrac{\sim m}{\therefore v}$

    The argument is valid using the disjunctive syllogism.

87. Label the statements.

        c: We are going to have a Caesar salad.

        e: We buy some eggs.

    In symbolic form the argument is:

    $c \rightarrow e$

    $\dfrac{\sim e}{\therefore \sim c}$

    The argument is valid using modus tollens.

88. Label the statements

        l: We serve lasagna.

        d: Eva will not come to our dinner party.

    In symbolic form the argument is:

$l \rightarrow d$

$\dfrac{\sim l}{\therefore \sim d}$

The argument is invalid using the fallacy of the inverse.

89. The argument is valid.

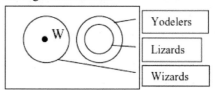

90. The argument is invalid.

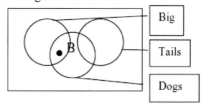

91. The argument is invalid.

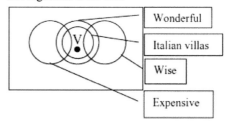

92. The argument is valid.

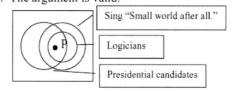

CHAPTER 3 TEST

1.   a.   The sentence "Look for the cat" is a command. It is not a statement.

      b.   The sentence "Clark Kent is afraid of the dark" is either true or false, so the sentence is a statement.

2.   a.   All trees are green.

      b.   Some apartments are available.

3.   a.   Since 5 is not less than 4, and 5 is not equal to 4, the statement is false.

      b.   Since $-2 = -2$, the statement is true.

4. a. False.

p	q	r	(p	∨	~q)	∧	(~r	∧	q)
T	F	T	T	T	T	F	F	F	F
			1	5	2	7	3	6	4

   b. True.

p	q	r	(r	∨	~p)	∨	[(p	∨	~q)	↔	(q	→	r)]
T	F	T	T	T	F	T	T	T	T	T	F	T	T
			1	7	2	11	3	8	4	10	5	9	6

5.

p	q	~	(p	∧	~q)	∧	(q	→	p)
T	T	T	T	F	F	T	T	T	T
T	F	F	T	T	T	T	F	T	T
F	T	T	F	F	F	T	T	F	F
F	F	T	F	F	T	T	F	T	F
		6	1	5	2	8	3	7	4

6.

p	q	r	(r	↔	~q)	∧	(p	→	q)
T	T	T	T	F	F	F	T	T	T
T	T	F	F	T	F	T	T	T	T
T	F	T	T	T	T	F	T	F	F
T	F	F	F	F	T	F	T	F	F
F	T	T	T	F	F	F	F	T	T
F	T	F	F	T	F	T	F	T	T
F	F	T	T	T	T	T	F	T	F
F	F	F	F	F	T	F	F	T	F
			1	5	2	7	3	6	4

7. Let *b* represent the statement "Elle ate breakfast." Let *l* represent the statement "Elle took a lunch break." In symbolic form, the original sentence is ~*b* ∧ ~*l*. An equivalent form is ~(*b* ∨ *l*). Thus an equivalent statement is "It is not the case that Elle ate breakfast or took a lunch break."

8. A tautology is a statement that is always true.

9. ~*p* ∨ *q*

10. a. The statement is of the form *"q, if p,"* where *p* is $|x| = |y|$ and *q* is $x = y$. This statement is false when *p* is true and *q* is false. It is true in all other cases. When $x = 1$ and $y = -1$, then $|x| = |y|$ is true but $x = y$ is false. So the statement is false.

    b. When $x = 2$, $y = 1$, and $z = -1$, the statement $x > y$ is true, but $xz > yz$ is false. So, *p* is true and *q* is false, and the statement is false.

11. a. *Converse:* If $x > 4$, then $x + 7 > 11$.

    b. *Inverse:* If $x + 7 \leq 11$, then $x \leq 4$.

    c. *Contrapositive:* If $x \leq 4$, then $x + 7 \leq 11$.

12. The symbolic form is:

$$p \to q$$
$$\underline{p \qquad}$$
$$\therefore q$$

13. The symbolic form is:

$$p \to q$$
$$\underline{q \to r}$$
$$\therefore p \to r$$

14. The symbolic form is:

$$p \to q$$
$$\underline{\sim q \qquad}$$
$$\therefore \sim p$$

15. The symbolic form is:

$$p \to q$$
$$\underline{\sim p \qquad}$$
$$\therefore \sim q$$

16.

p	q	First premise $(p \wedge \sim q) \wedge (\sim p \rightarrow q)$	Second premise p	Conclusion $\sim q$
T	T	F	T	F
T	F	T	T	T
F	T	F	F	F
F	F	F	F	T

The premises are true in row 2. Because the conclusion is true and the premises are both true, the argument is valid.

17.

p	q	r	First premise r	Second premise $p \rightarrow \sim r$	Third premise $\sim p \rightarrow q$	Conclusion $p \wedge q$
T	T	T	T	F	T	T
T	T	F	F	T	T	T
T	F	T	T	F	T	F
T	F	F	F	T	T	F
F	T	T	T	T	T	F
F	T	F	F	T	T	F
F	F	T	T	T	F	F
F	F	F	F	T	F	F

The premises are true in row 5. Because the conclusion is false and the premises are true, we know the argument is invalid.

18. Label the statements.
        t: We wish to win the talent contest.
        p: We must practice.
In symbolic form the argument is:

$t \rightarrow p$

$\dfrac{\sim t}{\therefore \sim p}$

The argument is invalid using the fallacy of the inverse.

19. Label the statements.
        a: Gina will take a job in Atlanta.
        k: Gina will take a job in Kansas City.
In symbolic form the argument is:

$a \vee k$

$\dfrac{\sim a}{\therefore k}$

The argument is valid using the disjunctive syllogism.

20. The argument is invalid.

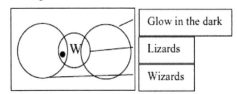

21. The argument is invalid.

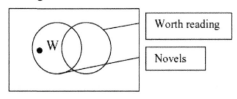

22. Label the statements.
        c: I cut my night class.
        p: I will go to the party.
In symbolic form the argument is:

$c \rightarrow p$

$\dfrac{p}{\therefore c}$

The argument is invalid using the fallacy of the converse.

Chapter 4: Apportionment and Voting

EXCURSION EXERCISES, SECTION 4.1

1. Answers may vary.

3. If each state has one representative, the denominator of the Huntington-Hill number $\dfrac{(P_A)^2}{a(a+1)}$ is 2 for all the states. Therefore, the Huntington-Hill number is simply half of the square of the population for each state, and the state with the highest Huntington-Hill number, which is the state that would receive the next representative, is the most populous state, Virginia.

EXERCISE SET 4.1

1. To calculate the standard divisor, divide the total population p by the number of items to apportion n.

3. The standard quota for a state is the whole number part of the quotient of the state's population divided by the standard divisor.

5. a. $\dfrac{1547-1215}{1215}=0.273$

   b. $\dfrac{1498-1195}{1195}=0.254$

   c. Salinas

7.

	Summer Hill Average	Seaside Average	Abs. Unfair
Summer Hill receives assoc.	$\dfrac{5289}{587+1}$ $=8.99$	$\dfrac{6215}{614}$ $=10.12$	1.13
Seaside receives assoc.	$\dfrac{5289}{587}$ $=9.01$	$\dfrac{6215}{614+1}$ $=10.11$	1.10

Relative unfairness if Summer Hill receives the new associate: $\dfrac{1.13}{8.99}=0.126$

Relative unfairness if Seaside receives the new associate: $\dfrac{1.10}{10.11}=0.109$

Using the apportionment principle, Seaside Mall should get the new associate.

9. a. $\dfrac{p}{n}=\dfrac{308,745,538}{435}\approx 709,760$

   There is one representative for every 709,760 citizens in the U.S.

   b. Underrepresented; the average constituency is greater than the standard divisor.

   c. Over-represented; the average constituency is less than the standard divisor.

11. a. $\dfrac{p}{n}=\dfrac{1781}{48}\approx 37.10$

   There is one new nurse for every 37.10 beds.

   b.

Hospital	Quot.	Q	SQ	beds
Sharp	$\dfrac{242}{37.1}$	6.52	6	7
Palomar	$\dfrac{356}{37.1}$	9.60	9	10
Tri-City	$\dfrac{308}{37.1}$	8.30	8	8
Del Raye	$\dfrac{190}{37.1}$	5.12	5	5
R. Verde	$\dfrac{275}{37.1}$	7.41	7	7
Bel Aire	$\dfrac{410}{37.1}$	11.05	11	11

   c. The modified standard divisor is 34.5.

Hospital	Quot.	Q	SQ	beds
Sharp	$\dfrac{242}{34.5}$	7.01	7	7
Palomar	$\dfrac{356}{34.5}$	10.32	10	10
Tri-City	$\dfrac{308}{34.5}$	8.93	8	8
Del Raye	$\dfrac{190}{34.5}$	5.51	5	5
R. Verde	$\dfrac{275}{34.5}$	7.97	7	7
Bel Aire	$\dfrac{410}{34.5}$	11.88	11	11

   d. They are identical.

13. The population paradox occurs when the population of one state is increasing faster than that of another state, yet the first state still loses a representative.

15. The Balinski-Young Impossibility Theorem states that any apportionment method will either violate the quota rule or will produce paradoxes such as the Alabama paradox.

17. a.   The standard divisor is 6.57.

Resort	A	B	C	D
Guest rooms	23	256	182	301
@ 115	4	39	27	45
@ 116	3	39	28	46

Yes. Resort A will lose a TV.

b.   The standard divisor is 6.51.

@ 117	4	39	28	46

No. No resort loses a TV.

c.   The standard divisor is 6.46.

@ 118	3	40	28	47

Yes. Resort A will lose a TV.

19. a.   The standard divisor is 61.9.

Office	Boston	Chicago
Employees	151	1210
Vice presidents	2	20

b.   The standard divisor is 62.3.

Office	Boston	Chicago	SF
Employees	151	1210	135
Quota	2.424	19.422	2.167
VPs	3	19	2

Yes. Chicago lost a vice president and Boston gained one.

21. a.

	avg. constit. in 5th grade	avg. constit. in 6th grade
Add to 5th grade	$\dfrac{604}{19+1}=30.2$	$\dfrac{698}{21}\approx 33.2$
Add to 6th grade	$\dfrac{604}{19}\approx 31.8$	$\dfrac{698}{21+1}\approx 31.7$

The absolute unfairness of the apportionment if the 5th grade gets the new teacher is 33.2–30.2; the absolute unfairness of the apportionment if the sixth grade gets the new teacher is 31.8 – 31.7.
The relative unfairness of the apportionment for each grade equals the absolute unfairness of the apportionment for the grade divided by the average constituency if the grade gets the new teacher. If the fifth grade gets the new teacher, the relative unfairness of the apportionment is
(33.2–30.2)/30.2 ≈ 0.0993.
If the sixth grade gets the new teacher, the relative unfairness of the apportionment is
(31.8 – 31.7)/31.7 ≈ 0.0032.
Because the smaller relative unfairness results from adding the teacher to the sixth

grade, sixth grade should get the teacher.

b.   The Huntington-Hill number for the fifth grade is

$$\frac{(S_5)^2}{t(t+1)}=\frac{(604)^2}{(19)(20)}\approx 960.$$

The Huntington-Hill number for the sixth grade is

$$\frac{(698)^2}{(21)(22)}\approx 1054.6.$$

The grade with the greater Huntington-Hill number, which is grade 6, receives the new teacher. This is the same result as using the relative unfairness of the apportionment.

23.

School	Computers	Students	H-H number
Rose	26	625	556
Lincoln	22	532	559
Midway	26	620	548
Valley	31	754	573

Because Valley School has the greatest Huntington-Hill number, it should be assigned the new computer.

25. New York's population is much larger than Louisiana's, so New York has many more representatives, and a percentage change in its population adds or subtracts more representatives than for smaller states.

27.

a.   The modified standard divisor is 88.

Div.	Quot.	Q	SQ	PCs
Lib. Arts	3455/88	39.26	39	39
Bus.	5780/88	65.68	66	66
Hum.	1896/88	21.55	22	22
Sci.	4678/88	53.16	53	53

This is the same result as the Hamilton method.

b.   Using the Jefferson method, the humanities division gets one less computer and the sciences division gets one more computer, compared with using the Webster method.

29. The Jefferson and Webster methods

31. The Huntington-Hill method

33. To use the Adams method, calculate each state's standard quota using the standard divisor and always round up to the next highest whole number. Modify the standard divisor until the sum of the states' calculated quotas equals the number of representatives to be apportioned. The Adams method violates the quota rule but is not susceptible to any paradox.

35. a. $\dfrac{P_A}{a+1}$

   b. $\dfrac{P_B}{b}$

   c. Using the formula:
   $$\dfrac{\dfrac{P_B}{b} - \dfrac{P_A}{a+1}}{\dfrac{P_A}{a+1}}$$

   d. Using the formula:
   $$\dfrac{\dfrac{P_A}{a} - \dfrac{P_B}{b+1}}{\dfrac{P_B}{b+1}}$$

   e. Comparing:
   $$\dfrac{\dfrac{P_A}{a} - \dfrac{P_B}{b+1}}{\dfrac{P_B}{b+1}} < \dfrac{\dfrac{P_B}{b} - \dfrac{P_A}{a+1}}{\dfrac{P_A}{a+1}}$$

   f. Simplify the complex fractions and write them in standard form:
   $$\dfrac{P_B(a+1)}{P_A b} - 1 < \dfrac{P_A(b+1)}{P_B a} - 1$$
   Add 1 to both sides.
   $$\dfrac{P_B(a+1)}{P_A b} < \dfrac{P_A(b+1)}{P_B a}$$
   Multiply both sides of the inequality by $P_A b P_B a$, the product of the denominators, and cancel common factors.
   $$(P_B)^2 a(a+1) < (P_A)^2 b(b+1)$$
   Regroup.
   $$\dfrac{(P_B)^2}{b(b+1)} < \dfrac{(P_A)^2}{a(a+1)}$$

EXCURSION EXERCISES, SECTION 4.2

1. First: lemon; second: raspberry; third: plain

3. No

5. Yes

7. No

EXERCISE SET 4.2

1. A majority means that a choice receives more than 50% of the votes. A plurality means that the choice with the most votes wins. It is possible to have a plurality without a majority if there are more than two choices.

3. If there are n choices in an election, each voter ranks the choices by giving n points to the voter's first choice, $n-1$ points to the voter's second choice, and so on, with the voter's least favorite choice receiving 1 point. The choice with the most points is the winner.

5. In the pairwise comparison voting method, each choice is compared one-on-one with each of the other choices. A choice receives 1 point for a win, 0.5 points for a tie, and 0 points for a loss. The choice with the greatest number of points is the winner.

7. No; no. By Arrow's Impossibility Theorem, no possible voting system involving three or more choices satisfies the fairness criteria.

9. a. Al Gore

   b. No, since 50,999,897 is less than 50% of the total votes cast: 50,456,002 + 50,999,897 + 2,882,955 = 104,338,854

   c. George Bush

11. a. 35

    b. 18

    c. Scooby Doo-13 votes

13. Stream Online: $8 \cdot 3 + (13 + 15) \cdot 2 + (7+7) = 94$
    Theater: $(15 + 7) \cdot 3 + (8 + 7) \cdot 2 + 13 = 109$
    DVD: $(13 + 7) \cdot 3 + 7 \cdot 2 + (15+8) = 97$
    This group of consumers prefers to go to the theater.

15. Mickey Mouse: $(10 \cdot 3) + (11 \cdot 2) + 14 = 66$
    Bugs Bunny: $(12 \cdot 3) + (14 \cdot 2) + 9 = 73$
    Scooby Doo: $(13 \cdot 3) + (10 \cdot 2) + 12 = 71$
    The students prefer Bugs Bunny.

17. Lee: $(41 \cdot 4) + (81 \cdot 3) + (80 \cdot 2) + 31 = 598$
    Brewer: $(84 \cdot 4) + 0 + (86 \cdot 2) + 63 = 571$
    Garcia: $(36 \cdot 4) + (121 \cdot 3) + (31 \cdot 2) + 45 = 614$
    Turley: $(72 \cdot 4) + (31 \cdot 3) + (36 \cdot 2) + 94 = 547$
    Elaine Garcia should be class president.

19. Red and white have the least first place votes (none) and is eliminated. The preference schedule is now:

green/yellow	3	1	3	1
red/blue	2	3	2	3
blue/white	1	2	1	2
votes	4	2	5	4

Red and blue now has no first-place votes, and is eliminated. The preference schedule is now:

green/yellow	2	1	2	1
blue/white	1	2	1	2
votes	4	2	5	4

Blue and white has 9 first-place votes, and green and yellow has 6. The team chooses blue and white uniforms.

21. Garcia has the least first-place votes (36) and is eliminated. The preference schedule is now:

Lee	1	2	1	2	3	2
Brewer	3	1	2	3	1	3
Turley	2	3	3	1	2	1
votes	36	53	41	27	31	45

Turley now has the least first-place votes (72) and is eliminated. The preference schedule becomes:

Lee	1	2	1	1	2	1
Brewer	2	1	2	2	1	2
votes	36	53	41	27	31	45

Lee has 149 first-place votes; Brewer has 84 first-place votes. Raymond Lee should be class president.

23. a. "Buy new computers for the club," which has 19 first-place votes.

b. Neither "throw an end-of-year party" nor "donate to charity" received any first-place votes, and are eliminated first. Of the remaining options, "establish a scholarship" received the fewest first-place votes (8) and is also eliminated. The final preference schedule is now:

convention	1	1	2	1	2
computers	2	2	1	2	1
votes	8	5	12	9	7

"Travel to a convention" has 22 first-place votes; "Buy new computers" has 19 first-place votes. The money should be spent to pay for several members to travel to a convention.

c. Scholarship:
$(8 \cdot 5) + (5 \cdot 4) + (21 \cdot 3) + (7 \cdot 2) + 0 = 137$
Convention:
$(14 \cdot 5) + (20 \cdot 4) + 0 + 0 + 7 = 157$
Computers:
$(19 \cdot 5) + 0 + (13 \cdot 3) + (9 \cdot 2) + 0 = 152$
Party: $0 + (16 \cdot 4) + 0 + (8 \cdot 2) + 17 = 97$
Charity: $0 + 0 + (7 \cdot 3) + (17 \cdot 2) + 17 = 72$
The money should be spent to pay for several members to travel to a convention.

d. Answers will vary.

25.

versus	X-Men	X2: X-Men United	X-Men: Days of Future Past	X-Men: First Class
X-Men	--	X-Men	X-Men: Days of Future Past	X-Men
X2: X-Men United	--	--	X-Men: Days of Future Past	X-Men: First Class
X-Men: Days of Future Past	--	--	--	X-Men: Days of Future Past
X-Men: First Class	--	--	--	--

Total votes = 2532

X-Men preferred over *X2: X-Men United* on 429 + 1137 + 384 = 1950 ballots, and *X2: X-Men United* preferred over *X-Men* on 2532 − 1950 = 582 ballots. *X-Men* win the match.

X-Men preferred on 813 ballots over *X-Men: Days of Future Past*, so *X-Men: Days of Future Past* preferred on 1719 ballots, so *X-Men: Days of Future Past* wins the matchup.

X-Men preferred on all ballots over the *X-Men: First Class*.

X2: X-Men United preferred over *X-Men: Days of Future Past* on 966 ballots, so *X-Men: Days of Future Past* preferred over *X2: X-Men United* on 1566 ballots, and is the winner of that matchup.

X2: X-Men United also preferred over *X-Men: First Class* on 966 ballots, so *X-Men: First Class* wins that matchup.

Finally, *X-Men: Days of Future Past* is favored on all ballots over the *X-Men: First Class*.

X-Men: Days of Future Past has three wins; *X-Men* have two, and *X-Men: First Class* has one. The online forum prefers *X-Men: Days of Future Past*.

27.

versus	Bulldog	Panther	Hornet	Bobcat
Bulldog	--	Panther	Hornet	Bobcat
Panther	--	--	Hornet	Bobcat
Hornet	--	--	--	Hornet
Bobcat	--	--	--	--

Total votes = 3150
Bulldog has 390 ballots versus 2760 for Panther. Bulldog has 1028 ballots versus 2122 for Hornet. Bulldog has 390 ballots versus 2760 for Bobcat. Panther has 1562 ballots versus 1588 for Hornet. Panther has 1449 ballots versus 1701 for Bobcat. Hornet has 1839 ballots versus 1311 for Bobcat. Hornet has three wins, so the Hornet mascot should be chosen.

29.

versus	red/white	green/yellow	red/blue	blue/white
red/white	--	red/white	red/white	blue/white
green/yellow	--	--	red/blue	blue/white
red/blue	--	--	--	blue/white
blue/white	--	--	--	--

Total votes = 15
Red/white has 9 ballots versus 6 for green/yellow; red/white has 10 ballots versus 5 for red/blue; red/white has 4 ballots versus 11 for blue/white; green/yellow has 6 ballots versus 9 for red/blue; green/yellow has 6 ballots versus 9 for blue/white; red/blue has 0 ballots versus 15 for blue/white.
Blue and white wins three matchups, red and white wins two matchups, and red and blue wins one matchup. The players prefer blue and white uniforms.

31.

versus	Mickey	Bugs	Scooby
Mickey	--	Bugs	Scooby
Bugs	--	--	Bugs
Scooby	--	--	--

No. Using plurality voting, Scooby Doo won the election, but Bugs Bunny wins both head-to-head contests.

33.

versus	Scholarship	Convention	Computer	Party	Charity
Scholarship	--	convention	scholarship	scholarship	scholarship
Convention	--	--	convention	convention	convention
Computers	--	--	--	computer	computer
Party	--	--	--	--	party
Charity	--	--	--	--	--

Yes. Using the Borda Count method, the members voted to pay to travel to a convention. Traveling to a convention also won all of its head-to-head contests, so the Condorcet criterion is satisfied.

35. No. Using the Borda Count method, Elaine Garcia won the election, although no candidate received a majority of the first-place votes.

37. a. Lorenz: $(2691 \cdot 3) + 0 + 2653 = 10{,}726$
Beasley: $(2416 \cdot 3) + (237 \cdot 2) + 2691 = 10{,}413$
Hyde: $(237 \cdot 3) + (5107 \cdot 2) + 0 = 10{,}925$
Stephen Hyde wins the election.

b. Stephen Hyde received the fewest number of first-place votes.

c. John Lorenz beats Marcia Beasley by $2691 - 2653$, and beats Stephen Hyde by the same vote, so Lorenz wins all head-to-head comparisons.

d. The candidate that wins all the head-to-head matches does not win the election.

e. Lorenz gets $(2691 \cdot 2) + 2653 = 8035$ points, while Hyde gets $(2653 \cdot 2) + 2691 = 7997$ points.

f. The candidate winning the original election (Stephen Hyde) did not remain the winner in a recount in which a losing candidate withdrew from the race.

39. Cynthia:
$0 + (16 \cdot 4) + (27 \cdot 3) + (22 \cdot 2) + 16 = 205$
Andrew:
$(16 \cdot 5) + (22 \cdot 4) + (10 \cdot 3) + (6 \cdot 2) + 27 = 237$
Jen:
$(10 \cdot 5) + (43 \cdot 4) + (6 \cdot 3) + 0 + 22 = 262$
Hector:
$(28 \cdot 5) + 0 + 0 + (43 \cdot 2) + 10 = 236$
Medin:
$(27 \cdot 5) + 0 + (38 \cdot 3) + (10 \cdot 2) + 6 = 275$
Medin is the president, Jen is the vice-president, Andrew is the secretary, and Hector is the treasurer.

41. a.

A	2	1	2	5	4	2	4	5
B	5	3	5	3	2	5	3	2
C	4	5	4	1	4	3	2	1
D	1	3	2	4	5	1	3	2
	26	42	19	33	24	8	24	33

A: $(26 \cdot 2) + (42 \cdot 1) + (19 \cdot 2) + (33 \cdot 5)$
$+ (24 \cdot 4) + (8 \cdot 2) + (24 \cdot 4) + (33 \cdot 5) = 670$
B: $(26 \cdot 5) + (42 \cdot 3) + (19 \cdot 5) + (33 \cdot 3)$
$+ (24 \cdot 2) + (8 \cdot 5) + (24 \cdot 3) + (33 \cdot 2) = 676$
C: $(26 \cdot 4) + (42 \cdot 5) + (19 \cdot 4) + (33 \cdot 1)$
$+ (24 \cdot 4) + (8 \cdot 3) + (24 \cdot 2) + (33 \cdot 1) = 624$
D: $(26 \cdot 1) + (42 \cdot 3) + (19 \cdot 2) + (33 \cdot 4)$
$+ (24 \cdot 5) + (8 \cdot 1) + (24 \cdot 3) + (33 \cdot 2) = 588$
The winner of the election is candidate B.

b. A: 90 votes; B: 53 votes; C: 42 votes; D: 24 votes. Candidate A wins a plurality contest.

43. Round Table is the best choice, as everyone finds it at least acceptable. Both Pizza Hut and Domino's are objectionable to more than one person.

EXCURSION EXERCISES, SECTION 4.3

1. The blocking coalitions are: {A, B}, {A, D}, {B, D}, {A, B, C}, {A, B, D}, {A, C, D}, {B, C, D}

3. A blocking coalition is one holding at least six votes. Using the first letter of each country in place of its name, the blocking coalitions are {B,F}, {B,G}, {B,I}, {F,G}, {F,I}, {F,N}, {G,I}, {G,N}, {I,N}, (B,F,G}, {B,F,I}, {B,F,L}, {B,F,N}, {B,G,I}, {B,G,N}, {B,G,L}, {B,I,L}, {B,I,N}, {F,G,I}, {F,G,L}, {F,G,N}, {F,I,L}, {F,I,N}, {F,L,N}, {G,I,L}, {G,I,N}, {G,L,N}, {I,L,N}, {B,F,G,I}, {B,F,G,L}, {B,F,G,N}, {B,F,I,L}, {B,F,I,N}, {B,F,L,N}, {B,G,I,N}, {B,G,I,L}, {B,G,L,N}, {B,I,L,N}, {F,G,I,L}, {Γ,G,I,N}, {Γ,G,L,N}, {F,I,L,N}, {G,I,L,N}, {B,F,G,I,L}, {B,F,G,I,N}, {B,F,G,L,N}, {B,F,I,L,N}, {B,G,I,L,N}, and {F,G,I,L,N}

5. Many answers are possible. One is {4: 4, 2, 1}, with winning coalitions {A,B},{A,C}, and {A, B, C}; there are no blocking coalitions, and A is the critical voter in all three winning coalitions. The Banzhaf power indices are

$$BPI(A) = \frac{3}{3} = 1; \quad BPI(B) = BPI(C) = \frac{0}{3} = 0$$

7. Many answers are possible. One is {4: 1, 1, 1, 1, 1}. Winning coalitions are {A,B,C,D}, {A,B,C,E}, {A,B,D,E}, {A,C,D,E}, {B,C,D,E}, and {A,B,C,D,E}. Each voter is critical in all coalitions except the last, where none is. Blocking coalitions comprise all 10 groups of two, all 10 groups of three, and the five above groups of four. Each voter is critical in the first type of group, and none are critical in the second and third types of groups. Each of the five voters is therefore critical eight times, and the BPI for each is $\frac{9}{9-5} = \frac{1}{5}$.

EXERCISE SET 4.3

1. a. 6
   b. 4
   c. 3
   d. 6
   e. No
   f. A and C
   g. $2^4 - 1 = 15$
   h. 6

3.

Winning coalition	Critical voters
{A, B}	A, B
{A, C}	A, C
{A, B, C}	A

$$BPI(A) = \frac{3}{5} = 0.6$$

$$BPI(B) = BPI(C) = \frac{1}{5} = 0.2$$

5.

Winning coalition	Critical voters
{A, B}	A, B
{A, B, C}	A, B
{A, B, D}	A, B
{A, C, D}	A, C, D
{A, B, C, D}	A

$$BPI(A) = \frac{5}{10} = 0.5$$

$$BPI(B) = \frac{3}{10} = 0.3$$

$$BPI(C) = \frac{1}{10} = 0.1 = BPI(D)$$

7.

Winning coalitions	Critical voters
{A, B}	A, B
{A, B, C}	A, B
{A, B, D}	A, B
{A, B, E}	A, B
{A, C, D}	A, C, D
{A, C, E}	A, C, E
{B, C, D}	B, C, D
{A, B, C, D}	none
{A, B, C, E}	A
{A, B, D, E}	A, B
{A, C, D, E}	A, C
{B, C, D, E}	B, C, D
{A, B, C, D, E}	none

$$BPI(A) = \frac{9}{25} = 0.36; \ BPI(B) = \frac{7}{25} = 0.28$$

$$BPI(C) = \frac{5}{25} = 0.20; \ BPI(D) = \frac{3}{25} = 0.12$$

$$BPI(E) = \frac{1}{25} = 0.04$$

9. No coalition without voter A can reach a quota. Voter A has a Banzhaf power index of 1; all the other voters have a BPI of 0.

11.

Winning Coalition	Critical Voters
{A, B}	A, B
{A, C}	A, C
{A, B, C}	A
{A, B, D}	A, B
{A, B, E}	A, B
{A, C, D}	A, C
{A, C, E}	A, C
{A, D, E}	A, D, E
{B, C, D}	B, C, D
{A, B, C, D}	none
{A, B, C, E}	A
{A, B, D, E}	A
{A, C, D, E}	A
{B, C, D, E}	B, D, C
{A, B, C, D, E}	none

$$BPI(A) = \frac{11}{25} = 0.44$$

$$BPI(B) = BPI(C) = \frac{5}{25} = 0.20$$

$$BPI(D) = \frac{3}{25} = 0.12$$

$$BPI(E) = \frac{1}{25} = 0.04$$

13. a. In Exercise 9, voter A never needs any of the other voters in any winning coalition, and is thus the dictator.

b. Veto power means that one or more voters

are critical in every winning coalition. In exercises 3 and 5, voter A is critical in each winning coalition. In Exercise 6, there is only one coalition, so each voter has veto power. In Exercise 9, voter A is a dictator and thus has veto power. In exercise 12, voter A is critical in all winning coalitions.

c. None

d. In Exercise 6, although voters have different weights, the only winning coalition comprises all the voters, so in practice each voter has one vote. In Exercise 8, each voter has one vote.

15. The winning coalitions are {D, T}, {D,P}, and {T, P}, with each voter critical to each. The director, teacher, and principal each have a Banzhaf power index of 0.33.

17. a. {12: 1, 1, 1, 1, 1, 1, 1, 1, 1, 1, 1, 1,}

b. Yes.

c. Yes, every juror has a veto, as the verdict must be unanimous.

d. Because this is a veto power system, divide the voter power of 1 by the quota of 12.

19. A is the dictator, and all of B, C, D, and E are dummies.

21. None.

23. a. The winning coalitions are {A, B}, {A, C}, {A, B, C}. A is critical to all three (and has veto power), and B and C are each critical in their own coalition with A. The BPI for A is 0.6, and for B and C the BPI is 0.2

b. With respect to the formation of winning coalitions, the coaches are the same.

25. a. There are four winning coalitions: {A, B}, {A, C}, {B, C}, and {A, B, C}. The critical voters are, respectively, A and B, A and C, B and C, and none of them. Each voter has a BPI of 0.33.

b. Despite the varied weights, this is a majority system. Any two of the three voters are needed for a quota.

27. a. 11 and 14

b. 15 and 16

c. No. D is the dummy for $q = 11$, but not for $q = 12$, for instance. The existence of a dummy depends on the combinations of voter weight as well as the quota.

29. a. The original voting system is {3:1,1,1,1,1,} with voters A, B, C, D, and E. The fraudulent voting system is {3:2,0,1,1,1}. The weight of A's vote is now 2 instead of 1. To determine *BPI(A)* in the fraudulent voting system, we need to determine the critical voters in each winning coalition.

Winning Coalitions and Critical Voters in the {3:2,0,1,1,1} Voting System				
Winning Coalition	**Critical Voters**		**Winning Coalition**	**Critical Voters**
{A, C}	A, C		{A, D, E}	A
{A, D}	A, D		{C, D, E}	C, D, E
{A, E}	A, E		{A, B, C, D}	A
{A, B, C}	A, C		{A, B, C, E}	A
{A, B, D}	A, D		{A, B, D, E}	A
{A, B, E}	A, E		{A, C, D, E}	None
{A, C, D}	A		{B, C, D, E}	C, D, E
{A, C, E}	A		{A, B, C, D, E}	None

A is a critical voter in 12 of the winning coalitions and B is never a critical voter. C, D, and E are each a critical voter in 4 of the winning coalitions. The number of times that any voter is a critical voter is $12 + 0 + 4 + 4 + 4 = 24$.

$$BPI(A) = \frac{12}{24} = 0.5 \qquad BPI(B) = \frac{0}{24} = 0 \qquad BPI(C) = BPI(D) = BPI(E) = \frac{4}{24} = 0.167$$

Thus *BPI(A)* in the fraudulent system is 0.5.

b. The original voting system is {3:1,1,1,1,1,} with voters A, B, C, D, and E. The fraudulent voting system is {3:1,1,2,1,1}. The weight of C's vote is now 2 instead of 1. To determine *BPI(C)* in the fraudulent voting system, we need to determine the critical voters in each winning coalition.

Winning Coalitions and Critical Voters in the {3:1,1,2,1,1} Voting System				
Winning Coalition	**Critical Voters**		**Winning Coalition**	**Critical Voters**
{A, C}	A, C		{B, C, D}	C
{B, C}	B, C		{B, C, E}	C
{C, D}	C, D		{B, D, E}	B, D, E
{C, E}	C, E		{C, D, E}	C
{A, B, C}	C		{A, B, C, D}	None
{A, B, D}	A, B, D		{A, B, C, E}	None
{A, B, E}	A, B, E		{A, B, D, E}	None
{A, C, D}	C		{A, C, D, E}	None
{A, C, E}	C		{B, C, D, E}	None
{A, D, E}	A, D, E		{A, B, C, D, E}	None

C is a critical voter in 10 of the winning coalitions. A, B, D, and E are each a critical voter in 4 of the winning coalitions. The number of times that any voter is a critical voter is $10 + 4 + 4 + 4 + 4 = 26$.

$$BPI(C) = \frac{10}{26} = 0.385 \qquad BPI(A) = BPI(B) = BPI(D) = BPI(E) = \frac{4}{26} = 0.154$$

Thus *BPI(C)* is about 0.385 in the fraudulent system.

CHAPTER 4 REVIEW EXERCISES

1. a. The standard divisor is $\frac{9606}{50} = 192.18$.

Div.	Enrol.	Quota	Proj.
Health	1280	6.66	7
Bus.	3425	17.82	18
Eng.	1968	10.24	10
Sci.	2936	15.28	15

Health division gets 7 projectors, the Business division gets 18, the Engineering division gets 10, and the Science division gets 15.

b. The modified standard divisor is 183.4.

Div.	Enrol.	Quota	Proj.
Health	1280	6.98	6
Bus.	3425	18.68	18
Eng.	1968	10.73	10
Sci.	2936	16.01	16

c. A modified standard divisor of 191 yields the following apportionment.

Div.	Enrol.	Quota	Proj.
Health	1280	6.70	7
Bus.	3425	17.93	18
Eng.	1968	10.30	10
Sci.	2936	15.37	15

2. a. The standard divisor is $\frac{110}{35} = 3.143$.

Airport	Agts.	Quota	Empl.
Newark	28	8.91	9
Cleve.	19	6.05	6
Chi.	34	10.82	11
Phila.	13	4.14	4
Det.	16	5.09	5

Newark gets 9 security employees, Cleveland 6, Chicago 11, Philadelphia 4, and Detroit 5.

   b. The modified standard divisor is 3.05.

Airport	Agts.	Quota	Empl.
Newark	28	9.18	9
Cleve.	19	6.23	6
Chicago	34	11.15	11
Phila.	13	4.26	4
Detroit	16	5.25	5

   c. The same modified standard divisor of 3.05 yields the Webster apportionment.

Airport	Agts.	Quota	Empl.
Newark	28	9.18	9
Cleve.	19	6.23	6
Chi.	34	11.15	11
Phila.	13	4.26	4
Det.	16	5.25	5

3. a. Relative unfairness if High Desert receives the new controller: $\frac{326-297}{297} \approx 0.098$

   b. Relative unfairness if Eastlake receives the new controller: $\frac{302-253}{253} \approx 0.194$.

   c. Using the apportionment principle, High Desert Airport should get the new controller.

4.

	Morena Valley Average	West Keyes Average	Abs. Unfair
Morena Valley receives professor	$\frac{1437}{38+1} = 36.85$	$\frac{1504}{46} = 32.70$	4.15
West Keyes receives professor	$\frac{1437}{38} = 37.82$	$\frac{1504}{46+1} = 32.00$	5.82

Relative unfairness if Morena Valley receives the new professor: $\frac{4.15}{36.85} \approx 0.113$

Relative unfairness if Eastlake receives the new controller: $\frac{5.82}{32.00} \approx 0.182$

Morena Valley should receive the new professor.

5. a. The standard divisor is 13.79. Office A has a quota is 1.38, B has a quota of 14.14, C has a quota of 22.34, and D has a quota of 29.15. Office A gets the next printer, and no office loses a printer. The Alabama paradox does not occur.

   b. The standard divisor is 13.59, and the quotas are as follows: A:1.40; B: 14.35; C: 22.66; and D: 29.58. Offices C and D get the new printers, and Office A falls back to 1 printer. This is an example of the Alabama paradox.

6. a. The standard divisor is 20.96.

Center	Quota	Autos
A	1.48	2
B	5.15	5
C	3.34	3
D	15.70	16
E	2.34	2

Regional center D gets the additional automobile. There is no Alabama paradox, as no center loses an automobile.

   b. The standard divisor is 20.24.

Center	Quota	Autos
A	1.53	2
B	5.34	5
C	3.46	4
D	16.25	16
E	2.42	2

Regional center C gets the additional automobile. There is no Alabama paradox, as no center loses an automobile.

7. a. The standard divisor is 151.4. Los Angeles gets 9 servers, and Newark gets 2.

   b. The standard divisor is now 148. LA gets 10, and Newark and KC get 1 each. Yes, this is the new states' paradox, since Newark loses a server without having lost people.

8. a. The standard divisor is 1408. The apportionment is A: 10, B: 3, C: 21.

   b. The standard divisor is 1428. The apportionment is A: 11, B: 2, C: 21. Yes, this is the population paradox. Although the population of region B has grown at a higher rate than the population region A, region B still loses an inspector in reapportionment.

9. Yes. See exercise 6b above.

10. No, it cannot occur.

11. a. The Huntington-Hill number is given by

    $\dfrac{(P_A)^2}{a(a+1)}$, where P_A is the population, and

    a is the current apportionment.

Bldg	A	B	C
Guard	25	43	18
Empl.	414	705	293
H-H	263.7	262.7	251.0

    Building A has the greatest Huntington-Hill number, and is assigned the next guard.

    b.

Bldg	A	B	C
Guard	26	43	18
Empl.	414	705	293
H-H	244.2	262.7	251.0

    Building B now has the greatest Huntington-Hill number, and is assigned the 88th guard.

12. a. Ortega, with 8 votes.

    b. No, he has 8 of 23 votes.

    c. Kelly: $(2 \cdot 8) + (3 \cdot 5) + (3 \cdot 4) + (4 \cdot 6)$
        $= 67$
        Ortega: $(4 \cdot 8) + (2 \cdot 5) + (1 \cdot 4) + (2 \cdot 6)$
        $= 58$
        Nisbet: $(3 \cdot 8) + (1 \cdot 5) + (4 \cdot 4) + (3 \cdot 6)$
        $= 63$
        Toyama: $(1 \cdot 8) + (4 \cdot 5) + (2 \cdot 4) + (1 \cdot 6)$
        $= 42$
        Crystal Kelly wins the essay contest.

13. a. Vail 23, Aspen 22, Powderhorn 18 The club chooses Vail.

    b. Aspen:
        $(5 \cdot 14) + (5 \cdot 8) + (3 \cdot 11) + (4 \cdot 18) + (3 \cdot 12)$
        $= 251$
        Copper Mtn.:
        $(1 \cdot 14) + (2 \cdot 8) + (4 \cdot 11) + (2 \cdot 18) + (2 \cdot 12)$
        $= 134$
        Powderhorn:
        $(3 \cdot 14) + (4 \cdot 8) + (1 \cdot 11) + (5 \cdot 18) + (1 \cdot 12)$
        $= 187$
        Telluride:
        $(2 \cdot 14) + (1 \cdot 8) + (2 \cdot 11) + (1 \cdot 18) + (4 \cdot 12)$
        $= 124$
        Vail:
        $(4 \cdot 14) + (3 \cdot 8) + (5 \cdot 11) + (3 \cdot 18) + (5 \cdot 12)$
        $= 249$
        Using the Borda Count method, the club chooses Aspen.

14. Schneider has no first-place votes, and is eliminated. The preference schedule is now:

Reynolds	2	2	1	3
Hernandez	1	3	3	2
Kim	3	1	2	1
votes	132	214	93	119

    Reynolds now has the fewest first-place votes, and is eliminated. The preference schedule is now:

Hernandez	1	2	2	2
Kim	2	1	1	1
votes	132	214	93	119

    Kim wins the election, 426 to 132.

15. Twix has no first-place votes, and is eliminated. The preference schedule is now:

Crunch	1	3	3	2	2
Snickers	2	1	2	3	1
Milky Way	3	2	1	1	3
votes	15	38	27	16	22

    Nestle Crunch has the fewest first-place votes, and is eliminated. The preference schedule is now:

Snickers	1	1	2	2	1
Milky Way	2	2	1	1	2
votes	15	38	27	16	22

    Snickers is the favorite candy bar, 75 to 43.

16.

versus	GR	LH	AK	JS
GR	--	GR	AK	GR
LH	--	--	AK	JS
AK	--	--	--	AK
JS	--	--	--	--

    Kim wins 3 matches to 2 for Reynolds and 1 for Schneider.

17.

versus	Crunch	Snick.	Milky	Twix
Crunch	--	Snick.	Milky	Twix
Snick.	--	--	Snick.	Snick.
Milky	--	--	--	Milky
Twix	--	--	--	--

    Snickers is the favorite candy bar, with 3 matches compared to 2 for Milky Way and 1 for Twix.

18. a. Cynthia:
        $(3 \cdot 11) + (2 \cdot 97) + (1 \cdot 112) = 339$
        Hannah:
        $(3 \cdot 112) + (1 \cdot 97) + (1 \cdot 11) = 444$
        Shannon:
        $(3 \cdot 97) + (2 \cdot 112) + (2 \cdot 11) = 537$
        Shannon is the homecoming queen.

b.

versus	Cynthia	Hannah	Shannon
Cynthia	--	Hannah	Shannon
Hannah	--	--	Hannah
Shannon	--	--	--

Hannah wins all head-to-head contests.

c. The candidate who wins all head-to-head contests should win the election of all candidates, but while Hannah wins all head-to-head contests, Shannon wins the overall election.

d. Hannah, with 112 votes, wins both the plurality and the majority of votes.

e. Hannah won the majority of first-place votes, but Shannon won the overall election.

19. a. The resulting preference schedule:

Hannah	1	2	2
Shannon	2	1	1
votes	112	97	11

Hannah wins the election, 112 to 108.

b. Cynthia L., a losing candidate, withdrew from the race and caused a change in the overall winner of the election.

20. The montonicity criterion has been violated. If a candidate wins an election, then, if the only change in voters' preferences is that supporters of another candidate change their votes to support the winner, then the result must remain the same. Margaret won the scholarship, but, when voters changed their preference from Terry to Margaret, Jean won the scholarship.

21. a. 18

b. 18

c. Yes

d. A and C

e. 15

f. 6

22. a. 35

b. 35

c. Yes

d. A

e. 31

f. 10

23. The winning coalitions are {A, B}, with A and B critical members; {A, C}, with A and C critical members; and {A, B, C}, with A as the critical member. $BPI(A) = \frac{3}{5} = 0.6$

$$BPI(B) = BPI(C) = \frac{1}{5} = 0.2$$

24. In a one-person, one-vote system, each voter has the same power as any other. As there are five voters in this system, each one has a BPI of $\frac{1}{5} = 0.2$.

25.

Winning coalitions	Critical voters
{A, B}	A, B
{A, C}	A, C
{A, B, C}	A
{A, B, D}	A, B
{A, C, D}	A, C
{B, C, D}	B, C, D
{A, B, C, D}	none

$BPI(A) = \frac{5}{12} \approx 0.42$

$BPI(B) = BPI(C) = \frac{3}{12} = 0.25$

$BPI(D) = \frac{1}{12} \approx 0.08$

26.

Winning Coalition	Critical Voters
{A, B}	A, B
{A, C}	A, C
{A, B, C}	A
{A, B, D}	A, B
{A, B,E}	A, B
{A, C, D}	A, C
{A, C, E,}	A, C
{A, D, E}	A, D, E
{A, B, C, D}	A
{A, B, C, E}	A
{A, B, D, E}	A
{A, C, D, E}	A
{A, B, C, D, E}	A

$BPI(A) \approx 0.62$;
$BPI(B) = BPI(C) \approx 0.14$;
$BPI(D) = BPI(E) \approx 0.05$

27. A is the dictator, and B, C, D, and E are dummies.

28. There is no dictator, and D is a dummy.

29.

Winning Coalition	Critical Voters
{A, B}	A, B
{A, C}	A, C
{A, D}	A, D
{A, B, C}	A
{A, B, D}	A
{A, C, D}	A
{A, B, C, D}	A
{A, E}	A, E
{B, C, D, E}	B, C, D, E

$BPI(A) = 0.5$;
$BPI(B) = BPI(C) = BPI(D) = BPI(E) = 0.125$

CHAPTER 4 TEST

1.

	Spring Valley Average	Summer-ville Average	Abs. Unfair
Spring Valley receives carrier	$\dfrac{67{,}530}{158+1}$ $= 424.72$	$\dfrac{59{,}950}{129}$ $= 418.22$	6.5
Summer-ville receives carrier	$\dfrac{67{,}530}{158}$ $= 427.41$	$\dfrac{59{,}950}{129+1}$ $= 415$	12.41

Relative unfairness if Spring Valley receives the new carrier: $\dfrac{6.5}{424.72} \approx 0.015$

Relative unfairness if Summerville receives the new carrier: $\dfrac{12.41}{415} \approx 0.030$

Spring Valley should receive the new carrier.

2. a. The standard divisor is 58.6.

Div.	Empl.	Quota	Comp.
Sales	1008	17.20	17
Adver.	234	3.99	4
Serv.	625	10.67	11
Manu.	3114	53.14	53

Sales get 17 computers, advertising 4, service 11, and manufacturing 53.

b. The modified standard divisor is 57.0.

Div.	Empl.	Quota	Comp.
Sales	1008	17.68	17
Adver.	234	4.11	4
Serv.	625	10.96	10
Manu.	3114	54.63	54

No, the quota rule is not violated; each division gets either the standard quota, or one more, computers.

3. a. The formula for the Huntington-Hill number is $\dfrac{(P_A)^2}{a(a+1)}$.

The Huntington-Hill number for Cedar Falls equals 77,792 and the Huntington-Hill number for Lake View equals 70,290.

b. Cedar Falls has the higher Huntington-Hill number, and should be assigned the new counselor.

4. a. 33

b. 26

c. No.

d. A and C

e. $2^5 - 1 = 31$

f. 10

5. a. Aquafina, which has 39 votes to 31 for Arrowhead and 30 for Evian.

b. No, Aquafina has 39 of 100 votes and needs 51 for a majority.

c. Arrowhead:
$(2\cdot 22)+(1\cdot 17)+(3\cdot 31)+(1\cdot 11)+(2\cdot 19)$
$= 203$
Evian:
$(1\cdot 22)+(2\cdot 17)+(2\cdot 31)+(3\cdot 11)+(3\cdot 19)$
$= 208$
Aquafina:
$(3\cdot 22)+(3\cdot 17)+(1\cdot 31)+(2\cdot 11)+(1\cdot 19)$
$= 189$

Evian is the preferred brand of water, using the Borda Count method.

6.

versus	NY	Dal.	LA	Atl.
NY	--	NY	NY	NY
Dal.	--	--	LA	Atl.
LA	--	--	--	LA
Atl.	--	--	--	--

New York wins 3 matches, LA wins 2, and Atlanta wins 1. New York should be selected.

7. a. Evening had the fewest first-place votes and is eliminated. The preference schedule is now:

Noon	2	2	2	2	1
PM	1	3	1	3	2
votes	12	16	9	5	13

Noon has the fewest first-place votes and is eliminated.

The preference schedule is now:

AM	2	1	2	1	2
PM	1	2	1	2	1
votes	12	16	9	5	13

Afternoon has 34 first-place votes and morning has 21. She should schedule the session for the afternoon.

b. Morning:
$(1 \cdot 12) + (4 \cdot 16) + (1 \cdot 9) + (3 \cdot 5) + (2 \cdot 13)$
$= 126$

Noon:
$(2 \cdot 12) + (3 \cdot 16) + (3 \cdot 9) + (2 \cdot 5) + (4 \cdot 13)$
$= 161$

Afternoon:
$(4 \cdot 12) + (2 \cdot 16) + (4 \cdot 9) + (1 \cdot 5) + (3 \cdot 13)$
$= 160$

Evening:
$(3 \cdot 12) + (1 \cdot 16) + (2 \cdot 9) + (4 \cdot 5) + (1 \cdot 13)$
$= 103$

Using the Borda Count method, the best time for the session is noon.

8. a. Proposal A, with 40 votes compared to 9 for proposal B and 39 for proposal C.

b. Proposal B, which wins 48 votes compared to 40 for proposal A.

c. Eliminating a losing choice changed the outcome of the election.

d.

versus	**A**	**B**	**C**
A	--	B	C
B	--	--	B
C	--	--	--

Proposal B wins all head-to-head contests.

e. Proposal B won all head-to-head contests but lost the vote when all the choices were on the ballot.

9.

Winning Coalition	Critical Voters
{A, B}	A, B
{A, C}	A, C
{A, B, C}	A
{A, B, D}	A, B
{A, C, D}	A, C
{B, C, D}	B, C, D
{A, B, C, D}	none

$BPI(\text{A}) = \dfrac{5}{12} \approx 0.42$

$BPI(\text{B}) = BPI(\text{C}) = \dfrac{3}{12} = 0.25$

$BPI(\text{D}) = \dfrac{1}{12} \approx 0.08$

10.

Winning Coalition	Critical Voters
{A, B}	A, B
{A, C}	A, C
{A, B, C}	A
{A, D}	A, D
{B, C, D}	B, C, D

$BPI(\text{A}) = \dfrac{4}{10} = 0.4$

$BPI(\text{B}) = BPI(\text{C}) = BPI(\text{D}) = \dfrac{2}{10} = 0.2$

Chapter 5: The Mathematics of Graphs

EXCURSION EXERCISES, SECTION 5.1

1. One answer is

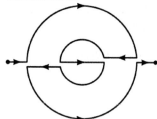

3. One answer is

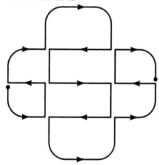

5. More than two vertices are of odd degree.

EXERCISE SET 5.1

1. The graph is shown below:

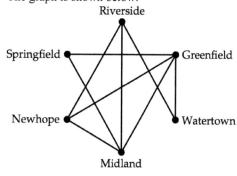

3. a. No. There is no line connecting John and Stacy.

   b. 3

   c. Ada

   d. A loop would correspond to a friend talking to himself or herself.

5. a. 6
   b. 7
   c. 6
   d. Yes
   e. No

7. a. 6
   b. 4
   c. 4
   d. Yes
   e. Yes

9. Equivalent

11. Not equivalent

13. The graph on the right has a vertex of degree 4 and the graph on the left does not.

15. a. Yes. D-A-E-B-D-C-E-D is an Euler circuit.

17. a. Not Eulerian. There are vertices of odd degree.

    b. Yes. A-E-A-D-E-D-C-E-C-B-E-B is an Euler walk.

19. a. Not Eulerian. There are vertices of odd degree.

    b. This graph does not have an Euler walk. More than two vertices are of odd degree.

21. a. Not Eulerian. There are vertices of odd degree.

    b. Yes. E-A-D-E-G-D-C-G-F-C-B-F-A-B-E-F is an Euler walk.

23. a. As a graph:

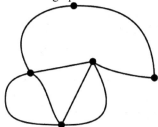

    b. It is possible to cross each bridge once and return to the starting point. Every vertex of the graph has an even degree.

25. Yes. There are exactly two vertices of odd degree, so an Euler walk is possible.

27. Yes, the hamster can travel through every tube without going through the same tube twice. Draw a graph of the Habitrail, with the vertices representing the cages and the edges representing the tubes and find an Euler circuit or an Euler walk. There is no Euler circuit; therefore the hamster cannot return to its starting point.

29. The graph is shown below:
    Yes. You can always return to the starting point.

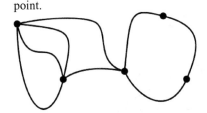

31. a. 2
    b. 4
    c. Wayne

33. a. AD
    b. EF, CF
    c. none
    d. BC, CD, AD

EXCURSION EXERCISES, SECTION 5.2

1. C-A-B-E-D-C, total weight 18
    D-E-B-A-C-D, total weight 18
    E-B-A-C-D-E, total weight 18

3. The Edge-Picking Algorithm generates the circuit A-F-C-B-D-E-A. It has the same weight as the circuit with the smallest total weight of 61 as found in exercise 2.

EXERCISE SET 5.2

1. The graph is connected and has at least three vertices and no multiple edges, so the theorem applies. Every vertex has a degree of 4 or more, so the graph is Hamiltonian. A Hamiltonian circuit is A-B-C-D-E-G-F-A.

3. The graph is connected and has at least three vertices and no multiple edges, so the theorem applies. Every vertex has a degree of 4 or more, so the graph is Hamiltonian. A Hamiltonian circuit is A-B-E-C-H-D-F-G-A.

5. One such route is Springfield-Greenfield-Watertown-Riverside-Newhope-Midland-Springfield.

7. Two Hamiltonian circuits are A-B-E-D-C-A, total weight 31, and A-D-E-B-C-A, total weight 32.

9. Two Hamiltonian circuits are A-D-C-E-B-F-A, total weight 114, and A-C-D-E-B-F-A, total weight 158.

11. A-D-B-C-F-E-A

13. A-C-E-B-D-A

15. A-D-B-F-E-C-A

17. A-C-E-B-D-A

19. As a graph:

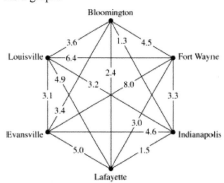

Louisville-Evansville-Bloomington-Indianapolis-Lafayette-Fort Wayne-Louisville

21. Louisville-Evansville-Bloomington-Indianapolis-Layfayette-Fort Wayne-Louisville

23. As a graph:

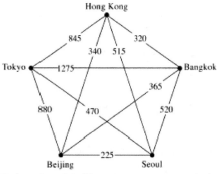

Tokyo-Seoul-Beijing-Hong Kong-Bangkok-Tokyo

25. Tokyo-Seoul-Beijing-Hong Kong-Bangkok-Tokyo

27. Represent the information as a graph:

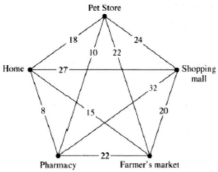

The Greedy Algorithm generates the route home-pharmacy-pet store-farmer's market-shopping mall-home. The Edge-Picking Algorithm generates the route home-pharmacy-pet store-shopping mall-farmer's market-home.

29. Represent the information as a graph:

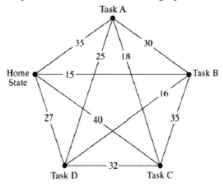

The Greedy Algorithm generates the sequence home state-task B-task D-task A-task C-home state. The Edge-Picking Algorithm gives the same sequence.

31. An Euler circuit would be most efficient. The officer needs to travel each street, or edge, once.

33. a.   There are many possible answers.

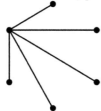

b.   There are many possible answers.

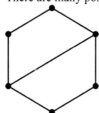

c.   There are many possible answers.

EXCURSION EXERCISES, SECTION 5.3

1.   As a graph:

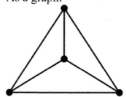

3.   a.   3

   b.   12 faces × 5 edges = 60. Each edge is shared by two faces, so divide by 2. There are 30 edges in the projected graph.

   c. $v + f = e + 2$
         $v + 12 = 30 + 2$
         $v = 20$. There are 20 vertices.

EXERCISE SET 5.3

1.   As a planar graph:

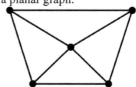

3.   As a planar graph:

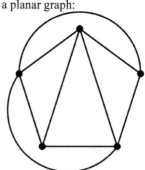

5.   As a planar graph:

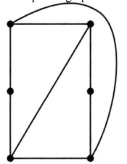

7.   As a planar graph:

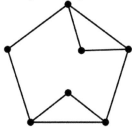

9.   Finding a subgraph:

The highlighted subgraph is the Utilities Graph.

11. The graph contains K_5.

13. Starting with the original graph:

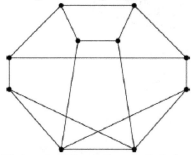

First contract the vertices on the left and right sides.

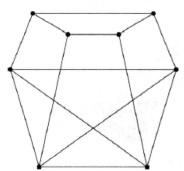

Then contract each diagonal pair of vertices at the top of the graph.

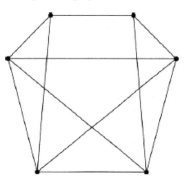

Finally contract the two vertices at the top of the graph.

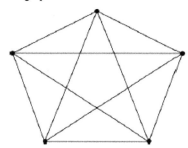

15. The graph contracts to the Utilities Graph.

17. 5 faces, 5 vertices, 8 edges
$$v + f = e + 2$$
$$5 + 5 = 8 + 2$$
$$10 = 10$$

19. 2 faces, 8 vertices, 8 edges
$$v + f = e + 2$$
$$8 + 2 = 8 + 2$$
$$10 = 10$$

21. 5 faces, 10 vertices, 13 edges
$$v + f = e + 2$$
$$10 + 5 = 13 + 2$$
$$15 = 15$$

23. $$v + f = e + 2$$
$$8 + f = 15 + 2$$
$$f = 9 \text{ faces}$$

25. The graph is shown below:

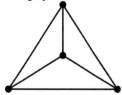

27. $v + f = e + 2$
    $v + 3 = v + 2$
    The above equation is never true, so Euler's Formula cannot be satisfied.

29. With the labeling shown in the first figure, A-B-C-D-E-A is a Hamiltonian Circuit, as shown in the second figure. Of the remaining five edges to be drawn, only two can be drawn inside the circuit without intersection, and it is impossible to draw the other three edges outside the circuit without intersections, as shown in the third figure. (Edge CE cannot be drawn,)

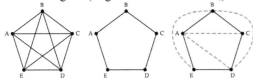

EXCURSION EXERCISES, SECTION 5.4

1. As graphs:

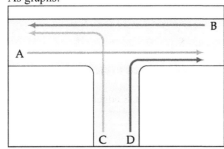

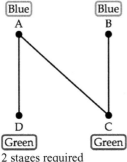

2 stages required

3. As graphs:

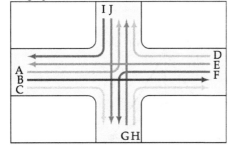

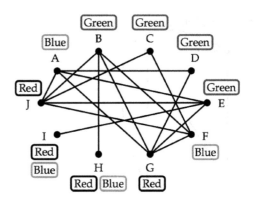

3 stages required

EXERCISE SET 5.4

1. Requires three colors:

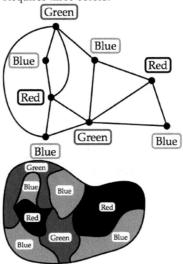

3. Requires three colors:

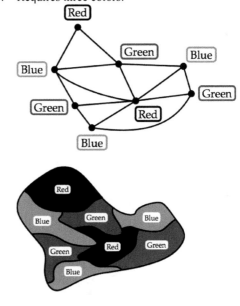

5. As a graph:

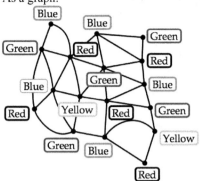

7.

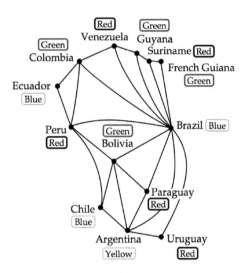

9. The graph is shown below:

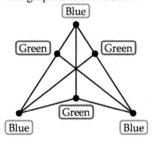

11. The graph is not 2-colorable because it has a circuit that consists of an odd number of vertices.

13. The graph is shown below:

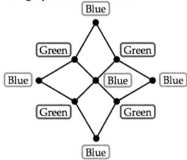

15. The graph is shown below:

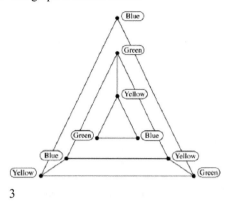

3

17. The graph is shown below:

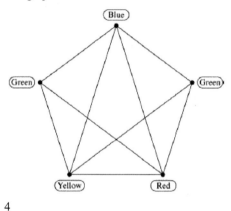

4

19. The graph is shown below:

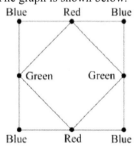

3

21. Represent the information as a graph:

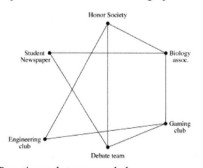

Two times slots are needed.

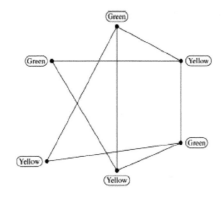

23. Represent the information as a graph:

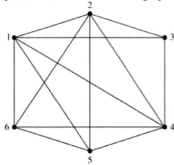

5 days are required.
Day 1: Group 1
Day 2: Group 2
Day 3: Groups 3 and 5
Day 4: Group 4
Day 5: Group 6

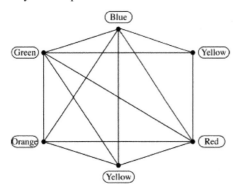

25. Represent the information as a graph:

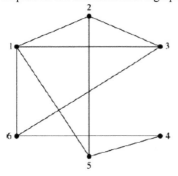

3 days are needed.

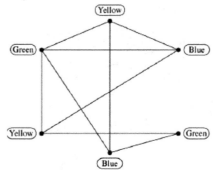

Day 1: Films 1 and 4
Day 2: Films 2 and 6
Day 3: Films 3 and 5

27. The chromatic number of the graph tells the engineers, the minimum number of transmitting channels needed.

29. The graph must be a disconnected graph with no edges.

31. The graph is shown below:

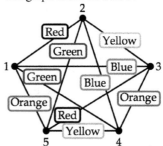

Day 1: 1 vs. 3, 2 vs. 4; Day 2: 1 vs. 4, 2 vs. 5; Day 3: 1 vs. 2, 3 vs. 5; Day 4: 2 vs. 3, 4 vs. 5; Day 5: 1 vs. 5, 3 vs. 4.

CHAPTER 5 REVIEW EXERCISES

1. a. 8

    b. 4

    c. All vertices have degree 4.

    d. Yes

2. a. 6

    b. 7

    c. 1, 1, 2, 2, 2, 2

    d. No

3. As a graph:

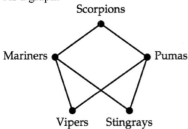

4. a. No

    b. 4

    c. 110, 405

    d. 105

5. Equivalent

6. Equivalent

7. a. E-A-B-C-D-B-E-C-A-D is an Euler walk.

    b. It is not possible to find an Euler circuit because there are vertices of odd degree.

8. a. It is not possible to find an Euler walk because there are more than 2 vertices of odd degree.

    b. It is not possible to find an Euler circuit because there are vertices of odd degree.

9. a. F-A-E-C-B-A-D-B-E-D-C-F is an Euler walk.

    b. F-A-E-C-B-A-D-B-E-D-C-F is an Euler circuit.

10. a. B-A-E-C-A-D-F-C-B-D-E is an Euler walk.

    b. It is not possible to find an Euler circuit because there are vertices of odd degree.

11. As a graph:

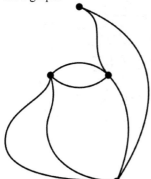

Yes. An Euler circuit exists.

12. Yes, it is possible to walk through each doorway exactly once, i.e. an Euler walk exists. No, it is not possible to do so and return to the starting point, i.e. an Euler circuit does not exist.

13. The graph is connected and has at least three vertices and no multiple edges, so the theorem applies. Every vertex has a degree of 3 or more, so the graph is Hamiltonian. A Hamiltonian circuit is A-B-C-E-D-A.

14. The graph is connected and has at least three vertices and no multiple edges, so the theorem applies. Every vertex has a degree of 4 or more, so the graph is Hamiltonian. A Hamiltonian circuit is A-D-F-B-C-E-A.

15. As a graph:

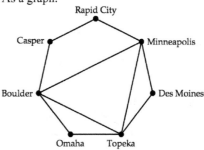

Casper-Rapid City-Minneapolis-Des Moines-Topeka-Omaha-Boulder-Casper

16. Casper-Boulder-Topeka-Minneapolis-Boulder-Omaha-Topeka-Des Moines-Minneapolis-Rapid City-Casper

17. A-D-F-E-B-C-A

18. A-B-E-C-D-A

19. A-D-F-E-C-B-A

20. A-B-E-D-C-A

21. As a graph:

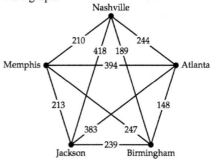

Memphis-Nashville-Birmingham-Atlanta-Jackson-Memphis

22. As a graph:

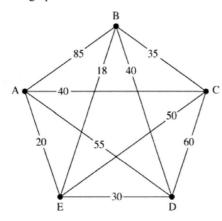

A-E-B-C-D-A

23. Redrawing:

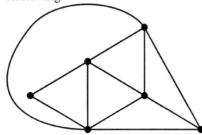

24. Redrawing:

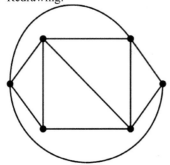

25. K_5 is a subgraph of the graph, so it is not planar.

26. The graph can be contracted into the Utilities Graph, so it is not planar.

27. 5 vertices, 8 edges, 5 faces
$$v + f = e + 2$$
$$5 + 5 = 8 + 2$$
$$10 = 10$$

28. 14 vertices, 16 edges, 4 faces
$$v + f = e + 2$$
$$14 + 4 = 16 + 2$$
$$18 = 18$$

29. As a graph:

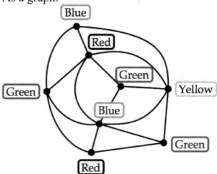

Requires four colors.

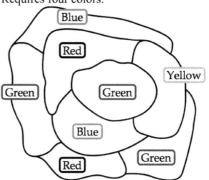

30. As a graph:

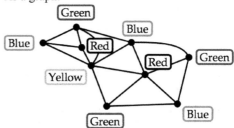

Requires four colors.

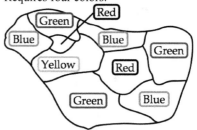

31. The graph has no circuits consisting of an odd number of vertices, so it is 2-colorable.

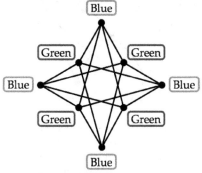

32. The graph has a circuit that consists of an odd number of vertices, so it is not 2-colorable.

33. The chromatic number is 3.

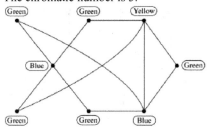

34. The chromatic number is 5.

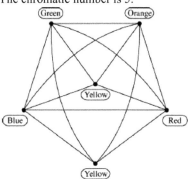

35. As a graph:

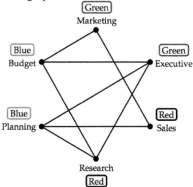

Three times slots are needed.
Time 1: Budget and Planning
Time 2: Marketing and Executive
Time 3: Sales and Research

CHAPTER 5 TEST

1. a. No

    b. Monique

    c. 0

    d. No

2. The graphs are equivalent. You can see this by making a list of all the edges in the each graph and comparing the lists.

3. a. No. The graph has vertices of odd degree, so it is not Eulerian.

    b. One possible answer: A-B-E-A-F-D-C-F-B-C-B-D is an Euler path.

4. Represent the information with a graph:

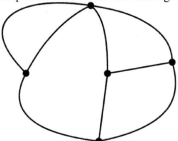

The path described corresponds to an Euler walk or an Euler circuit. The graph has more than 2 vertices of odd degree, so neither an Euler walk nor an Euler circuit exists. Thus such a walk is impossible.

5. a. A graph with more than three vertices and no multiple edges is Hamiltonian if every vertex is of degree at least n/2, where n is the number of vertices. The graph is connected and has more than three vertices and no multiple edges, so the theorem applies. All of the vertices are of degree at least 4, so the theorem is satisfied and the graph is Hamiltonian.

    b. A-G-C-D-F-B-E-A is a Hamiltonian circuit.

6. a. As a graph:

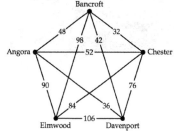

b. Angora-Elmwood-Chester-Bancroft-
    Davenport-Angora.
    The total cost is
    $90 + 84 + 32 + 42 + 36 = \$284$.

c. Look for an Euler circuit. One exists
    because each vertex has degree 4, an even
    number.

7. A-E-D-B-C-F-A

8. a. Redrawing:

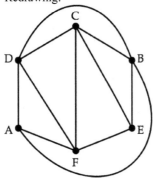

b. Contracting vertices B and H and vertices
    A and G gives the Utilities Graph, so the
    graph is not planar.

9. a. 6

    b. 7

    c. 12 vertices, 7 faces, 17 edges
        $v + f = e + 2$
        $12 + 7 = 17 + 2$
        $19 = 19$

10. a. As a graph:

    b. 3

    c. Coloring:

11. Coloring:

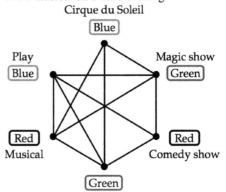

12. Three evenings: Cirque du Soleil and a play on
    one evening; a magic show and a tribute band
    concert on another evening; and a comedy show
    and a musical on a third evening.

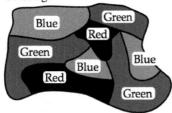

Chapter 6: Numeration Systems and Number Theory

EXCURSION EXERCISES, SECTION 6.1

1.

		¹S		
²C	³O	L	U	M	⁴N
	N		M		I
⁵T	E	N			N
	S		⁶Z	E	R	O

3. $357 = 3 \times 100 + 5 \times 10 + 7$

三
百
五
十
七

5. a. 4 numerals: 4, 5, 2, and 8.

   b. 7 numerals:

四千五百二十 and 八

EXERCISES 6.1

1. $46 = 40 + 6$
∩∩∩∩||||||

3. $103 = 100 + 3$
911|

5. $2568 = 2000 + 500 + 60 + 8$

7. $23,402 = 20,000 + 3,000 + 400 + 2$

9. $65,800 = 60,000 + 5,000 + 800$

11. $1,405,203$
$= 1,000,000 + 400,000 + 5,000 + 200 + 3$

13. $2 \times 1,000 + 100 + 3 \times 10 + 4 \times 1 = 2134$

15. $8 \times 100 + 4 \times 10 + 5 = 845$

17. $1 \times 1000 + 2 \times 100 + 3 \times 10 + 2 = 1232$

19. $2 \times 100,000 + 2 \times 10,000 + 1,000 + 10 + 1$
$= 221,011$

21. $6 \times 10,000 + 5 \times 1000 + 7 \times 100 + 6 \times 10 + 9$
$= 65,769$

23. $5 \times 1,000,000 + 1 \times 100,000 \times 2 \times 10,000$
$+ 2 \times 1000 + 4 \times 100 + 6$
$= 5,122,406$

25.

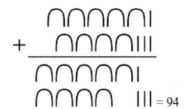

27.

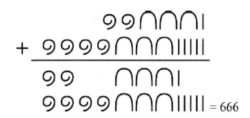

29.

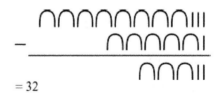
$= 32$

31.

Borrow 1 heel bone and change it to ten strokes.
Borrow 1 scroll and change it to 10 heel bones.
∩∩∩∩||||||| $= 56$

33. $CLXI = 100 + 50 + 10 + 1 = 161$

35. $DCL = 500 + 100 + 50 = 650$

37. $MCDIX = 1000 + (500 - 100) + (10 - 1) = 1409$

39. $MCCXL = 1,000 + 100 + 100 + (50 - 10)$
$= 1240$

41. $DCCCXL = 500 + 100 + 100 + 100 + (50 - 10)$
$= 840$

71

43. $\overline{\text{IX}}$XLIV
$$= (10-1)\times 1000 + (50-10) + (5-1)$$
$$= 9044$$

45. $\overline{\text{XI}}$CDLXI
$$= 11\times 1000 + (500-100) + 50 + 10 + 1$$
$$= 11{,}461$$

47. $39 = 10 + 10 + 10 + (10-1) = \text{XXXIX}$

49. $157 = 100 + 50 + 5 + 1 + 1 = \text{CLVII}$

51. $542 = 500 + (50-10) + 2 = \text{DXLII}$

53. $1197 = 1000 + 100 + (100-10) + 5 + 1 + 1$
$= \text{MCXCVII}$

55. 787
$= 500 + 100 + 100 + 50 + 10 + 10 + 10 + 5 + 2$
$= \text{DCCLXXXVII}$

57. 683
$= 500 + 100 + 50 + 10 + 10 + 10 + 3$
$= \text{DCLXXXIII}$

59. $6898 = 6\times 1000 + 500 + 100 + 100 + 100$
$$+ (100-10) + 5 + 3$$
$$= \overline{\text{VI}}\text{DCCCXCVIII}$$

61.
√	1	47
	2	94
√	4	188
	5	235

$5\times 47 = 235$

63.
	1	63
	2	126
	4	252
√	8	504
	8	504

$8\times 63 = 504$

65.
√	1	29
√	2	58
√	4	116
	7	203

$7\times 29 = 203$

67.
√	1	35
	2	70
	4	140
	8	280
√	16	560
	17	595

$17\times 35 = 595$

69.
√	1	108
√	2	216
√	4	432
	8	864
√	16	1728
	23	2484

$23\times 108 = 2484$

71. a. Answers will vary. For example, it may be easier to perform addition and subtraction in the Egyptian numeration system.

b. Answers will vary. For example, less symbols are needed to express some numbers in the Roman numeration system.

73. Reports will vary. The Ionic Greek numeration system gained popularity in Greco-Roman times, starting around 100 B.C. It consisted of 24 Greek letters and three letters taken from the Phoenician alphabet. The first nine letters were used to represent the counting numbers from 1 to 9. The next nine letters were used to represent the multiples of 10 from 10 to 90. The third set of letters was used to represent the multiples of 100 from 100 to 900. For example, α (alpha) represented 1, λ (lambda) represented 30, and φ (phi) represented 500. In the Ionic Greek numeration system, the value of a numeral was the sum of each of the individual numerals that comprised it. For instance, 531 was represented by $\alpha\lambda\varphi$ (500 + 30 + 1). A prime symbol was used to multiply the numeral by 1,000. One of the advantages of the Ionic Greek system was that large numbers could be written using very few symbols. A disadvantage of the Ionic Greek system was that it was difficult to perform arithmetic using Ionic numerals.

EXCURSION EXERCISES, SECTION 6.2

1. Rewriting:
$$\begin{bmatrix} 724 \\ -351 \\ \hline \end{bmatrix} \rightarrow \begin{bmatrix} 724 \\ +648 \\ \hline 1372 \end{bmatrix} \rightarrow \begin{bmatrix} 372 \\ + 1 \\ \hline 373 \end{bmatrix}$$
$724 - 351 = 373$

3. Rewriting:
$$\begin{bmatrix} 91{,}572 \\ -07{,}824 \\ \hline \end{bmatrix} \rightarrow \begin{bmatrix} 91{,}572 \\ +92{,}175 \\ \hline 183{,}747 \end{bmatrix} \rightarrow \begin{bmatrix} 83{,}747 \\ + 1 \\ \hline 83{,}748 \end{bmatrix}$$
$91{,}572 - 7824 = 83{,}748$

5. Rewriting:

$$\begin{bmatrix} 3,156,782 \\ -0,875,236 \end{bmatrix} \rightarrow \begin{bmatrix} 3,156,782 \\ +9,124,763 \\ \hline 12,281,545 \end{bmatrix} \rightarrow \begin{bmatrix} 2,281,545 \\ + \qquad 1 \\ \hline 2,281,546 \end{bmatrix}$$

$$3,156,782 - 875,236 = 2,281,546$$

7. The nines complement and the end-around carry procedure produce a correct answer because applying these processes is equivalent to adding and subtracting the same number from a given difference.
Example:

$$641 - 235 = 641 + 999 - 235 - 999$$
$$= 641 + (999 - 235) - 999$$
$$= (641 + 764) - 999$$
$$= 1405 - 1000 + 1$$
$$= 406$$

Adding the nines complement of 235 to 641 produces a result that is 999 larger than $641 - 235$. Applying the end carry-around procedure decreases the previous sum by 999. Thus adding the nines complement and performing the end-around carry nullify each other and produce the correct difference.

EXERCISES 6.2

1. $48 = (4 \times 10^1) + (8 \times 10^0)$

3. $420 = (4 \times 10^2) + (2 \times 10^1) + (0 \times 10^0)$

5. 6803
$= (6 \times 10^3) + (8 \times 10^2) + (0 \times 10^1) + (3 \times 10^0)$

7. $10,208$
$= (1 \times 10^4) + (0 \times 10^3) + (2 \times 10^2) + (0 \times 10^1) + (8 \times 10^0)$

9. $400 + 50 + 6 = 456$

11. $5000 + 70 + 6 = 5076$

13. $30,000 + 5000 + 400 + 7 = 35,407$

15. $600,000 + 80,000 + 3000 + 40 = 683,040$

17. Expanding:
$$35 = (3 \times 10) + 5$$
$$\underline{+41 = (4 \times 10) + 1}$$
$$= (7 \times 10) + 6 = 76$$

19. Expanding:
$$257 = (2 \times 100) + (5 \times 10) + 7$$
$$\underline{+138 = (1 \times 100) + (3 \times 10) + 8}$$
$$= (3 \times 100) + (8 \times 10) + 15$$
$$= (3 \times 100) + (9 \times 10) + 5 = 395$$

21. Expanding:
$$1023 = (1 \times 1000) + (0 \times 100) + (2 \times 10) + 3$$
$$\underline{+1458 = (1 \times 1000) + (4 \times 100) + (5 \times 10) + 8}$$
$$= (2 \times 1000) + (4 \times 100) + (7 \times 10) + 11$$
$$= (2 \times 1000) + (4 \times 100) + (8 \times 10) + 1$$
$$= 2481$$

23. Write 60 as $5 \times 10 + 10$.
$$62 = (5 \times 10) + 12$$
$$\underline{-35 = (3 \times 10) + 5}$$
$$= (2 \times 10) + 7 = 27$$

25. Write 700 as $6 \times 100 + 10 \times 10$.
$$4725 = (4 \times 1000) + (6 \times 100) + (12 \times 10) + 5$$
$$\underline{-1362 = (1 \times 1000) + (3 \times 100) + (6 \times 10) + 2}$$
$$= (3 \times 1000) + (3 \times 100) + (6 \times 10) + 3$$
$$= 2481$$

27. Write 3000 as $2 \times 1000 + 10 \times 100$.
$$23,168 = 2 \cdot 10^4 + 2 \cdot 10^3 + 11 \cdot 10^2 + 6 \cdot 10 + 8$$
$$\underline{-12,857 = 1 \cdot 10^4 + 2 \cdot 10^3 + 8 \cdot 10^2 + 5 \cdot 10 + 7}$$
$$= 1 \cdot 10^4 + 0 \cdot 10^3 + 3 \cdot 10^2 + 1 \cdot 10 + 1$$
$$= 10,311$$

29. $20 + 3 = 23$

31. $(1 \times 60) + 30 + 7 = 97$

33. Expanding:
$$(20 \times 60^2) + (2 \times 60) + 10 + 3$$
$$= (20 \times 3600) + (2 \times 60) + 10 + 3$$
$$= 72,133$$

35. Expanding:
$$(10 \times 60^3) + (3 \times 60^2) + (11 \times 60) + 6$$
$$= (10 \times 216,000) + (3 \times 3600) + (660) + 6$$
$$= 2,171,466$$

37. $42 = 40 + 2$

39. $128 = 2 \times 60 + 8$

41. $5678 = (1 \times 60^2) + (34 \times 60^1) + (38 \times 60^0)$

43. $10{,}584 = (2 \times 60^2) + (56 \times 60^1) + (24 \times 60^0)$

45. $21{,}345 = (5 \times 60^2) + (55 \times 60^1) + (45 \times 60^0)$

47. Converting:

$$45$$
$$\underline{+23}$$
$$68 = 60 + 8$$

49. Converting:

$$33 \cdot 60 + 42$$
$$\underline{+32 \cdot 60 + 21}$$
$$65 \cdot 60 + 63 = 66 \cdot 60 + 3 = 1 \cdot 60^2 + 6 \cdot 60 + 3$$

51. Converting:

$$10 \cdot 60^2 + 31 \cdot 60 + 43$$
$$\underline{+1 \cdot 60^2 + 21 \cdot 60 + 32}$$
$$11 \cdot 60^2 + 52 \cdot 60 + 75 = 66 \cdot 60 + 3$$
$$= 11 \cdot 60^2 + 53 \cdot 60 + 15$$

53. $9 \times 20 + 14 \times 1 = 194$

55. $5 \times 360 + 0 \times 20 + 3 \times 1 = 1803$

57. $2 \times 7200 + 0 \times 360 + 4 \times 20 + 12 \times 1 = 14{,}492$

59. $5 \times 7200 + 0 \times 360 + 5 \times 20 + 3 \times 1 = 36{,}103$

61. $137 = (6 \times 20^1) + (17 \times 20^0)$

63. $948 = (2 \times 360) + (11 \times 20^1) + (8 \times 20^0)$

65. $1693 = (4 \times 360) + (12 \times 20^1) + (13 \times 20^0)$

67. $7432 = (1 \times 7200) + (0 \times 360) + (11 \times 20^1)$
$\quad + (12 \times 20^0)$

69. a. Answers will vary.

    b. Answers will vary.

71. a. $(2 \times 3^1) + (1 \times 3^0) = 7$

    b. $(2 \times 3^2) + (0 \times 3^1) + (2 \times 3^0) = 20$

    c. $(1 \times 3^3) + (0 \times 3^2) + (2 \times 3^1) + (2 \times 3^0) = 35$

    d. $37 = (1 \times 3^3) + (1 \times 3^2) + (0 \times 3^1) + (1 \times 3^0)$
        UUZU

    e. $87 = (1 \times 3^4) + (0 \times 3^3) + (0 \times 3^2) + (2 \times 3^1)$
        $+ (0 \times 3^0)$ or UZZTZ

    f. $144 = (1 \times 3^4) + (2 \times 3^3) + (1 \times 3^2) + (0 \times 3^1) + (0 \times 3^0)$ or UTUZZ

EXCURSION EXERCISES, SECTION 6.3

1. The cards are in numeric order from smallest, $00001_{\text{two}} = 1$, to largest $11111_{\text{two}} = 31$.

    The first implementation of the sorting procedure places all odd numerals behind the even numerals. Thus the base two numerals of the form xxxx1 are behind the base two numerals of the form xxxx0, where x represents either a 0 or a 1.

    The second implementation of the sorting procedure places the base two numerals of the form xxx1x behind the base two numerals of the form xxx0x. Thus from front to back, the base two numerals are now in the order xxx00, xxx01, xxx10, xxx11.

    The third implementation of the sorting procedure places the base two numerals of the form xx1xx behind the base two numerals of the form xx0xx. Thus the order of the cards from front to back is now given by the base two numerals xx000, xx001, xx010, xx011, xx100, x101, xx110, xx111.

    The fourth implementation of the sorting procedure places the base two numerals of the form of the form x1xxx behind the base numerals of the form x0xxx. Thus the order of the cards from front to back is now given by the base two numerals x0000, x0001, x0010, x0011, x0100, x0101, x0110, x0111, x1000, x1001, x1010, x1011, x1100, x1101, x1110, x1111.

The fifth implementation of the sorting procedure places the base two numerals of the form 1xxxx behind the base two numerals of the form 0xxxx. Thus the order of the cards, from front to back, is given by the base two numerals 00001, 00010, 00011, 00100, 00101, 00110, 00111, 01000, 01001, 01010, 01011, 01100, 01101, 01110, 01111, 10000, 10001, 10010, 10011, 10100, 10101, 10110, 10111, 11000, 11001, 11010, 11011, 11100, 11101, 11110, 11111. In terms of base ten numerals, the cards are now arranged in the order 1, 2, 3, 4, …, 31.

3. A base three number system requires three numerals, such as 0, 1, and 2. A notch in a card can be used to represent a 0 and a hole in a card can be used to represent a 1, as in base two, but there is no convenient method that can be used to represent the numeral 2.

EXERCISE SET 6.3

1. Converting:
$$243_{\text{five}} = 2 \times 5^2 + 4 \times 5^1 + 3 \times 5^0$$
$$= 2 \times 25 + 4 \times 5 + 3 \times 1$$
$$= 73$$

3. Converting:
$$67_{\text{nine}} = 6 \times 9^1 + 7 \times 9^0$$
$$= 6 \times 9 + 7 \times 1$$
$$= 61$$

5. Converting:
$$3154_{\text{six}} = 3 \times 6^3 + 1 \times 6^2 + 5 \times 6^1 + 4 \times 6^0$$
$$= 3 \times 216 + 1 \times 36 + 5 \times 6 + 4$$
$$= 718$$

7. Converting:
$$13211_{\text{four}} = 1 \times 4^4 + 3 \times 4^3 + 2 \times 4^2 + 1 \times 4^1 + 1 \times 4^0$$
$$= 1 \times 256 + 3 \times 64 + 2 \times 16 + 1 \times 4 + 1$$
$$= 485$$

9. Converting:
$$B5_{\text{sixteen}} = 11 \times 16^1 + 5 \times 16^0$$
$$= 11 \times 16 + 5$$
$$= 181$$

11.
5	267	
5	53	2
5	10	3
	2	0

$$267 = 2032_{\text{five}}$$

13.
6	1932	
6	322	0
6	53	4
6	8	5
	1	2

$$1932 = 12540_{\text{six}}$$

15.
9	15306	
9	1700	6
9	188	8
9	20	8
	2	2

$$15306 = 22886_{\text{nine}}$$

17.
2	4060	
2	2030	0
2	1015	0
2	507	1
2	253	1
2	126	1
2	63	0
2	31	1
2	15	1
2	7	1
2	3	1
	1	1

$$4060 = 11111101110_{\text{two}}$$

19.
12	283	
12	23	7
	1	11 (B in base twelve)

$$283 = 1B7_{\text{twelve}}$$

21. Expanding:
$$1101_{\text{two}} = 1 \times 2^3 + 1 \times 2^2 + 0 \times 2^1 + 1 \times 2^0$$
$$= 1 \times 8 + 1 \times 4 + 0 \times 2 + 1 \times 1$$
$$= 13$$

23. Expanding:
$$11011_{\text{two}}$$
$$= 1 \times 2^4 + 1 \times 2^3 + 0 \times 2^2 + 1 \times 2^1 + 1 \times 2^0$$
$$= 1 \times 16 + 1 \times 8 + 0 \times 4 + 1 \times 2 + 1 \times 1$$
$$= 27$$

25. Expanding:
$$1100100_{\text{two}}$$
$$= 1 \times 2^6 + 1 \times 2^5 + 0 \times 2^4 + 0 \times 2^3$$
$$\quad + 1 \times 2^2 + 0 \times 2^1 + 0 \times 2^0$$
$$= 1 \times 64 + 1 \times 32 + 0 \times 16 + 1 \times 8$$
$$\quad + 1 \times 4 + 0 \times 2 + 0 \times 1$$
$$= 100$$

27. Expanding:

10001011_{two}

$$= 1\times 2^7 + 0\times 2^6 + 0\times 2^5 + 0\times 2^4$$
$$+ 1\times 2^3 + 0\times 2^2 + 1\times 2^1 + 1\times 2^0$$
$$= 1\times 128 + 0\times 64 + 0\times 32 + 0\times 16$$
$$+ 1\times 8 + 0\times 4 + 1\times 2 + 1\times 1$$
$$= 139$$

29.
```
1     0     1     0     0     1
   double dabble double double dabble
     = 2   = 5   = 10  = 20  = 41
```
$101001_{two} = 41$

31.
```
1    0    1    1    0    1    0
  double dabble dabble double dabble double
    = 2  = 5  = 14 = 22 = 45 = 90
```
$1011010_{two} = 90$

33.
```
1     0     1     0     0     1
   double dabble double double dabble
     = 2   = 5   = 10  = 20  = 41
1     1     0     1     0
   dabble dabble double dabble double
     = 83  = 167 = 334 = 669 = 1338
```
$10100111010_{two} = 1338$

35. First, convert the number to base ten.

$$34_{six} = 3\times 6^1 + 4\times 6^0$$
$$= 3\times 6 + 4\times 1$$
$$= 22$$

Now, change 22 into base eight.
```
8|22
   2  6
```
$34_{six} = 28_{eight}$

37. First, convert the number to base ten.

$$878_{nine} = 8\times 9^2 + 7\times 9^1 + 8\times 9^0$$
$$= 8\times 81 + 7\times 9 + 8\times 1$$
$$= 719$$

Now, change 719 into base four.
```
4|719
4|179  3
4|4    3
4|11   0
   2   3
```
$878_{nine} = 23033_{four}$

39. First, convert the number to base ten.

$$1110_{two} = 1\times 2^3 + 1\times 2^2 + 1\times 2^1 + 0\times 2^0$$
$$= 14$$

Now, change 14 into base five.
```
5|14
   2  4
```
$1110_{two} = 24_{five}$

41. First, convert the number to base ten.

$$3440_{eight} = 3\times 8^3 + 4\times 8^2 + 4\times 8^1 + 0\times 8^0$$
$$= 3\times 512 + 4\times 64 + 4\times 8 + 0\times 1$$
$$= 1824$$

Now, change 1824 into base nine.
```
9|1824
9|202  6
9|22   4
   2   4
```
$3440_{eight} = 2446_{nine}$

43. First, convert the number to base ten.

$$56_{sixteen} = 5\times 16^1 + 6\times 16^0$$
$$= 5\times 16 + 6\times 1$$
$$= 86$$

Now change 86 into base eight.
```
8|86
8|10  6
   1  2
```
$56_{sixteen} = 128_{eight}$

45. First, convert the number to base ten.

$$A4_{twelve} = 10\times 12^1 + 4\times 12^0$$
$$= 10\times 12 + 4\times 1$$
$$= 124$$

Now, change 124 into base sixteen.
```
16|124
    7   12 (C in base sixteen)
```
$A4_{twelve} = 7C_{sixteen}$

47. Use the tables given to write the 3-digit binary representation of each digit.

3	5	2	(base eight)
011	101	010	(base two)

$352_{eight} = 11101010_{two}$

49. Starting at the right, divide the numeral into groups of 3 digits, adding zeroes as needed. Replace each 3-digit group with its base eight symbol.

011	001	010	(base two)
3	1	2	(base eight)

$11001010_{two} = 312_{eight}$

51. Starting at the right, divide into groups of 4 digits, adding zeroes in front if needed. Replace each 4-digit group with its base sixteen symbol. You may write base ten first, if that is easier.

0001	0101	0001	(base two)
1	5	1	(base ten)
1	5	1	(base sixteen)

$101010001_{two} = 151_{sixteen}$

53. Use the tables given to write the 4-digit binary representation of each digit.

B	E	F	3	(base sixteen)
11	14	15	3	(base ten)
1011	1110	1111	0011	(base two)

$BEF3_{sixteen} = 1011111011110011_{two}$

55. Use the tables given to write the 4-digit binary representation of each digit or change to base ten digits and replace each one with a 4-digit binary equivalent.

B	A	5	C	F	(base 16)
11	10	5	12	15	(base 10)
1011	1010	0101	1100	1111	(base 2)

$BA5CF_{sixteen} = 10111010010111001111_{two}$

57. Answers will vary. Start at the left with the first number that is not a zero and move to the right. Every time you pass by a zero, triple your current number. Every time you pass by a 1, whipple. Whippling is accomplished by tripling your current number and adding 1. Every time you pass by a 2, zipple. Zippling is accomplished by tripling your current number and adding 2.

59. The following chart shows the American Standard Code for Information Interchange (ASCII) character set. Many modern computers make use of ASCII to represent internally the characters that can be typed on a computer keyboard. The chart for coding ASCII characters uses the hexadecimal numeration system. For instance, the capital letter A is shown in row 4 and column 1. Thus the ASCII code for the capital letter A is $41_{sixteen}$, which is 65 in base ten and 01000001 as an 8-bit binary numeral. The lower-case letter m is shown in row 6 and column D. The ASCII code for a lower-case m is $6D_{sixteen}$, which is 109 in base ten and 01101101 as an 8-bit binary numeral. *(Source:* http://www.aelius.com/products/asciitable.phtml)

	0	1	2	3	4	5	6	7	8	9	A	B	C	D	E	F
0	NUL	SOH	STX	ETX	EOT	ENQ	ACK	BEL	BS	HT	LF	VT	FF	CR	SO	SI
1	DLE	DC1	DC2	DC3	DC4	NAK	SYN	ETB	CAN	EM	SUB	ESC	FS	GS	RS	US
2	SPC	!	"	#	$	%	&	'	(	)	*	+	,	−	.	/
3	0	1	2	3	4	5	6	7	8	9	:	;	<	=	>	?
4	@	A	B	C	D	E	F	G	H	I	J	K	L	M	N	O
5	P	Q	R	S	T	U	V	W	X	Y	Z	[	\	]	^	_
6	`	a	b	c	d	e	f	g	h	i	j	k	l	m	n	o
7	p	q	r	s	t	u	v	w	x	y	z	{	|	}	~	DEL

61. Converting:

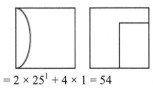

$= 2 \times 25^1 + 4 \times 1 = 54$

63. The numeral for 15 is the numeral for 3 rotated 90° counterclockwise.

65. To represent 8, overlay the symbol for 5 on the symbol for 3.

67. The first symbol is the symbol for 10, since it is a 2 rotated. It is on the left so it is to be multiplied by 25. The second symbol is the symbol for 6, since it is the overlay of 5 and 1. The number is 256.

69. a and b. Answers will vary.

EXCURSION EXERCISES, SECTION 6.4

1. The ones complement of the subtrahend is
    0110_{two}.

    $1110_{two} + 0110_{two} = 10100_{two}$
    $1110_{two} - 1001_{two} = 101_{two}$

3. The ones complement of the subtrahend is
    110100010_{two}.

    $101001010_{two} + 110100010_{two} = 1011101100_{two}$
    $101001010_{two} - 1011101_{two} = 11101101_{two}$

5. The ones complement of the subtrahend is
    1110110000_{two}.

    $1111101011_{two} + 1110110000_{two}$
    $= 11110011011_{two}$
    $1111101011_{two} - 1001111_{two}$
    $= 1110011100_{two}$

EXERCISE SET 6.4

1. Adding:

    $$\begin{array}{cccc} & & 1 & \\ & 2 & 0 & 4_{five} \\ + & 1 & 2 & 3_{five} \\ \hline & 3 & 3 & 2_{five} \end{array}$$

3. Adding:

    $$\begin{array}{cccc} 1 & & 1 & \\ 5 & 6 & 2 & 5_{seven} \\ + & 6 & 3 & 4_{seven} \\ \hline 6 & 5 & 6 & 2_{seven} \end{array}$$

5. Adding:

    $$\begin{array}{ccccccc} 1 & 1 & & 1 & 1 & 1 & \\ & 1 & 1 & 0 & 1 & 0 & 1_{two} \\ + & & 1 & 0 & 0 & 1 & 1_{two} \\ \hline 1 & 0 & 0 & 1 & 1 & 0 & 0 & 0_{two} \end{array}$$

7. Adding:

    $$\begin{array}{cccc} 1 & 1 & 1 & \\ & 8 & B & 5_{twelve} \\ + & 5 & 7 & 8_{twelve} \\ \hline 1 & 2 & 7 & 1_{twelve} \end{array}$$

9. Adding:

    $$\begin{array}{ccccc} 1 & 1 & 1 & \\ C & 4 & 8 & 9_{sixteen} \\ + & B & A & D_{sixteen} \\ \hline D & 0 & 3 & 6_{sixteen} \end{array}$$

11. Adding:

    $$\begin{array}{cccc} 1 & 1 & 1 & \\ & 4 & 3 & 5_{six} \\ + & 2 & 4 & 5_{six} \\ \hline 1 & 1 & 2 & 4_{six} \end{array}$$

13. Borrow 5 fives from the twenty-fives column.

    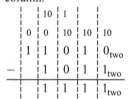

    $$\begin{array}{ccc} 3 & 13 & \\ 4 & 3 & 4_{five} \\ - 1 & 4 & 3_{five} \\ \hline 2 & 4 & 1_{five} \end{array}$$

15. Borrow 8 sixty-fours and 8 eights.

    $$\begin{array}{cccc} & 12 & 12 & \\ 6 & 13 & & \\ 7 & 3 & 2 & 5_{eight} \\ - & 5 & 6 & 3_{eight} \\ \hline 6 & 5 & 4 & 2_{eight} \end{array}$$

17. Borrow 2 ones from the twos column. Borrow 2
    twos by first borrowing 2 fours from the eights
    column. Then borrow 2 eights from the sixteens
    column.

    $$\begin{array}{ccccc} & 10 & 1 & & \\ 0 & 0 & 10 & 10 & 10 \\ 1 & 1 & 0 & 1 & 0_{two} \\ - & 1 & 0 & 1 & 1_{two} \\ \hline & 1 & 1 & 1 & 1_{two} \end{array}$$

19. Borrow 2 ones from the twos column by first
    borrowing 2 twos from the fours column. Then
    borrow 2 eights from the sixteens column. Next
    borrow 2 sixteens from the thirty-twos column
    by first borrowing 2 thirty-twos from the sixty-
    fours column. Finally, borrow 2 sixty-fours
    from the one hundred twenty-eights column.

    $$\begin{array}{cccccccc} & 10 & 1 & 10 & & & 1 & \\ 0 & 0 & 10 & 0 & 10 & 0 & 10 & 10 \\ 1 & 1 & 0 & 1 & 0 & 1 & 0 & 0_{two} \\ - & 1 & 0 & 1 & 1 & 0 & 1 & 1_{two} \\ \hline & 1 & 1 & 1 & 1 & 0 & 0 & 1_{two} \end{array}$$

21. Borrow 12 ones from the twelves column for a
    total of 19 or 17_{twelve} ones.

    $$\begin{array}{cccc} & & 9 & 17 \\ 4 & 3 & A & 7_{twelve} \\ - & 2 & 8 & 9_{twelve} \\ \hline 4 & 1 & 1 & A_{twelve} \end{array}$$

    So, $43A7_{twelve} - 289_{twelve} = 411A_{twelve}$

23. Borrow 9 ones from the nines column and 9 nines from the eighty-ones column.

$$
\begin{array}{ccc}
 & 15 & 12 \\
6 & 16 & \\
7 \quad 6 & 2_{nine} \\
- \quad 3 \quad 6 & 7_{nine} \\
\hline
3 \quad 8 & 4_{nine}
\end{array}
$$

25. $3 \times 5 = 15 = 23_{six}$
$3 \times 4 + 2 = 14 = 22_{six}$

$$
\begin{array}{ccc}
2 & 2 & \\
1 & 4 & 5_{six} \\
\times & & 3_{six} \\
\hline
5 & 2 & 3_{six}
\end{array}
$$

27. $2 \times 2 = 4 = 11_{three}$
$2 \times 1 + 1 = 3 = 10_{three}$
$2 \times 2 + 1 = 5 = 12_{three}$

$$
\begin{array}{ccccc}
1 & 1 & 1 & \\
 & 2 & 1 & 2_{three} \\
\times & & & 2_{three} \\
\hline
1 & 2 & 0 & 1_{three}
\end{array}
$$

29. $5 \times 4 = 20 = 24_{eight}$
$5 \times 5 + 2 = 27 = 3_{eight}$
$5 \times 7 + 2 = 37 = 45_{eight}$

$$
\begin{array}{ccccc}
2 & 3 & 2 & \\
7 & 3 & 5 & 4_{eight} \\
\times & & & 5_{eight} \\
\hline
4 & 5 & 2 & 3 & 4_{eight}
\end{array}
$$

31. Multiplying:

$$
\begin{array}{ccccccc}
1 & 0 & 1 & 0 & 1 & 0_{two} \\
\times & & & & 1 & 0_{two} \\
\hline
0 & 0 & 0 & 0 & 0 & 0 \\
1 & 0 & 1 & 0 & 1 & 0 \\
\hline
1 & 0 & 1 & 0 & 1 & 0 & 0_{two}
\end{array}
$$

33. $5 \times 3 = 15 = 17_{eight}$
$5 \times 5 + 1 = 26 = 32_{eight}$
$5 \times 4 + 3 = 23 = 27_{eight}$
$2 \times 5 = 10 = 12_{eight}$
$2 \times 4 + 1 = 9 = 11_{eight}$

$$
\begin{array}{cccccc}
 & 1 & 1 & & \\
 & & 4 & 5 & 3_{eight} \\
\times & & & 2 & 5_{eight} \\
\hline
 & 2 & 7 & 2 & 7 \\
1 & 1 & 2 & 6 & \\
\hline
1 & 4 & 2 & 0 & 7_{eight}
\end{array}
$$

35. $2 \times 3 = 6 = 12_{four}$
$2 \times 2 + 1 = 5 = 11_{four}$
$2 \times 3 + 1 = 7 = 13_{four}$
$3 \times 3 = 9 = 21_{four}$
$3 \times 2 + 2 = 8 = 20_{four}$
$3 \times 3 + 2 = 11 = 23_{four}$
$3 \times 1 + 2 = 5 = 11_{four}$

$$
\begin{array}{cccccc}
 & 2 & 2 & 2 & \\
 & 1 & 3 & 2 & 3_{four} \\
\times & & 1 & 3 & 2_{four} \\
\hline
 & 3 & 3 & 1 & 2 \\
1 & 1 & 3 & 0 & 1 \\
1 & 3 & 2 & 3 & \\
\hline
3 & 2 & 1 & 2 & 2 & 2_{four}
\end{array}
$$

37. $5_{sixteen} \times D_{sixteen} = 5 \times 13 = 65 = 41_{sixteen}$
$5_{sixteen} \times A_{sixteen} + 4_{sixteen} = 5 \times 10 + 4 = 54$
$\qquad\qquad = 36_{sixteen}$
$5_{sixteen} \times B_{sixteen} + 3_{sixteen} = 5 \times 11 + 3 = 58$
$\qquad\qquad = 3A_{sixteen}$

$$
\begin{array}{cccc}
3 & & 4 & \\
B & A & D_{sixteen} \\
\times & & 5_{sixteen} \\
\hline
3 & A & 6 & 1_{sixteen}
\end{array}
$$

39. Dividing:

$$
\begin{array}{r}
3 \quad 3_{four} \\
2_{four}\overline{)1 \quad 3 \quad 2_{four}} \\
\underline{1 \quad 2} \\
1 \quad 2 \\
\underline{1 \quad 2} \\
0_{four}
\end{array}
$$

Quotient: 33_{four}

41. Dividing:

$$
\begin{array}{r}
3 \quad 3_{four} \\
3_{four}\overline{)2 \quad 3 \quad 1_{four}} \\
\underline{2 \quad 1} \\
2 \quad 1 \\
\underline{2 \quad 1} \\
0_{four}
\end{array}
$$

Quotient: 33_{four}

43. Dividing:

$$
\begin{array}{r}
1\ 2\ 2\ 3_{\text{six}} \\
4_{\text{six}}\overline{)5\ 3\ 4\ 1_{\text{six}}} \\
\underline{4} \\
1\ 3 \\
\underline{1\ 2} \\
1\ 4 \\
\underline{1\ 2} \\
2\ 1 \\
\underline{2\ 0} \\
1_{\text{six}}
\end{array}
$$

Quotient: 1223_{six} Remainder: 1_{six}

45. Dividing:

$$
\begin{array}{r}
1\ 1\ 1\ 0_{\text{two}} \\
11_{\text{two}}\overline{)1\ 0\ 1\ 0\ 1\ 0_{\text{two}}} \\
\underline{1\ 1} \\
1\ 0\ 0 \\
\underline{1\ 1} \\
1\ 1 \\
\underline{1\ 1} \\
0_{\text{two}}
\end{array}
$$

47. Dividing:

$$
\begin{array}{r}
A\ 8_{\text{twelve}} \\
5_{\text{twelve}}\overline{)4\ 5\ 7_{\text{twelve}}} \\
\underline{4\ 2} \\
3\ 7 \\
\underline{3\ 4} \\
3_{\text{twelve}}
\end{array}
$$

Quotient: $A8_{\text{twelve}}$ Remainder: 3_{twelve}

49. Dividing:

$$
\begin{array}{r}
1\ 4_{\text{five}} \\
12_{\text{five}}\overline{)2\ 3\ 4_{\text{five}}} \\
\underline{1\ 2} \\
1\ 1\ 4 \\
\underline{1\ 0\ 3} \\
1\ 1_{\text{five}}
\end{array}
$$

Quotient: 14_{five} Remainder: 11_{five}

51. Expand both numerals.

$$
\begin{aligned}
143_x &= 1\cdot x^2 + 4\cdot x^1 + 3\cdot x^0 \\
&= x^2 + 4x + 3 \\
10200_{\text{three}} &= 1\cdot 3^4 + 0\cdot 3^3 + 2\cdot 3^2 + 0\cdot 3^1 + 0\cdot 3^0 \\
&= 99
\end{aligned}
$$

Set them equal to each other and solve the quadratic equation.

$$
\begin{aligned}
x^2 + 4x + 3 &= 99 \\
x^2 + 4x - 96 &= 0 \\
(x+12)(x-8) &= 0 \\
x = -12,\ &x = 8
\end{aligned}
$$

Since the base must be positive, $x = 8$.

53. a. $384 + 245 = 629$

b.   Convert 384 and 245 to base two.

$2\lfloor 384$		$2\lfloor 245$	
$2\lfloor 192$	0	$2\lfloor 122$	1
$2\lfloor 96$	0	$2\lfloor 61$	0
$2\lfloor 48$	0	$2\lfloor 30$	1
$2\lfloor 24$	0	$2\lfloor 15$	0
$2\lfloor 12$	0	$2\lfloor 7$	1
$2\lfloor 6$	0	$2\lfloor 3$	1
$2\lfloor 3$	0	1	1
$2\lfloor 1$	1		
1	1		

$384 = 110000000_{\text{two}}$
$245 = 11110101_{\text{two}}$

c.   Add the results in base two.

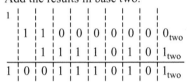

$$
\begin{array}{r}
1 \\
1\ 1\ 0\ 0\ 0\ 0\ 0\ 0\ 0_{\text{two}} \\
1\ 1\ 1\ 1\ 0\ 1\ 0\ 1_{\text{two}} \\
\hline
1\ 0\ 0\ 1\ 1\ 1\ 0\ 1\ 0\ 1_{\text{two}}
\end{array}
$$

d.   Convert the answer from part c. to base ten.
1001110101_{two}

$$
\begin{aligned}
&= 1\cdot 2^9 + 0\cdot 2^8 + 0\cdot 2^7 + 1\cdot 2^6 + 1\cdot 2^5 \\
&\quad + 1\cdot 2^4 + 0\cdot 2^3 + 1\cdot 2^2 + 0\cdot 2^1 + 1\cdot 2^0 \\
&= 629
\end{aligned}
$$

e.   The answers are the same.

55. a. $247 \times 26 = 6422$

b.   Convert 247 and 26 to base two.

$2\lfloor 247$		$2\lfloor 26$	
$2\lfloor 123$	1	$2\lfloor 13$	0
$2\lfloor 61$	1	$2\lfloor 6$	1
$2\lfloor 30$	1	$2\lfloor 3$	1
$2\lfloor 15$	1	1	1
$2\lfloor 7$	1		
$2\lfloor 3$	1		
1	1		

$247 = 11110111_{\text{two}}$
$26 = 11010_{\text{two}}$

c.   Multiply the results in base two.

$$
\begin{array}{r}
1\ 1\ 1\ 1\ 0\ 1\ 1\ 1_{\text{two}} \\
\times1\ 1\ 0\ 1\ 0_{\text{two}} \\
\hline
0\ 0\ 0\ 0\ 0\ 0\ 0\ 0 \\
1\ 1\ 1\ 1\ 0\ 1\ 1\ 1 \\
0\ 0\ 0\ 0\ 0\ 0\ 0\ 0 \\
1\ 1\ 1\ 1\ 0\ 1\ 1\ 1 \\
1\ 1\ 1\ 1\ 0\ 1\ 1\ 1 \\
\hline
1\ 1\ 0\ 0\ 1\ 0\ 0\ 0\ 1\ 0\ 1\ 1\ 0_{\text{two}}
\end{array}
$$

d. Convert the answer from part c to base ten.
$1100100010110_{\text{two}}$
$$= 1 \cdot 2^{12} + 1 \cdot 2^{11} + 0 \cdot 2^{10} + 0 \cdot 2^{9} + 1 \cdot 2^{8}$$
$$+ 0 \cdot 2^{7} + 0 \cdot 2^{6} + 0 \cdot 2^{5} + 1 \cdot 2^{4} + 0 \cdot 2^{3}$$
$$+ 1 \cdot 2^{2} + 1 \cdot 2^{1} + 0 \cdot 2^{0}$$
$$= 6422$$

e. The answers are the same.

57. The product of the ones digits is $4 \times 4 = 16$. Since the answer has a ones digit of 2, the base is either 14 or 7. Checking both shows that the base is 7.

59. In the base one numeration system, 0 would be the only numeral and the place values would be $1^0, 1^1, 1^2, 1^3, \ldots$, each of which equals 1. Thus 0 is the only number you could write using a base one numeration system.

61. M = 1, A = 4, S = 3, and O = 0

EXCURSION EXERCISES, SECTION 6.5

1. Factorial notation represents the product of every integer from 1 to the number. So, the first integer, $1,000,001! + 2$ can be written as $2m$, where m is some integer, because $1,000,001!$ has a factor of 2 . This is true of the other integers in the prime desert. The integer of the form $1,000,001! + I$ can be expressed as I times some number, so it is composite.

3. a. $21! + 2, 21! + 3, 21! + 4, \ldots, 21! + 21$

   b. $500,001! + 2, 500,001! + 3, \ldots,$
      $500,001! + 500,001$

   c. $7,000,000,001! + 2, 7,000,000,001! + 3,$
      $\ldots, 7,000,000,001! + 7,000,000,001$

EXERCISE SET 6.5

1. $20 = 1 \times 20 = 2 \times 10 = 4 \times 5$
   The divisors are 1, 2, 4, 5, 10, and 20.

3. $65 = 1 \times 65 = 5 \times 13$
   The divisors are 1, 5, 13, and 65.

5. $41 = 1 \times 41$. The divisors are 1 and 41.

7. $110 = 1 \times 110 = 2 \times 55 = 5 \times 22 = 10 \times 11$
   The divisors are 1, 2, 5, 10, 11, 22, 55, 110.

9. $385 = 1 \times 385 = 5 \times 77 = 7 \times 55 = 11 \times 35$
   The divisors are 1, 5, 7, 11, 35, 55, 77, 385.

11. The divisors of 21 are 1, 3, 7, and 21. Thus 21 is a composite number.

13. The only divisors of 37 are 1 and 37. Thus 37 is a prime number.

15. The only divisors of 101 are 1 and 101. Thus 101 is a prime number.

17. The only divisors of 79 are 1 and 79. Thus 79 is a prime number.

19. The divisors of 203 are 1, 7, 29, and 203. Thus 203 is a composite number.

21. 2: Yes. 210 is even.
    3: Yes. $2 + 1 + 0 = 3$, which is divisible by 3.
    4: No. 10 is not divisible by 4.
    5: Yes. 210 ends in a 0.
    6: Yes. 210 is divisible by both 2 and 3.
    8: No. 210 is not divisible by 8.
    9: No. $2 + 1 + 0 = 3$ and 3, which is not divisible by 9.
    10: Yes. 210 ends in a 0.

23. 2: No. 51 is not even
    3: Yes. $5 + 1 = 6$ and 6 is divisible by 3.
    4: No. 51 is not divisible by 4.
    5: No. 51 does not end in a 0 or 5.
    6: No. 51 is not divisible by both 2 and 3.
    8: No. 51 is not divisible by 8
    9: No. $5 + 1 = 6$ and 6 is not divisible by 9.
    10: No, because it does not end in a 0.

25. 2: Yes. 2568 is even.
    3: Yes. $2+5+6+8 = 21$, which is divisible by 3.
    4 : Yes. 68 is divisible by 4.
    5: No. 2568 does not end in a 0 or 5.
    6: Yes. 2568 is divisible by both 2 and 3.
    8: Yes. 568 is divisible by 8.
    9: No. $2+5+6+8=21$, which is not divisible by 9
    10: No. 2568 does not end in a 0.

27. 2: Yes. 4190 is even.
    3: No. $4+1+9+0=14$, which is not divisible by 3
    4: No. 90 is not divisible by 4.
    5: Yes. 4190 ends in a 0.
    6: No. 4190 is not divisible by both 2 and 3.
    8: No. 190 is not divisible by 8.
    9: No. $4+1+9+0=14$ and 14 is not divisible by 9
    10: Yes. 4190 ends in a 0.

29.

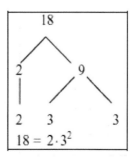

$$18 = 2 \cdot 3^2$$

31.

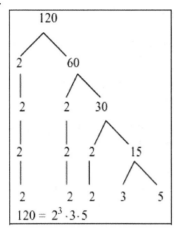

$$120 = 2^3 \cdot 3 \cdot 5$$

33.

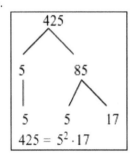

$$425 = 5^2 \cdot 17$$

35.

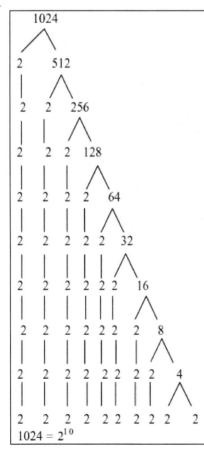

$$1024 = 2^{10}$$

37.

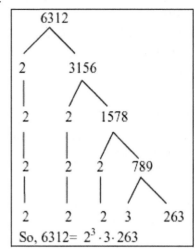

So, $6312 = 2^3 \cdot 3 \cdot 263$

39.

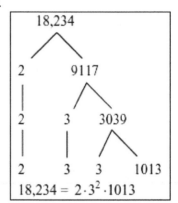

$$18{,}234 = 2 \cdot 3^2 \cdot 1013$$

41. Cross out all multiples of 2, 3, 5, 7, 11, and 13.
The rest of the numbers (bold) are prime.

2 3 4 **5** 6 **7** 8 9 10
11 12 **13** 14 15 16 **17** 18 **19** 20
21 22 **23** 24 25 26 27 28 **29** 30
31 32 33 34 35 36 **37** 38 39 40
41 42 **43** 44 45 46 **47** 48 49 50
51 52 **53** 54 55 56 57 58 **59** 60
61 62 63 64 65 66 **67** 68 69 70
71 72 **73** 74 75 76 77 78 **79** 80
81 82 **83** 84 85 86 87 88 **89** 90
91 92 93 94 95 96 **97** 98 99 100
101 102 **103** 104 105 106 **107** 108 **109** 110
111 112 **113** 114 115 16 117 118 119 120
121 122 123 124 125 126 **127** 128 129 130
131 132 133 134 135 136 **137** 138 **139** 140
141 142 143 144 145 146 147 148 **149** 150
151 152 153 154 155 156 **157** 158 159 160
161 162 **163** 164 165 166 **167** 168 169 170
171 172 **173** 174 175 176 177 178 **179** 180
181 182 183 184 185 186 187 188 189 190
191 192 **193** 194 195 196 **197** 198 **199** 200

43. 3 and 5, 5 and 7, 11 and 13, 17 and 19,
29 and 31, 41 and 43, 59 and 61, 71 and 73,
101 and 103, 107 and 109, 137 and 139,
149 and 151, 179 and 181, 191 and 193,
197 and 199.

45. 311 and 313, or 347 and 349.

47. a. $24 = 5 + 19$ also $24 = 11 + 13$

   b. $50 = 19 + 31$ also $50 = 3 + 47$

   c. $86 = 3 + 83$ also $86 = 7 + 79$ and others

   d. $144 = 5 + 139$ also $144 = 71 + 73$

   e. $210 = 11 + 199$ also $210 = 103 + 107$

   f. $264 = 7 + 257$ also $264 = 13 + 251$ and others

49. The double of the ones digit is 4.
$18 - 4 = 14$, which is divisible by 7.
Thus 182 is divisible by 7.

51. The double of the ones digit is 2.
$100 - 2 = 98$, which is divisible by 7.
Thus 1001 is divisible by 7.

53. The double of the ones digit is 2.
$1156 - 2 = 1154$
The double of the ones digit is 8.
$115 - 8 = 107$, which is not divisible by 7.
Thus 11,561 is not divisible by 7.

55. The double of the ones digit is
12 $20,431 - 12 = 20,419$.
The double of the ones digit is 18.
$2041 - 18 = 2023$
The double of the ones digit is 6. $202 - 6 = 196$.
The double of the ones digit is 12. $19 - 12 = 7$,
which is clearly divisible by 7.
Thus 204,316 is divisible by 7.

57. Four times the ones digit is 4. $9 + 4 = 13$, which
is divisible by 13.
Thus 91 is divisible by 13.

59. Four times the ones digit is 20.
$20 + 188 = 208$.
Four times the ones digit is 32.
$32 + 20 = 52$, which is divisible by 13.
Thus 1885 is divisible by 13.

61. Four times the ones digit is 28.
$28 + 1450 = 1478$.
Four times the ones digit is 32.
$32 + 147 = 179$.
Four times the ones digit is 36.
$36 + 17 = 53$, which is not divisible by 13
Thus 14,507 is not divisible by 13.

63. Four times the ones digit is 4.
$4 + 1335 = 1339$.
Four times the ones digit is 36.
$36 + 133 = 169$, which is divisible by 13.
Thus 13,351 is divisible by 13.

65. a. If $n = 1$, the two values are 2 and 0.
If $n = 2$, the values are 3 and 1. If $n = 3$,
the values are 7 and 5. Thus $n = 3$ is the
smallest value.

   b. The first few values are given in part a).
If $n = 4$, the values are 25 and 23.
Thus $n = 4$ is the smallest value.

67. To determine whether a number is divisible by
17, multiply the ones digit of the given number
by 5. Find the difference between this result and
the number formed by omitting the ones digit
from the given number. Keep repeating this
procedure until you obtain a small final
difference. If the final difference is divisible by
17, then the given number is divisible by 17. If
the final difference is not divisible by 17, then
the given number is not divisible by 17.

69. $60 = 6 \cdot 10 = 3 \cdot 2 \cdot 2 \cdot 5 = 2^2 \cdot 3^1 \cdot 5^1$
Adding one to each exponent gives the numbers
3, 2 and 2. Their product is $3 \cdot 2 \cdot 2 = 12$. There
are 12 distinct divisors of 60.

71. $297 = 3^3 \cdot 11$
Adding one to each exponent gives 4 and 2.
Their product is 8. There are 8 distinct divisors
of 297.

73. $360 = 36 \cdot 10 = 2 \cdot 3 \cdot 2 \cdot 3 \cdot 2 \cdot 5 = 2^3 \cdot 3^2 \cdot 5^1$
Adding one to each exponent gives the numbers
4, 3 and 2. Their product is $4 \cdot 3 \cdot 2 = 24$. There
are 24 distinct divisors of 360.

75. a. N is a product of all the prime numbers.
Thus $N - 1$, which must have at least one
prime factor, must share at least one prime
factor with the number N.

   b. The distributive property of multiplication
over addition (subtraction).

77. a. Definition of a divisor

   b. If both a and j were greater than $\sqrt{n}$, we
would have $aj > \sqrt{n}\sqrt{n} = n = aj$. This
gives us $aj > aj$, which is a contradiction.

   c. $aj = n$, where a and j are both natural
numbers, so both a and j are by definition
divisors of n.

79. a – f. Each statement is a conjecture
(as of July 2016).

EXCURSION EXERCISES 6.6

1. $200 = 2^3 \cdot 5^2$.
   The sum of all powers of 2 is
   $1 + 2^1 + 2^2 + 2^3 = 15$
   The sum of all powers of 5 is $1 + 5^1 + 5^2 = 31$.
   The sum of the proper factors of 200 is
   $(15)(31) - 200 = 265$.
   Since $265 > 200$, 200 is abundant.

3. $325 = 5^2 \cdot 13$.
   The sum of all powers of 5 is $1 + 5^1 + 5^2 = 31$.
   The sum of all powers of 13 is $1 + 13^1 = 14$
   The sum of the proper factors of 325 is
   $(31)(14) - 325 = 109$
   Since $109 < 325$, 325 is deficient.

5. The only divisors of a prime number are 1 and the number itself. Thus the only proper factor of a prime number is one. Hence every prime number is deficient.

EXERCISE SET 6.6

1. The proper factors of 18 are 1, 2, 3, 6, 9. Their sum is $21 > 18$, so 18 is abundant.

3. The proper factors of 91 are 1, 7, 13. Their sum is $21 < 91$, so 91 is deficient.

5. The only proper factor of 19 is 1. Since $1 < 19$, 19 is deficient.

7. The proper factors of 204 are 1, 2, 3, 4, 6, 12, 17, 34, 68, 51, 102. Their sum is $300 > 204$, so 204 is abundant.

9. The proper factors of 610 are 1, 2, 5, 10, 61, 122, 305. Their sum is $506 < 610$, so 610 is deficient.

11. The proper factors of 291 are 1, 3, 97. Their sum is $101 < 291$, so 291 is deficient.

13. The proper factors of 176 are 1, 2, 4, 8, 11, 16, 22, 44, 88. Their sum is $196 > 176$, so 176 is abundant.

15. The proper factors of 260 are 1, 2, 4, 5, 10, 13, 20, 26, 52, 65, 130. Their sum $328 > 260$, so 260 is abundant.

17. $2^3 - 1 = 8 - 1 = 7$ which is prime.

19. $2^7 - 1 = 128 - 1 = 127$ which is prime.

21. Euclid's theorem says that for n prime and a prime Mersenne number $2^n - 1$, $2^{n-1}(2^n - 1)$ is a perfect number. Letting $n = 127$ yields the perfect number $2^{127-1}(2^{127} - 1) = 2^{126}(2^{127} - 1)$.

23. The number of digits in the number, b^x is the greatest integer of $(x \log b) + 1$.
    We have $b = 2$ and $x = 17$.
    $(x \log b) + 1 = (17 \log 2) + 1 \approx 6.117$
    The greatest integer is 6, so the number of digits in $2^{17} - 1$ is 6.

25. The number of digits in the number, b^x is the greatest integer of $(x \log b) + 1$.
    We have $b = 2$ and $x = 1,398,269$.
    $(x \log b) + 1 = (1,398,269 \log 2) + 1$
    ≈ 420921.91
    The greatest integer is 420,921, so the number of digits in $2^{1398269} - 1$ is 420,921.

27. The number of digits in the number, b^x is the greatest integer of $(x \log b) + 1$.
    We have $b = 2$ and $x = 6,972,593$.
    $(x \log b) + 1 = (6,972,593 \log 2) + 1$
    $\approx 2,098,960.6$
    The greatest integer is 2,098,960, so the number of digits in $2^{6972593} - 1$ is 2,098,960.

29. The number of digits in the number, b^x is the greatest integer of $(x \log b) + 1$.
    We have $b = 2$ and $x = 37,156,667$.
    $(x \log b) + 1 = (37,156,667 \, log \, 2) + 1$
    $\approx 11,185,272.31$
    The greatest integer is 11,185,272, so the number of digits in $2^{37156667} - 1$ is 11,185,272.

31. Substituting, we have $9^5 + 15^5 = z^5$ Evaluating, we have $59,049 + 759,375 = 818,424 = z^5$
    However, the fifth root of 818,424 is approximately 15.226, which is not a natural number.
    So, $x = 9$, $y = 15$ and $n = 5$ do not yield a solution to the equation, $x^n + y^n = z^n$

33. a. False. For instance, if $n = 11$, then
       $2^{11} - 1 = 2047 = 23 \cdot 89$

    b. False. Fermat's Last Theorem was the last of Fermat's theories (conjectures) that other mathematicians were able to establish.

    c. True.

    d. Conjecture

35. a. Fermat's Little Theorem states $a^n - a$ is divisible by n if n is prime.
       Substituting gives $12^7 - 12 = 35,831,796$, which is divisible by 7.

    b. Fermat's Little Theorem states $a^n - a$ is divisible by n if n is prime.
       Substituting gives $8^{11} - 8 = 8,589,934,584$, which is divisible by 11.

37. $8128 = 1^3 + 3^3 + 5^3 + \ldots + 13^3 + 15^3$

39. a.   The divisors of 6 are 1, 2, 3, 6.
$$\frac{1}{1}+\frac{1}{2}+\frac{1}{3}+\frac{1}{6}=\frac{6}{6}+\frac{3}{6}+\frac{2}{6}+\frac{1}{6}=\frac{12}{6}=2$$

   b.   The divisors of 28 are 1, 2, 4, 7, 14, 28.
$$\frac{1}{1}+\frac{1}{2}+\frac{1}{4}+\frac{1}{7}+\frac{1}{14}+\frac{1}{28}$$
$$=\frac{28}{28}+\frac{14}{28}+\frac{7}{28}+\frac{4}{28}+\frac{2}{28}+\frac{1}{28}$$
$$=\frac{56}{28}=2$$

41. The first five Fermat numbers formed using $n = 0, 1, 2, 3,$ and 4 are all prime numbers. In 1732, Euler discovered that the sixth Fermat number 4,294,967,297, formed using $n = 5$, is not a prime number because it is divisible by 641.

43. 70

CHAPTER 6 REVIEW EXERCISES

1. 4,506,325
$= 4,000,000 + 500,000 + 6000 + 300 + 20 + 5$

2. 3,124,043
$= 3,000,000 + 100,000 + 20,000 + 4000 + 40 + 3$

3. $2 \times 100,000 + 3 \times 1000 + 2 \times 10,000 + 1 \times 10 + 3$
$= 223,013$

4. $2 \times 100,000 + 3 \times 100 + 1 \times 1000 + 4 + 5 \times 10$
$+ 2 \times 10,000$
$= 221,354$

5. CCCXLIX
$= 100 + 100 + 100 + (50 - 10) + (10 - 1)$
$= 349$

6. DCCLXXIV
$= 500 + 100 + 100 + 50 + 10 + 10 + (5 - 1)$
$= 774$

7. Rewriting:
$\overline{\text{IX}}\text{DCXL} = (10,000 - 1000) + 500$
$+ 100 + (50 - 10)$
$= 9640$

8. Rewriting:
$\overline{\text{XCII}}\text{CDXLIV} = (100,000 - 10,000)$
$+ 2000 + (500 - 100)$
$+ (50 - 10) + (5 - 1)$
$= 92,444$

9. $567 = 500 + 60 + 7 = \text{DLXVII}$

10. $823 = 800 + 20 + 3 = \text{DCCCXXIII}$

11. $2489 = 2000 + 400 + 80 + 9 = \text{MMCDLXXXIX}$

12. $1335 = 1000 + 300 + 30 + 5 = \text{MCCCXXXV}$

13. $432 = 400 + 30 + 2$
$= (4 \times 10^2) + (3 \times 10^1) + (2 \times 10^0)$

14. 456,327
$= 400,000 + 50,000 + 6000 + 300 + 20 + 7$
$= (4 \times 10^5) + (5 \times 10^4) + (6 \times 10^3) + (3 \times 10^2)$
$+ (2 \times 10^1) + (7 \times 10^0)$

15. $5,000,000 + 30,000 + 8000 + 200 + 4$
$= 5,038,204$

16. $300,000 + 80,000 + 7000 + 900 + 60 = 387,960$

17. $(13 \times 60) + (21 \times 1) = 801$

18. $(26 \times 60) + (43 \times 1) = 1603$

19. $(21 \times 60^2) + (14 \times 60) + (1 \times 1) = 76,441$

20. $(24 \times 60^2) + (16 \times 60) + (33 \times 1) = 87,393$

21. $721 = (12 \times 60) + (1 \times 1)$

22. $1080 = (18 \times 60) + (0 \times 1)$

23. $12,543 = (3 \times 3600) + (29 \times 60) + (3 \times 1)$

24. $19,281 = (5 \times 3600) + (21 \times 60) + (21 \times 1)$

25. $(9 \times 20) + (14 \times 1) = 194$

26. $(13 \times 20) + (7 \times 1) = 267$

27. $(6 \times 360) + (0 \times 20) + (18 \times 1) = 2178$

28. $(18 \times 360) + (5 \times 20) + (0 \times 1) = 6580$

29. $522 = (1 \times 360) + (8 \times 20) + (2 \times 1)$

30. $346 = (17 \times 20) + (6 \times 1)$

31. $1862 = (5 \times 360) + (3 \times 20) + (2 \times 1)$

32. $1987 = (5 \times 360) + (9 \times 20) + (7 \times 1)$

33. $45_{six} = (4 \times 6^1) + (5 \times 6^0) = 29$

34. $172_{nine} = (1 \times 9^2) + (7 \times 9^1) + (2 \times 9^0) = 146$

35. $E3_{sixteen} = (14 \times 16^1) + (3 \times 16^0) = 227$

36. $1BA_{twelve} = (1 \times 12^2) + (11 \times 12^1) + (10 \times 12^0)$
$= 286$

37. $3\underline{|45}$
    $3\underline{|15} \quad 0$
    $3\underline{|5} \quad \ 0$
    $\quad 1$
    $45 = 1200_{three}$

38. $7\underline{|123}$
    $7\underline{|17} \quad 4$
    $\quad 2 \quad 3$
    $123 = 234_{seven}$

39. $11\underline{|862}$
    $11\underline{|78} \quad 4$
    $\quad 7 \quad 1$
    $862 = 714_{eleven}$

40. $12\underline{|3021}$
    $12\underline{|251} \quad 9$
    $12\underline{|20} \quad 11 \text{ (B in base twelve)}$
    $\quad 1 \quad 8$
    $3021 = 18B9_{twelve}$

41. $346_{nine} = 3 \times 9^2 + 4 \times 9^1 + 6 \times 9^0$
$= 285$
$= 1 \times 6^3 + 1 \times 6^2 + 5 \times 6^1 + 3 \times 6^0$
$= 1153_{six}$

42. $1532_{six} = (1 \times 6^3) + (5 \times 6^2) + (3 \times 6^1) + (2 \times 6^0)$
$= 416$
$= (6 \times 8^2) + (4 \times 8^1) + (0 \times 8^0)$
$= 640_{eight}$

43. $275_{twelve} = (2 \times 12^2) + (7 \times 12^1) + (5 \times 12^0)$
$= 377$
$= (4 \times 9^2) + (5 \times 9^1) + (8 \times 9^0)$
$= 458_{nine}$

44. $67A_{sixteen} = (6 \times 16^2) + (7 \times 16^1) + (10 \times 16^0)$
$= 1658$
$= (11 \times 12^2) + (6 \times 12^1) + (2 \times 12^0)$
$= B62_{twelve}$

45. $\begin{array}{cc} 011 & 100 \\ 3 & 4 \end{array}$ (base 2)
    (base 8)
    $11100_{two} = 34_{eight}$

46. $\begin{array}{ccc} 001 & 010 & 100 \\ 1 & 2 & 4 \end{array}$ (base 2)
    (base 8)
    $1010100_{two} = 124_{eight}$

47. $\begin{array}{ccc} 0011 & 1000 & 1101 \\ 3 & 8 & D \end{array}$ (base 2)
    (base 16)
    $1110001101_{two} = 38D_{sixteen}$

48. $\begin{array}{ccc} 0111 & 0101 & 0100 \\ 7 & 5 & 4 \end{array}$ (base 2)
    (base 16)
    $11101010100_{two} = 754_{sixteen}$

49. $\begin{array}{cc} 2 & 5 \\ 010 & 101 \end{array}$ (base 8)
    (base 2)
    $25_{eight} = 10101_{two}$

50. $\begin{array}{cccc} 1 & 4 & 7 & 2 \\ 001 & 100 & 111 & 010 \end{array}$ (base 8)
    (base 2)
    $1472_{eight} = 1100111010_{two}$

51. $\begin{array}{cc} 4 & A \\ 0100 & 1010 \end{array}$ (base 16)
    (base 2)
    $4A_{sixteen} = 1001010_{two}$

52. $\begin{array}{ccc} C & 7 & 2 \\ 1100 & 0111 & 0010 \end{array}$ (base 16)
    (base 2)
    $C72_{sixteen} = 110001110010_{two}$

53.
1	1	0	0	1	1
dabble	double	double	dabble	dabble
$=3$	$=6$	$=12$	$=25$	$=51$

0	1	0
double	dabble	double
$=102$	$=205$	$=410$

$110011010_{two} = 410$

54.
1	0	0	0	1	0
double	double	double	dabble	double
$=2$	$=4$	$=8$	$=17$	$=34$

1	0	1
dabble	double	dabble
$=69$	$=138$	$=277$

$100010101_{two} = 277$

55.
1	0	0	0	0	0
double	double	double	double	double
$=2$	$=4$	$=8$	$=16$	$=32$

1	0	0	0	1
dabble	double	double	double	dabble
$=65$	$=130$	$=260$	$=520$	$=1041$

$10000010001_{two} = 1041$

56.
1	1	0	0	1	0
dabble	double	double	dabble	double
$=3$	$=6$	$=12$	$=25$	$=50$

1	0	0	0	0
dabble	double	double	double	double
$=101$	$=202$	$=404$	$=808$	$=1616$

$11001010000_{two} = 1616$

57. Adding:

$$\begin{array}{r} 1\ \ 1 \\ 2\ \ 3\ \ 5_{six} \\ +\ 1\ \ 4\ \ 4_{six} \\ \hline 4\ \ 2\ \ 3_{six} \end{array}$$

58. Adding:

$$\begin{array}{r} 1\ \ 1 \\ 6\ \ 7\ \ 3_{eight} \\ +\ 3\ \ 4\ \ 5_{eight} \\ \hline 1\ \ 2\ \ 4\ \ 0_{eight} \end{array}$$

59. Subtracting:

$$\begin{array}{r} 6\ \ 12 \\ 6\ \ 7\ \ 2_{nine} \\ -\ 1\ \ 3\ \ 5_{nine} \\ \hline 5\ \ 3\ \ 6_{nine} \end{array}$$

60. Subtracting:

$$\begin{array}{r} 2\ \ 12 \\ 1\ \ 3\ \ 3\ \ 2_{four} \\ -\ 2\ \ 1\ \ 3_{four} \\ \hline 1\ \ 1\ \ 1\ \ 3_{four} \end{array}$$

61. Multiplying:

$$\begin{array}{r} 2\ \ 1 \\ 5\ \ 4\ \ 2_{eight} \\ \times\ 2\ \ 5_{eight} \\ \hline 3\ \ 3\ \ 5\ \ 2 \\ 1\ \ 3\ \ 0\ \ 4 \\ \hline 1\ \ 6\ \ 4\ \ 1\ \ 2_{eight} \end{array}$$

62. Multiplying:

$$\begin{array}{r} 2\ \ 1 \\ 3\ \ 4\ \ 2\ \ 1_{five} \\ \times\ 4\ \ 3_{five} \\ \hline 2\ \ 1\ \ 3\ \ 1\ \ 3 \\ 3\ \ 0\ \ 2\ \ 3\ \ 4 \\ \hline 3\ \ 2\ \ 4\ \ 2\ \ 0\ \ 3_{five} \end{array}$$

63. Dividing:

$$\begin{array}{r} 1\ 1\ 1\ 0\ 0_{two} \\ 11_{two}\overline{)1\ 0\ 1\ 0\ 1\ 0\ 1_{two}} \\ \underline{1\ 1} \\ 1\ 0\ 0 \\ \underline{1\ 1} \\ 1\ 1 \\ \underline{1\ 1} \\ 0\ 0\ 1_{four} \end{array}$$

Quotient: 11100_{two} Remainder: 1_{two}

64. Dividing:

$$\begin{array}{r} 2\ \ 1_{four} \\ 12_{four}\overline{)3\ 2\ 1_{four}} \\ \underline{3\ 0} \\ 2\ 1 \\ \underline{1\ 2} \\ 3_{four} \end{array}$$

Quotient: 21_{four} Remainder: 3_{four}

65. 2: No. 1485 is not even.
3: Yes. $1+4+8+5 = 18$, which is divisible by 3.
4: No. 85 is not divisible by 4.
5: Yes. 1485 ends in 5.
6: No. 1485 is not divisible by both 2 and 3.
8: No. 485 is not divisible by 8.
9: Yes. $1+4+8+5 = 18$, which is divisible by 9.
10: No. 1485 does not end in 0.
11: Yes. $4+5 = 9$ and $1+8 = 9$ with $9-9 = 0$, which is divisible by 11.

66. 2: Yes. 4268 is even.
    3: No. $4 + 2 + 6 + 8 = 20$, which is not divisible by 3.
    4: Yes. 68 is divisible by 4.
    5: No. 4268 does not end in 0 or 5.
    6: No. 4268 is not divisible by both 2 and 3.
    8: No. 268 is not divisible by 8.
    9: No. $4 + 2 + 6 + 8 = 20$, which is not divisible by 9.
    10: No. 4268 does not end in 0.
    11: Yes. $2+8 = 10$ and $4+6 =10$ with $10-10 = 0$, which is divisible by 11.

67. 501 is divisible by 3, so it is a composite number.

68. 781 is divisible by 11, so it is a composite number.

69. 689 is divisible by 13, so it is a composite number.

70. 1003 is divisible by 17, so it is a composite number.

71. $45 = 3 \cdot 3 \cdot 5 = 3^2 \cdot 5$

72. $54 = 2 \cdot 3 \cdot 3 \cdot 3 = 2 \cdot 3^3$

73. $153 = 3 \cdot 3 \cdot 17 = 3^2 \cdot 17$

74. $285 = 3 \cdot 5 - 19$

75. The proper factors of 28 are 1, 2, 4, 7, and 14. Their sum is $1 + 2 + 4 + 7 + 14 = 28$, so 28 is a perfect number.

76. The proper factors of 81 are 1, 3, 9, and 27. Their sum is $1 + 3 + 9 + 27 = 40 < 81$, so 81 is a deficient number.

77. The proper factors of 144 are 1, 2, 3, 4, 6, 8, 9, 12, 16, 18, 24, 36, 48, and 72. Their sum is $1 + 2 + 3 + 4 + 6 + 8 + 9 + 12 + 16 + 18 + 24 + 36 + 48 + 72 = 259 > 144$, so 144 is an abundant number.

78. The proper factors of 200 are 1, 2, 4, 5, 8, 10, 20, 25, 40, 50, and 100. Their sum is $1 + 2 + 4 + 5 + 8 + 10 + 20 + 25 + 40 + 50 + 100 = 265 > 200$, so 200 is an abundant number.

79. Let $n = 61$ and use Euclid's procedure.
    $2^{60}(2^{61} - 1)$

80. Let $n = 1279$ and use Euclid's procedure.
    $2^{1278}(2^{1279} - 1)$

81.
	1	46
	2	92
	4	184
✓	8	368
	8	368

$8 \times 46 = 368$

82.
		1	57
✓		1	57
		2	114
		4	228
✓		8	456
		9	513

$9 \times 57 = 513$

83.
	1	83
✓	2	166
✓	4	332
✓	8	664
	14	1162

$14 \times 83 = 1162$

84.
	1	83
✓	1	83
	2	286
✓	4	572
	8	1144
✓	16	2288
	21	3003

$21 \times 143 = 3003$

85. $132{,}049 \log 2 + 1 \approx 39{,}751.7$
    The number has 39,751 digits.

86. $2{,}976{,}221 \log 2 + 1 \approx 895{,}932.8$
    The number has 895,932 digits.

87. zero

88. No. Evaluating, we have
    $$8 + 4913 = z^3$$
    $$4921 = z^3$$
    However, the third root of 4921 is approximately 17.009, which is not a natural number.

CHAPTER 6 TEST

1. $3124 = 3000 + 100 + 20 + 4$
   𐌙𐌙𐌙𐌙𐌙∩∩||||

2. $4000 + 200 + 60 + 3 = 4263$

3. $1000 + (500 - 100) + (50 - 10) + 5 + 1 + 1$
   $= 1447$

4. $2609 = 2000 + 600 + 9$ MMDCIX

5. $67{,}485$
   $= 60{,}000 + 7000 + 400 + 80 + 5$
   $= (6 \times 10^4) + (7 \times 10^3) + (4 \times 10^2) + (8 \times 10^1)$
   $+ (5 \times 10^0)$

6. $500,000 + 30,000 + 200 + 80 + 4 = 530,284$

7. $(10 \times 60^2) + (21 \times 60) + (14 \times 1) = 37,274$

8. $9675 = (2 \times 3600) + (41 \times 60) + (15 \times 1)$

    ▼▼ 《《《▼ 《▼▼▼▼▼

9. $(3 \times 360) + (11 \times 20) + (5 \times 1) = 1305$

10. $502 = (1 \times 360) + (7 \times 20) + (2 \times 1)$

    •

    ▬▬ • •

    •   •

11. Converting:
    $$3542_{six} = (3 \times 6^3) + (5 \times 6^2) + (4 \times 6^1) + (2 \times 6^0)$$
    $$= 854$$

12. a. Converting:
    $$2148 = 4 \times 8^3 + 1 \times 8^2 + 4 \times 8^1 + 4 \times 8^0$$
    $$= 4144_{eight}$$

    b. Converting:
    $$2148 = 1 \times 12^3 + 2 \times 12^2 + 11 \times 12^1 + 0 \times 12^0$$
    $$= 12B0_{twelve}$$

13. $4_{eight} = 100_{two}$
    $5_{eight} = 101_{two}$
    $6_{eight} = 110_{two}$
    $7_{eight} = 111_{two}$
    $4567_{eight} = 100101110111_{two}$

14. $\begin{array}{cccl} 1010 & 1011 & 0111 & \text{(base 2)} \\ A & B & 7 & \text{(base 16)} \end{array}$
    $101010110111_{two} = AB7_{sixteen}$

15. Adding:

    $\begin{array}{cccc} 1 & 1 & & \\ & & 3 & 4_{five} \\ + & & 2 & 3_{five} \\ \hline & 1 & 1 & 2_{five} \end{array}$

16. Subtracting:

    $\begin{array}{cccc} & 5 & 10 & \\ 4 & 6 & 2_{eight} \\ - & 1 & 4 & 7_{eight} \\ \hline & 3 & 1 & 3_{eight} \end{array}$

17. Multiplying:

    $\begin{array}{ccccccc} 1 & 0 & 1 & 1 & 1 & 0_{two} \\ \times & & & 1 & 0 & 1_{two} \\ \hline & 1 & 0 & 1 & 1 & 1 & 0 \\ 0 & 0 & 0 & 0 & 0 & 0 \\ 1 & 0 & 1 & 1 & 1 & 0 \\ \hline 1 & 1 & 1 & 0 & 0 & 1 & 1 & 0_{two} \end{array}$

18. Dividing:

    $\begin{array}{r} 6 \quad 1_{seven} \\ 5_{seven} \overline{)4 \quad 3 \quad 1_{seven}} \\ \underline{4 \quad 2} \\ 1 \quad 1 \\ \underline{5} \\ 3_{seven} \end{array}$

    Quotient: 61_{seven} Remainder: 3_{seven}

19. $230 = 2 \cdot 5 \cdot 23$

20. 1001 is divisible by 7, so it is a composite number.

21. a. The number is not even. No.

    b. The digits add up to 45, which is divisible by 3. Yes.

    c. The ones digit does not end in a 5 or 0. No.

22. a. The last two digits are divisible by 4. Yes.

    b. The number is even; however, the digits add up to 43, which is not divisible by 3, so the number is not divisible by 3. No.

    c. The digits add up to 18 and 25, and their difference is not divisible by 11. No.

23. The proper factors of 96 are 1, 2, 3, 4, 6, 8, 12, 16, 24, 32, and 48. Their sum is $1 + 2 + 3 + 4 + 6 + 8 + 12 + 16 + 24 + 32 + 48 = 156 > 96$, so 96 is an abundant number.

24. Let $n = 17$. Then Euclid's Procedure gives $2^{16}(2^{17} - 1)$.

25. Since Fermat's Last Theorem has recently been proven, we know that z is not a natural number.

Chapter 7: Measurement and Geometry

EXERCISE SET 7.1

1. $6 \text{ ft} = 6 \; \cancel{\text{ft}} \times \dfrac{12 \text{ in.}}{1 \; \cancel{\text{ft}}}$

    $= 6 \times 12 \text{ in.}$

    $= 70 \text{ in.}$

3. $5 \text{ yd} = 5 \; \cancel{\text{yd}} \times \dfrac{3 \; \cancel{\text{ft}}}{1 \; \cancel{\text{yd}}} \times \dfrac{12 \text{ in.}}{1 \; \cancel{\text{ft}}}$

    $= 5 \times 3 \times 12 \text{ in.}$

    $= 180 \text{ in.}$

5. $64 \text{ oz} = 64 \; \cancel{\text{oz}} \times \dfrac{1 \text{ lb}}{16 \; \cancel{\text{oz}}}$

    $= \dfrac{64 \text{ lb}}{16}$

    $= 4 \text{ lb}$

7. $1\frac{1}{2} \text{ lb} = \dfrac{3}{2} \; \cancel{\text{lb}} \times \dfrac{16 \text{ oz}}{1 \; \cancel{\text{lb}}}$

    $= \dfrac{48 \text{ oz}}{2}$

    $= 24 \text{ oz}$

9. $2\frac{1}{2} \text{ c} = \dfrac{5}{2} \; \cancel{\text{c}} \times \dfrac{8 \text{ fl oz}}{1 \; \cancel{\text{c}}}$

    $= \dfrac{40 \text{ ft oz}}{2}$

    $= 20 \text{ fl oz}$

11. $7 \text{ gal} = 7 \; \cancel{\text{gal}} \times \dfrac{4 \; \cancel{\text{qt}}}{1 \; \cancel{\text{gal}}} \times \dfrac{2 \text{ pt}}{1 \; \cancel{\text{qt}}}$

    $= 7 \times 4 \times 2 \text{ pt}$

    $= 56 \text{ pt}$

13. $62 \text{ cm} = 620 \text{ mm}$

15. $3.21 \text{ m} = 321 \text{ cm}$

17. $7421 \text{ g} = 7.421 \text{ kg}$

19. $0.45 \text{ g} = 4.5 \text{ dg}$

21. $7.5 \text{ ml} = 0.0075 \text{ L}$

23. $0.435 \text{ L} = 435 \text{ ml} = 435 \text{ cm}^3$

25. $145 \text{ lb} \approx 145 \; \cancel{\text{lb}} \times \dfrac{1 \text{ kg}}{2.2 \; \cancel{\text{lb}}} = \dfrac{145 \text{ kg}}{2.2} \approx 65.91 \text{ kg}$

     $145 \text{ lb} \approx 65.91 \text{ kg}$

27. $\dfrac{30 \text{ mi}}{\text{h}} \approx \dfrac{30 \; \cancel{\text{mi}}}{\text{h}} \times \dfrac{1.61 \text{ km}}{1 \; \cancel{\text{mi}}}$

     $= \dfrac{30 \times 1.61 \text{ km}}{\text{h}}$

     $= 48.3 \text{ km/h}$

     $30 \text{ mi/h} \approx 48.3 \text{ km/h}$

29. $\dfrac{\$32.99}{\text{gal}} \approx \dfrac{\$32.99}{\cancel{\text{gal}}} \times \dfrac{1 \; \cancel{\text{gal}}}{3.79 \text{ L}}$

     $= \dfrac{\$32.99}{3.79 \text{ L}} \approx \dfrac{\$8.704485}{\text{L}}$

     $\approx \dfrac{\$8.70}{\text{L}}$

     $\$32.99/\text{gal} \approx \$8.70/\text{L}$

31. $35 \text{ mm} = 3.5 \text{ cm}$

     $= 3.5 \; \cancel{\text{cm}} \times \dfrac{1 \text{ in.}}{2.54 \; \cancel{\text{cm}}}$

     $= \dfrac{3.5 \text{ in.}}{2.54} \approx 1.37795 \text{ in.}$

     $\approx 1.38 \text{ in.}$

     $35 \text{ mm} \approx 1.38 \text{ in.}$

33. $\dfrac{\$10}{\text{kg}} \approx \dfrac{\$10}{\cancel{\text{kg}}} \times \dfrac{1 \; \cancel{\text{kg}}}{2.2 \text{ lb}}$

     $= \dfrac{\$10}{2.2 \text{ lb}} \approx \dfrac{\$4.545454}{\text{lb}}$

     $\approx \dfrac{\$4.55}{\text{lb}}$

     $\$10/\text{kg} \approx \$4.55/\text{lb}$

35. $2.3 \text{ T} = 2.3 \times \dfrac{10^{12}}{10^{24}} \text{ Y} = 2.3 \times 10^{-12} \text{ Y}$

37. $0.65 \text{ Z} = 0.65 \times \dfrac{10^{21}}{10^{9}} \text{ G}$

     $= 0.65 \times 10^{12} \text{ G}$

     $= 6.5 \times 10^{11} \text{ G}$

39. $4.01 \text{ G} = 4.01 \times \dfrac{10^{9}}{10^{18}} \text{ E} = 4.01 \times 10^{-9} \text{ E}$

41. $\dfrac{3 \times 10^8 \text{ m}}{\text{s}} = \dfrac{3 \times 10^8 \; \cancel{\text{m}}}{\text{s}} \times \dfrac{1 \text{ Zm}}{10^{21} \; \cancel{\text{m}}}$

     $= \dfrac{3 \times 10^{-13} \text{ Zm}}{\text{s}}$

     $3 \times 10^8 \text{ m/s} = 3 \times 10^{-13} \text{ Zm/s}$

43. $1 \text{ ps} = 1 \times 10^{-12} \text{ s} = 0.000000000001 \text{ s}$

EXCURSION EXERCISES, SECTION 7.2

1. Find the measure of the angle in each sector:
Pepperoni:	$0.43 \times 360° \approx 155°$
Sausage:	$0.19 \times 360° \approx 68°$
Mushrooms:	$0.14 \times 360° \approx 50°$
Vegetables:	$0.13 \times 360° \approx 47°$
Other:	$0.07 \times 360° \approx 25°$
Onions:	$0.04 \times 360° \approx 14°$

EXERCISE SET 7.2

1. The three names for the angle are $\angle O$, $\angle AOB$, and $\angle BOA$.

3. Solving:
    $$x + 62° = 90°$$
    $$x = 28°$$
    A 28° angle

5. Solving:
    $$x + 162° = 180°$$
    $$x = 18°$$
    An 18° angle

7. An acute angle, since complementary angles are two angles whose sum is 90°.

9. An obtuse angle, since the sum of supplementary angles are 180°, and an acute angle is less than 90°.

11. Solving:
    $$AB + BC + CD = AD$$
    $$12 + BC + 9 = 35$$
    $$BC = 14 \text{ cm}$$

13. Solving:
    $$QR + RS = QS$$
    $$QR + 3QR = QS$$
    $$7 + 21 = QS$$
    $$QS = 28 \text{ ft}$$

15. Solving:
    $$m\angle LOM + m\angle MON = m\angle LON$$
    $$53° + m\angle MON = 139°$$
    $$m\angle MON = 86°$$

17. Solving:
    $$x + 74° = 145°$$
    $$x = 71°$$

19. Solving:
    $$x + 2x = 90°$$
    $$3x = 90°$$
    $$x = 30°$$

21. Solving:
    $$a + 53° = 180°$$
    $$a = 127°$$

23. Solving:
    $$a + 76° + 168° = 360°$$
    $$a = 116°$$

25. Solving:
    $$3x + 4x + 2x = 180°$$
    $$9x = 180°$$
    $$x = 20°$$

27. Solving:
    $$5x + x + 20° + 2x = 180°$$
    $$8x = 160°$$
    $$x = 20°$$

29. $\angle a$ is complementary to $\angle b$'s supplementary angle.
    $$m\angle a + m\angle x = 90°$$
    $$51° + m\angle x = 90°$$
    $$m\angle x = 39°$$
    $$m\angle b + m\angle x = 180°$$
    $$m\angle b + 39° = 180°$$
    $$m\angle b = 141°$$

31. Solving:
    $$x + 74° = 180°$$
    $$x = 106°$$

33. The two angles are vertical angles, so their measures are equal.
    $$5x = 3x + 22°$$
    $$2x = 22°$$
    $$x = 11°$$

35. $\angle a$ is a corresponding angle with the 38° angle.
    Therefore, $m\angle a = 38°$.
    $$m\angle a + m\angle b = 180°$$
    $$38° + m\angle b = 180°$$
    $$m\angle b = 142°$$

37. $\angle a$ is an alternate interior angle with the 47°
    angle. Therefore, $m\angle a = 47°$.
    $$m\angle a + m\angle b = 180°$$
    $$47° + m\angle b = 180°$$
    $$m\angle b = 133°$$

39. Solving:
    $$5x + 4x = 180°$$
    $$9x = 180°$$
    $$x = 20°$$

41. Solving:
    $$2x + x + 39° = 180°$$
    $$3x = 141°$$
    $$x = 47°$$

43. The three lines form a triangle. The sum of the
    interior angles of the triangle is
    $$m\angle b + (180° - m\angle a) + (180° - m\angle x) = 180°$$
    $$70° + (180° - 95°) + (180° - m\angle x) = 180°$$
    $$70° + 265° - m\angle x = 180°$$
    $$-m\angle x = -155°$$
    $$m\angle x = 155°$$
    $\angle y$ and $\angle b$ are vertical angles:
    $$m\angle y = m\angle b = 70°$$

45. $\angle a$ and $\angle y$ are vertical angles $m\angle a = \angle y = 45°$
    The three lines form a triangle. The sum of the
    interior angles of the triangle is
    $$m\angle a + 90° + (180° - m\angle b) = 180°$$
    $$45° + 90° - m\angle b = 0°$$
    $$-m\angle b = -135°$$
    $$m\angle b = 135°$$

47. Solving:
    $$90° + 30° + x = 180°$$
    $$x = 60°$$

49. Solving:
    $$42° + 103° + x = 180°$$
    $$x = 35°$$

51. False, the sum of the measures of the interior
    angles of a triangle must be 180°. Two obtuse
    angles would be greater than 180°.

53. True, an obtuse angle is greater than 90°, and
    the sum of the measures of the interior angles of
    the triangle is 180°. Therefore, the remaining
    angles in the triangle would be less than 90°.

55. Solving:
    $$m\angle x + m\angle y + m\angle z$$
    $$= (180° - m\angle c) + (180° - m\angle a) + (180° - m\angle c)$$
    $$= 180° + 180° + 180° - m\angle a - m\angle b - m\angle c$$
    $$= 540° - (m\angle a + m\angle b + m\angle c)$$
    The sum of the interior angles of a triangle is
    180°, so
    $$= 540 - (180°)$$
    $$= 360°$$

57. $\angle AOC$ and $\angle BOC$ are supplementary angles;
    therefore, $m\angle AOC + m\angle BOC = 180°$. Because
    $m\angle AOC = m\angle BOC$, by substitution, $m\angle AOC +$
    $m\angle AOC = 180°$. Therefore, $2(m\angle AOC) = 180°$,
    and $m\angle AOC = 90°$. Hence $\overline{AB} \perp \overline{CD}$.

EXCURSION EXERCISES, SECTION 7.3

1.

Number of seconds elapsed	Length (in inches)	Width (in inches)	Perimeter (in inches)	Area (in square inches)
0	5	4	18	20
1	5.5	3.8	18.6	20.9
2	6	3.6	19.2	21.6
3	6.5	3.4	19.8	22.1
4	7	3.2	20.4	22.4
5	7.5	3.0	21	22.5
6	8	2.8	21.6	22.4
7	8.5	2.6	22.2	22.1
8	9	2.4	22.8	21.6
9	9.5	2.2	23.4	20.9
10	10	2.0	24	20

3. increased

5. decreased

EXERCISE SET 7.3

1. a. Perimeter is measured in linear units, not
      square units.

   b. Area is measured in square units, not linear
      units.

3. a. $P = 2l + 2w = 2(5) + 2(10) = 30$ m

   b. $A = lw = (5)(10) = 50$ m^2

5. a. $P = 4s = 4(10) = 40$ km

   b. $A = s^2 = (10)^2 = 100$ km^2

7. a. $P = 2(b) + 2(s) = 2(12) + 2(8) = 40$ ft

    b. $A = bh = (12)(6) = 72$ ft^2

9. a. $C = 2\pi r = 2\pi(4) = 8\pi$ cm ≈ 25.13 cm

    b. $A = \pi r^2 = \pi(4)^2 = 16\pi$ cm$^2 \approx 50.27$ cm^2

11. a. $C = 2\pi r = 2\pi(5.5) = 11\pi$ mi ≈ 34.56 mi

    b. $A = \pi r^2 = \pi(5.5)^2 = 30.25\pi$ mi$^2 \approx 95.03$ mi^2

13. a. $C = \pi d = \pi(17) = 17\pi$ ft ≈ 53.41 ft

    b. $r = \dfrac{d}{2} = \dfrac{17}{2} = 8.5$ ft

      $A = \pi r^2 = \pi(8.5)^2 = 72.25\pi$ ft$^2 \approx 226.98$ ft^2

15. $P = 5s = 5(4) = 20$ in.

17. $P = 2b + 2s = 2(3) + 2(2) = 10$ mi

19. $P = 2l + 2w = 2(68) + 2(42) = 220$ in.
    15 ft $= (15)(12) = 180$ in.
    1 package is not enough.
    2 packages $= 2(180) = 360$ in.
    Two packages are needed.

21. The perimeter of the square equals $4s$. the circumference of the circle equals πs. Since $4 > \pi$, the perimeter of the square is greater.

23. $A = s^2 = (12)^2 = 144$ m^2

25. $P = 4s = 36$ in.
    $s = 9$ in.

27. $A = lw$

      $(l)(8) = 312$ ft^2

      $l = 39$ ft

29. Comparing:
    $10 \times 5 = 50$
    $10 \times 10 = 100$
    $10 \times 15 = 150$
    $10 \times 20 = 200$
    The smallest unit that will provide 175 ft^2 of floor space is 10 feet by 20 feet.

31. $A = \dfrac{1}{2}bh = \dfrac{1}{2}(21)(13) = 136.5$ ft^2

33. $A = \dfrac{1}{2}h(b_1 + b_2) = \dfrac{1}{2}(10)(23 + 9) = 160$ km^2

35. $A = \dfrac{1}{2}h(b_1 + b_2) = \dfrac{1}{2}(10)(10 + 12) = 110$ ft^2
    110 ft$^2 \div 55$ ft^2 per quart $= 2$ quarts

37. $A = lw$
    Wall 1 $= (10)(8) = 80$ ft^2
    Wall 2 $= (12)(8) = 96$ ft^2
    80 ft^2 + 96 ft^2 = 176 ft^2 total area
    176 ft$^2 \div 40$ ft^2 per roll = 4.4, which is 5 rolls
    $5 \times \$96 = \$480$

39. $A = lw = (8)(9) = 72$ ft^2
    $A = lw = (12)(12) = 144$ ft^2
    $72 + 144 = 216$ ft^2 total area
    216 ft$^2 \div 9$ ft^2 per yard$^2 = 24$ yd^2
    24 yd$^2 \times \$38 = \$912$

41. Length of total region $= 25 + 4 = 29$ m
    Width of total region $= 15 + 4 = 19$ m
    Total Area $= lw = (29)(19) = 551$ m^2
    Area of Plot $= lw = (25)(15) = 375$ m^2
    Area of Walkway $= 551 - 375 = 176$ m^2

43. $C = \pi d = \pi(4.2) \approx 13.19$ ft

45. $C = \pi d = \pi(16) \approx 50.265$ in.
    50.265 in. per revolution $\times$ 240 revolutions per minute = 12,064 in. per minute

47. $C = \pi d = \pi(16) \approx 50.265$ in.
    50.265 in $\times$ 15 revolutions ≈ 753.98 in.
    753.98 in. $\div 12 \approx 62.83$ ft

49. $A = \pi r^2 = \pi(50)^2 = 2{,}500\pi$ ft^2

51. $A = \pi r^2$
    Area of Small Pizza $= \pi(6)^2 \approx 113.10$ in^2
    Area of Large Pizza $= \pi(12)^2 \approx 452.39$ in^2
    $452.39 - 113.10 = 339.29$ in^2
    $2(113.10) = 226.10$ in^2, the area of the large pizza is more than twice that of the small pizza; it is four times the area of the small pizza.

53. $A = \dfrac{1}{2}h(b_1 + b_2)$

    $A = \dfrac{1}{2}(2.75)(9 + 7.8) = 23.1$

    $A = \dfrac{1}{2}(2.75)(10.3 + 9) = 26.5375$

    $A = \dfrac{1}{2}(2.75)(12 + 10.3) = 30.6625$

    $A = \dfrac{1}{2}(2.75)(11.3 + 12) = 32.0375$

    $A = \dfrac{1}{2}(2.75)(11.1 + 11.3) = 30.8$

    $A = \dfrac{1}{2}(2.75)(9.8 + 11.1) = 28.7375$

    $A = \dfrac{1}{2}(2.75)(9.2 + 9.8) = 26.125$

    $A = \dfrac{1}{2}(2.75)(8.4 + 9.2) = 24.2$

    A total = sum of all totals = 222.2 mi^2

55. Length of Rectangle = $4r$
    Width of Rectangle = $2r$
    Area of Rectangle = $(4r)(2r) = 8r^2$
    Area of Circles = $2\pi r^2$
    Area of shaded region = $8r^2 - 2\pi r^2$

57. $A = lw$
    Double the length and the width.
    $A = (2l)(2w) = 4(lw)$
    The area is four times that of the original rectangle.

59. a. $s = \dfrac{a+b+c}{2} = \dfrac{4.4+5.7+6.2}{2} = 8.15$

    $A = \sqrt{s(s-a)(s-b)(s-c)}$
    $\quad = \sqrt{8.15(8.15-4.4)(8.15-5.7)(8.15-6.2)}$
    $\quad = 12.1 \text{ in}^2$

    b. $s = \dfrac{a+b+c}{2} = \dfrac{8.3+8.3+8.3}{2} = 12.45$

    $A = \sqrt{s(s-a)(s-b)(s-c)}$
    $\quad = \sqrt{12.45(12.45-8.3)^3}$
    $\quad = 29.8 \text{ cm}^2$

    c. $P = 3+4+5 = 12$

    $s = \dfrac{3+4+5}{2} = 6$

    $A = \sqrt{6(6-5)(6-4)(6-3)} = 6$

    The lengths of the sides of the triangle are 3 in., 4 in., and 5 in.

EXCURSION EXERCISES, SECTION 7.4

1. Genus is the number of holes in a figure.
   a. 1
   b. 8
   c. 0
   d. 3

3. Spatula: The spatula has a genus of 4 and all the other figures have a genus of 0.

5. a. Genus of 0: 1, 2, 3, 5, 7
      Genus of 1: 4, 6, 9
      Genus of 2: 8

   b. Genus of 0: 6, 9
      Genus of 1: 1 4 5 7 8
      Genus of 2: 2 3

EXERCISE SET 7.4

1. $\dfrac{7}{14} = \dfrac{1}{2}$

3. $\dfrac{6}{8} = \dfrac{3}{4}$

5. $\dfrac{5}{9} = \dfrac{4}{x}$
   $5x = 36$
   $x = 7.2 \text{ cm}$

7. $\dfrac{3}{5} = \dfrac{2}{x}$
   $3x = 10$
   $x \approx 3.3 \text{ m}$

9. $\dfrac{4}{8} = \dfrac{x}{6}$
   $8x = 24$
   $x = 3 \text{ m}$
   $P = 3+4+5 = 12 \text{ m}$

11. $\dfrac{4}{12} = \dfrac{x}{15}$
    $12x = 60$
    $x = 5 \text{ in.}$
    $P = 3+4+5 = 12 \text{ in.}$

13. $\dfrac{15}{40} = \dfrac{x}{20}$
    $40x = 300$
    $x = 7.5 \text{ cm}$
    $A = \dfrac{1}{2}bh = \dfrac{1}{2}(15)(7.5) = 56.3 \text{ cm}^2$

15. $\dfrac{24}{8} = \dfrac{x}{6}$
    $8x = 144$
    $x = 18 \text{ ft}$

17. $\dfrac{8}{4} = \dfrac{x}{8}$
    $4x = 64$
    $x = 16 \text{ m}$

19. 5 ft 9 in. = 5.75 ft
    $\dfrac{12}{30} = \dfrac{5.75}{x}$
    $12x = 172.5$
    $x = 14.375 = 14\dfrac{3}{8} \text{ ft}$

21. $\dfrac{8}{20} = \dfrac{6}{x}$
    $8x = 120$
    $x = 15 \text{ m}$

23. Solving:

$$\frac{BC}{AC} = \frac{CE - DE}{CE}$$
$$AC = 3 + 3 = 6 \text{ ft}$$
$$\frac{3}{6} = \frac{x - 4}{x}$$
$$3x = 6x - 24$$
$$3x = 24$$
$$x = 8 \text{ ft}$$

25. $$\frac{24}{12} = \frac{39 - x}{x}$$
$$24x = 468 - 12x$$
$$36x = 468$$
$$x = 13 \text{ cm}$$

27. $$\frac{8}{20} = \frac{14}{x}$$
$$8x = 280$$
$$x = 35 \text{ m}$$

29. Yes, SAS Theorem.

31. Yes, SSS Theorem.

33. Yes, ASA Theorem.

35. No.

37. Yes, SAS Theorem.

39. No.

41. No.

43. Using the Pythagorean Theorem:

$$a^2 + b^2 = c^2$$
$$3^2 + 4^2 = c^2$$
$$25 = c^2$$
$$5 = c$$

The side is 5 in.

45. Using the Pythagorean Theorem:

$$a^2 + b^2 = c^2$$
$$5^2 + 7^2 = c^2$$
$$74 = c^2$$
$$8.6 \approx c$$

The side is 8.6 cm.

47. Using the Pythagorean Theorem:

$$a^2 + b^2 = c^2$$
$$10^2 + b^2 = 15^2$$
$$b^2 = 225 - 100 = 125$$
$$b \approx 11.2$$

The side is 11.2 ft.

49. Using the Pythagorean Theorem:

$$a^2 + b^2 = c^2$$
$$4^2 + b^2 = 6^2$$
$$b^2 = 36 - 16 = 20$$
$$b \approx 4.5$$

The side is 4.5 cm.

51. Using the Pythagorean Theorem:

$$a^2 + b^2 = c^2$$
$$9^2 + 9^2 = c^2$$
$$162 = c^2$$
$$12.7 \approx c$$

The side is 12.7 yd.

53. Using the Pythagorean Theorem:

$$a^2 + b^2 = c^2$$
$$3^2 + 8^2 = c^2$$
$$73 = c^2$$
$$8.5 \approx c$$

The holes are 8.5 cm apart.

55. Using the Pythagorean Theorem:

$$a^2 + b^2 = c^2$$
$$5^2 + 9^2 = c^2$$
$$106 = c^2$$
$$10.3 \approx c$$

The hypotenuse has length 10.3 cm. Adding: 5 + 9 + 10.3 = 24.3. The perimeter is 24.3 cm.

57. a. The sum of the three angles in any triangle is 180°. If two angles in one triangle are equal to two angles in another triangle, then the third angles are 180 less the sum of the two known angles. There is only one value equal to that difference, so the remaining angles in each triangle are equal. This is always true.

    b. Isosceles triangles are triangles with two sides of equal length. There are obtuse, right, and acute isosceles triangles. A right isosceles triangle would not be similar to an acute isosceles triangle. This statement is only sometimes true.

    c. Equilateral triangles have three sides congruent as well as all three angles congruent. This means that all three angles in any equilateral triangle must measure 60°. All equilateral triangles have all angles equal in measure, so all equilateral triangles are similar.

    d. The sum of the three angles in any triangle is 180°. A right triangle contains one right angle, so the right angles in all right triangles are equal to each other. If second angles in two right triangles are also equal in measure, then the third angles are 180 less the sum of the two known angles. There is only one value equal to that difference, so the remaining angles in each triangle are equal. This is always true.

EXCURSION EXERCISES, SECTION 7.5

1. Find the volume of cylinder:

    $V = \pi r^2 h$

    $V = \pi(2)^2 10 = 40\pi$

    Let h be height of water displacement. Set

    $40\pi = lwh$

    $40\pi = (20)(30)h$

    $40\pi = 600h$

    $h = \dfrac{\pi}{15} \approx 0.21$ cm

3. Solving:

    $D = \dfrac{w}{V} = \dfrac{15}{V}$ $h = 0.42$

    $V = lwh = (12)(12)(0.42) = 60.48$

    $D = \dfrac{15}{60.48} \approx 0.25 \; \dfrac{\text{lbs}}{\text{in}^3}$

EXERCISE SET 7.5

1. $V = lwh = (14)(10)(6) = 840 \text{ in}^3$

3. $V = \dfrac{1}{3} s^2 h = \dfrac{1}{3}(3)^2(5) = 15 \text{ ft}^3$

5. Solving:

    $V = \dfrac{1}{2} d = \dfrac{1}{2}(3) = \dfrac{3}{2}$

    $V = \dfrac{4}{3}\pi r^3 = \dfrac{4}{3}\pi \left(\dfrac{3}{2}\right)^3 = \left(\dfrac{4}{3}\right)\left(\dfrac{27}{8}\right)\pi$

    $V = 4.5\pi \approx 14.14 \text{ cm}^3$

7. Solving:

    $l = 4$ $w = 5$ $h = 3$

    $S = 2lw + 2lh + 2wh$

    $S = 2(4)(5) + 2(4)(3) + 2(5)(3)$

    $S = 94 \text{ m}^2$

9. Solving:

    $S = 4$ $l = 5$

    $S = s^2 + 2sl$

    $= 4^2 + 2(4)(5)$

    $\quad 16 + 40$

    $S = 56 \text{ m}^2$

11. Solving:

    $r = 6$ $h = 2$

    $S = 2\pi r^2 + 2\pi rh$

    $S = 2\pi(6)^2 + 2\pi(6)(2)$

    $S = 96\pi \text{ in}^2 \approx 301.59 \text{ in}^2$

13. $V = lwh = (6.8)(2.5)(2) = 34 \text{ m}^3$

15. $V = s^3 = (2.5)^3 = 15.625 \text{ in}^3$

17. Solving:

    $V = \dfrac{4}{3}\pi r^3$ $r = 3$

    $V = \dfrac{4}{3}\pi(3)^3 = 36\pi \text{ ft}^3$

19. Solving:

    $d = 24$ $r = \dfrac{1}{2}(24) = 12$ $h = 18$

    $V = \pi r^2 h$

    $V = \pi(12)^2(18) = 2592\pi \approx 8143.01 \text{ cm}^3$

21. Solving:

    $r = 5$ $h = 9$

    $V = \dfrac{1}{3}\pi r^2 h$

    $V = \dfrac{1}{3}\pi(5)^2(9) = 75\pi \text{ in}^3$

23. Solving:

    $s = 6$ $h = 10$

    $V = \dfrac{1}{3}s^2 h = \dfrac{1}{3}(6)^2(10) = 120 \text{ in}^3$

25. Since the length of the side of the cube equals the radius of the sphere, then $s = r$.

    Volume of cube $= s^3$

    $\qquad\qquad\qquad = r^3$

    Volume of sphere $= \dfrac{4}{3}\pi r^3$

    Sphere has a greater volume, because $\dfrac{4}{3}\pi > 1$.

27. Solving:

    $C = 3.5$ $h = 8$

    If $C = 2\pi r$, then $r = \dfrac{C}{2\pi} = \dfrac{3.5}{2\pi} \approx 0.557$

    $V = \pi r^2 h = \pi(0.557)^2(8) \approx 7.80 \text{ ft}^3$

29. $l = 1000$ $w = 110$ $h = 43$

    $V = lwh = (1000)(110)(43) = 4{,}730{,}000 \text{ ft}^3$

    $(4{,}730{,}000 \text{ ft}^3)(7.48 \text{ gal/ft}^3) = 35{,}380{,}400 \text{ gal}$

31. Solving:

    $s = 3.4$

    $S = 6s^2 = 6\,(3.4)^2 = 69.36 \text{ m}^2$

33. Solving:
$$r = \frac{1}{2}(15) = 7.5$$
$$S = 4\pi r^2 = 4\pi(7.5)^2 = 225\pi \text{ cm}^2$$

35. Solving:
$$r = 4 \qquad h = 12$$
$$S = 2\pi r^2 + 2\pi rh = 2\pi(4)^2 + 2\pi(4)(12)$$
$$S = 32\pi + 96\pi = 128\pi \approx 402.12 \text{ in}^2$$

37. Solving:
$$r = 1.5 \qquad l = 2.5$$
$$S = \pi r^2 + \pi rl$$
$$S = \pi(1.5)^2 + \pi(1.5)(2.5)$$
$$S = 2.25\pi + 3.75\pi = 6\pi \text{ ft}^2$$

39. Solving:
$$s = 9 \qquad l = 12$$
$$S = s^2 + 2sl = (9)^2 + 2(9)(12)$$
$$S = 297 \text{ in}^2$$

41. Solving:
$$l = 7 \qquad h = 3 \qquad V = 52.5$$
$$V = lwh$$
$$w = \frac{V}{lh}$$
$$w = \frac{52.5}{7(3)} = \frac{52.5}{21} = 2.5 \text{ ft}$$

43. Solving:
$$r = 12 \qquad h = 30$$
$$S = 2\pi r^2 + 2\pi rh$$
$$S = 2\pi(12)^2 + 2\pi(12)(30)$$
$$S = 1008\pi \approx 3166.73 \text{ ft}^2$$
Each can covers 300 ft².
$$\frac{S}{300} = \frac{3166.73}{300} \approx 10.56$$
11 cans are needed.

45. Solving:
Pyramid:
$$S = s^2 + 2sl = (5)^2 + 2(5)(8) = 105 \text{ cm}^2$$
Cone:
$$r = 2.5$$
$$S = \pi r^2 + \pi rl$$
$$S = \pi(2.5)^2 + \pi(2.5)(8) = 26.25\pi \approx 82.47 \text{ cm}^2$$
Difference = $105 - 82.47 = 22.53 \text{ cm}^2$

47. Break figure into 2 rectangular prisms.
Prism 1 = $1.5 \times 1.5 \times 2$
$$V_1 = (1.5)(1.5)(2) = 4.5$$
Prism 2 = $0.5 \times 0.5 \times 2$
$$V_2 = (0.5)(0.5)(2) = 0.5$$
$$V_{\text{total}} = V_1 + V_2$$
$$V_{\text{total}} = 4.5 + 0.5 = 5 \text{ m}^3$$

49. $V_{\text{total}} = V_1 + V_2$
$$V_1 = \pi(3)^2(2) = 18\pi$$
$$V_2 = \pi(1)^2(4) = 4\pi$$
$$V_{\text{total}} = V_1 + V_2 = 18\pi + 4\pi$$
$$V_{\text{total}} = 22\pi \approx 69.12 \text{ in}^3$$

51. $V_{\text{total}} = V_{\text{prism}} + V_{\text{pyramid}}$
Rectangular Prism:
$$V_{\text{prism}} = lwh = (8)(8)(2) = 128 \text{ m}^3$$
Square Pyramid:
$$h = 5 - 2 = 3 \text{ in.}$$
$$V_{\text{pyramid}} = \frac{1}{3}s^2 h = \frac{1}{3}(8)^2(3) = 64 \text{ m}^3$$
$$V_{\text{total}} = 128 + 64 = 192 \text{ m}^3$$

53. Solving:
$$S_{\text{total}} = S_{\text{bottom}} + 4S_{\text{rect.face}} + 4S_{\text{triangle face}}$$
$$S_{\text{bottom}} = (8)(8) = 64$$
$$S_{\text{rect.face}} = (2)(8) = 16$$
$$S_{\text{triangle face}} = \frac{1}{2}(8)(5) = 20$$
$$S_{\text{total}} = 64 + 4(16) + 4(20) = 208 \text{ in}^2$$

55. Solving:
$$S_{\text{total}} = S_{\text{cylinder}} + S_{\text{prism}} - 2S_{\text{circle}}$$
$$S_{\text{cylinder}} = 2\pi(1)^2 + 2\pi(1)(2) = 6\pi$$
$$S_{\text{prism}} = 2(8)(2) + 2(8)(2) + 2(8)(8) = 192$$
$$S_{\text{circle}} = \pi(1)^2 = \pi$$
$$S_{\text{total}} = (6\pi + 192) - 2(\pi) = 204.57 \text{ cm}^2$$

57. Solving:
$$V = 6(2.5)(5) + 4(1)(5) = 95 \text{ m}^3$$
$$V = (95 \text{ m}^3)\left(\frac{1000 \text{ L}}{1 \text{ m}^3}\right) = 95{,}000 \text{ L}$$

59. Solving:
$$V_s = \frac{4}{3}r_s^3$$
$$V_l = \frac{4}{3}r_l^3$$
$$r_l = 3r_s$$
$$V_l = \frac{4}{3}(3r_s)^3 = 27V_s$$
Value of large sphere is 27 times the value of the small sphere or $4860.

61. a. Drawings will vary. For example:

b. Drawings will vary. For example:

63. a. Always true. The cross section is a right triangle whose slant height is the hypotenuse.

b. Never true. Height can never exceed slant height.

c. Sometimes true. A face can be any isosceles triangle.

65. a. For example, make a cut perpendicular to the top and bottom faces and parallel to two of the sides.

b. For example, beginning at an edge that is perpendicular to the bottom face, cut at an angle through the bottom face.

c. For example, beginning at the top face at a distance d from the vertex, cut at an angle to the bottom face, ending at a distance greater than d from the opposite vertex.

d. For example, beginning at the top face at a distance d from a vertex, cut across the cube to a point just above the opposite vertex.

EXCURSION EXERCISES, SECTION 7.6

1. – 8. Drawings should be similar to that shown in the text. Approximate values will vary. As produced by a calculator, $\sin 35° = 0.5736$, $\cos 35° = 0.8192$ and $\tan 35° = 0.7002$.

EXERCISE SET 7.6

1. a. $\sin A = \dfrac{\text{opp}}{\text{hyp}} = \dfrac{a}{c}$

b. $\sin B = \dfrac{\text{opp}}{\text{hyp}} = \dfrac{b}{c}$

c. $\cos A = \dfrac{\text{adj}}{\text{hyp}} = \dfrac{b}{c}$

d. $\cos B = \dfrac{\text{adj}}{\text{hyp}} = \dfrac{a}{c}$

e. $\tan A = \dfrac{\text{opp}}{\text{adj}} = \dfrac{a}{b}$

f. $\tan A = \dfrac{\text{opp}}{\text{adj}} = \dfrac{b}{a}$

3. Use the Pythagorean Theorem to find the unknown side:
$$a^2 + b^2 = c^2$$
$$12^2 + 5^2 = c^2$$
$$169 = c^2$$
$$13 = c$$
$$\sin \theta = \frac{5}{13}, \quad \cos \theta = \frac{12}{13}, \quad \tan \theta = \frac{5}{12}$$

5. Use the Pythagorean Theorem to find the unknown side:
$$a^2 + b^2 = c^2$$
$$a^2 + 24^2 = 25^2$$
$$a^2 = 49$$
$$a = 7$$
$$\sin \theta = \frac{24}{25}, \quad \cos \theta = \frac{7}{25}, \quad \tan \theta = \frac{24}{7}$$

7. Use the Pythagorean Theorem to find the unknown side:
$$a^2 + b^2 = c^2$$
$$8^2 + 7^2 = c^2$$
$$113 = c^2$$
$$\sqrt{113} = c$$
$$\sin \theta = \frac{8}{\sqrt{113}}, \quad \cos \theta = \frac{7}{\sqrt{113}}, \quad \tan \theta = \frac{8}{7}$$

9. Use the Pythagorean Theorem to find the unknown side:
$$a^2 + b^2 = c^2$$
$$a^2 + 40^2 = 80^2$$
$$a^2 = 4800$$
$$a = \sqrt{4800} = 40\sqrt{3}$$
$$\sin \theta = \frac{1}{2}, \quad \cos \theta = \frac{40\sqrt{3}}{80} = \frac{\sqrt{3}}{2},$$
$$\tan \theta = \frac{40}{40\sqrt{3}} = \frac{1}{\sqrt{3}}$$

11. 0.6820

13. 1.4281

15. 0.9971

17. 1.9970

19. 0.8878

21. 0.8453

23. 0.8508

25. 0.6833

27. 38.6°

29. 41.1°

31. 21.3°

33. 38.0°

35. 72.5°

37. $0.6°$

39. $66.1°$

41. $29.5°$

43.

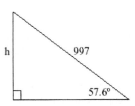

$$\sin 57.6° = \frac{h}{997}$$
$$h = 997 \sin 57.6° \approx 841.79$$
The balloon is 841.8 ft. off the ground.

45.

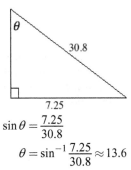

$$\sin \theta = \frac{7.25}{30.8}$$
$$\theta = \sin^{-1} \frac{7.25}{30.8} \approx 13.6$$
The angle is $13.6°$

47.

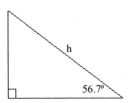

$$\cos 56.7° = \frac{16}{h}$$
$$h = \frac{16}{\cos 56.7°} \approx 29.14$$
The wire is 29.1 ft. long.

49.

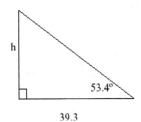

$$\tan 53.4° = \frac{h}{39.3}$$
$$h = 39.3 \tan 53.4° \approx 52.92$$
The height is 52.9 ft.

51.

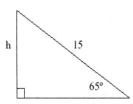

$$\sin 65 = \frac{h}{15}$$
$$h = 15 \sin 65° \approx 13.59$$
The ladder reaches 13.6 ft. up the side.

53.

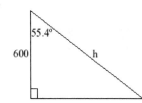

$$\cos 55.4° = \frac{600}{h}$$
$$h = \frac{600}{\cos 55.4°} \approx 1056.63$$
The wire is 1056.6 ft. long.

55.

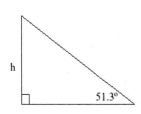

$$\tan 51.3° = \frac{h}{23.7}$$
$$h = 23.7 \tan 51.3° \approx 29.58$$
The tree is 29.6 yd. tall.

57. $\theta = \dfrac{12 \text{ cm}}{3 \text{ cm}} \text{ radians} = 4 \text{ radians}$

59. $\theta = \dfrac{6 \text{ in.}}{9 \text{ in.}} \text{ radian} = \dfrac{2}{3} \text{ radian}$

61. $\left(\dfrac{180}{\pi}\right)^{\circ}$

63. Converting:

$$45° \cdot \left(\frac{\pi \text{ radians}}{180°} \right) = \frac{45\pi}{180} = \frac{\pi}{4} \text{ radian}$$
$$\approx 0.7854 \text{ radian}$$

65. Converting:

$$315° \cdot \left(\frac{\pi \text{ radians}}{180°} \right) = \frac{315\pi}{180} = \frac{7\pi}{4} \text{ radians}$$
$$\approx 5.4978 \text{ radians}$$

67. Converting:

$$210° \cdot \left(\frac{\pi \text{ radians}}{180°} \right) = \frac{210\pi}{180} = \frac{7\pi}{6} \text{ radians}$$
$$\approx 3.6652 \text{ radians}$$

69. Converting:

$$\frac{\pi}{3} \text{ radians} \cdot \left(\frac{180°}{\pi \text{ radians}} \right) = \left(\frac{180\pi}{3\pi} \right)^{\circ} = 60°$$

71. Converting:

$$\frac{4\pi}{3} \text{ radians} \cdot \left(\frac{180°}{\pi \text{ radians}} \right) = \left(\frac{720\pi}{3\pi} \right)^{\circ} = 240°$$

73. Converting:

$$3 \text{ radians} \cdot \left(\frac{180°}{\pi \text{ radians}} \right) = \left(\frac{540}{\pi} \right)^{\circ} \approx 171.8873°$$

EXCURSION EXERCISES, SECTION 7.7

1. a.

The geodesic from A to B dips slightly below circle C.

  b.

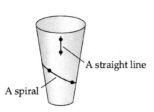

3. A piece of tape placed directly over circle E will not lie flat. Thus, by the inverse of the tape test theorem, circle E is not a geodesic of the figure.

5. c. Godthaab, Greenland

7. a. A piece of tape placed directly over the blue circle will lie flat. A piece of tape placed directly over either of the red circles will not lie flat.

  b. A piece of tape placed directly over the blue parabola will lie flat. A piece of tape placed directly over either red parabolas will not lie flat.

EXERCISE SET 7.7

1. a. Through a given point not on a given line, exactly one line can be drawn parallel to the given line.

  b. Through a given point not on a given line, there are at least two lines parallel to the given line.

  c. Through a given point not on a given line, there exist no lines parallel to the given line.

3. Carl Friedrich Gauss

5. Imaginary Geometry

7. A geodesic is a curve on a surface such that for any two points of the curve the portion of the curve between the points is the shortest path on the surface that joins these points.

9. An infinite saddle surface

11. Calculating:

$$m\angle A = 150°, \quad m\angle B = 120°, \quad m\angle C = 90°$$
$$S = (m\angle A + m\angle B + m\angle C - 180°)\left(\frac{\pi}{180°} \right) r^2$$
$$S = (150° + 120° + 90° - 180°)\left(\frac{\pi}{180°} \right)(1)^2$$
$$S = \pi \text{ units}^2$$

13. Using the formulas:

$$d_E(P,Q) = \sqrt{(x_2 - x_1)^2 + (y_2 - y_1)^2}$$
$$= \sqrt{(4 - (-3))^2 + (1 - 1)^2}$$
$$= \sqrt{49} = 7 \text{ blocks}$$
$$d_C(P,Q) = |x_2 - x_1| + |y_2 - y_1|$$
$$= |4 - (-3)| + |1 - 1|$$
$$= |7| + |0| = 7 \text{ blocks}$$

15. Using the formulas:

$$d_E(P,Q) = \sqrt{(x_2 - x_1)^2 + (y_2 - y_1)^2}$$
$$= \sqrt{(-3 - 2)^2 + (5 - (-3))^2}$$
$$= \sqrt{25 + 64} = \sqrt{89} \approx 9.4 \text{ blocks}$$
$$d_C(P,Q) = |x_2 - x_1| + |y_2 - y_1|$$
$$= |-3 - 2| + |5 - (-3)|$$
$$= |-5| + |8| = 13 \text{ blocks}$$

17. Using the formulas:
$$d_E(P,Q) = \sqrt{(x_2 - x_1)^2 + (y_2 - y_1)^2}$$
$$= \sqrt{(5-(-1))^2 + (-2-4)^2}$$
$$= \sqrt{36+36} = \sqrt{72} \approx 8.5 \text{ blocks}$$
$$d_C(P,Q) = |x_2 - x_1| + |y_2 - y_1|$$
$$= |5-(-1)| + |-2-4|$$
$$= |6| + |-6| = 12 \text{ blocks}$$

19. Using the formulas:
$$d_E(P,Q) = \sqrt{(x_2 - x_1)^2 + (y_2 - y_1)^2}$$
$$= \sqrt{(3-2)^2 + (-6-0)^2}$$
$$= \sqrt{1+36} = \sqrt{37} \approx 6.1 \text{ blocks}$$
$$d_C(P,Q) = |x_2 - x_1| + |y_2 - y_1|$$
$$= |3-2| + |-6-0|$$
$$= |1| + |-6| = 7 \text{ blocks}$$

21. Using the formula:
$$d_C(P,Q) = \frac{1+|m|}{\sqrt{1+m^2}} d_E(P,Q)$$
$$= \frac{1+\left|\frac{3}{4}\right|}{\sqrt{1+\left(\frac{3}{4}\right)^2}} \cdot 5 = \frac{\frac{7}{4}}{\sqrt{\frac{25}{16}}} \cdot 5$$
$$= \frac{\frac{7}{4}}{\frac{5}{4}} \cdot 5 = \frac{7}{4} \cdot \frac{4}{5} \cdot 5 = 7 \text{ blocks}$$

23. Using the formula:
$$d_C(P,Q) = \frac{1+|m|}{\sqrt{1+m^2}} d_E(P,Q)$$
$$= \frac{1+\left|-\frac{2}{3}\right|}{\sqrt{1+\left(-\frac{2}{3}\right)^2}} \cdot \sqrt{13} = \frac{\frac{5}{3}}{\sqrt{\frac{13}{9}}} \cdot \sqrt{13}$$
$$= \frac{\frac{5}{3}}{\frac{\sqrt{13}}{3}} \cdot \sqrt{13} = \frac{5}{3} \cdot \frac{3}{\sqrt{13}} \cdot \sqrt{13}$$
$$= 5 \text{ blocks}$$

25. Using the formula:
$$d_C(P,Q) = \frac{1+|m|}{\sqrt{1+m^2}} d_E(P,Q)$$
$$= \frac{1+\left|\frac{1}{4}\right|}{\sqrt{1+\left(\frac{1}{4}\right)^2}} \cdot \sqrt{17} = \frac{\frac{5}{4}}{\sqrt{\frac{17}{16}}} \cdot \sqrt{17}$$
$$= \frac{\frac{5}{4}}{\frac{\sqrt{17}}{4}} \cdot \sqrt{17} = \frac{5}{4} \cdot \frac{4}{\sqrt{17}} \cdot \sqrt{17}$$
$$= 5 \text{ blocks}$$

27. A city distance may be associated with more than one Euclidean distance. For example, if $P = (0,0)$ and $Q = (2,0)$, then the city distance between the points is 2 blocks and the Euclidean distance is also 2 blocks. However, if $P = (0,0)$ and $Q = (1,1)$, then the city distance between the points is still 2 blocks, but the Euclidean distance is $\sqrt{2}$ blocks.

29.

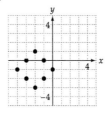

31.

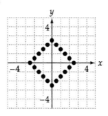

33. $4n$

35. a.

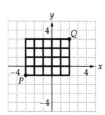

b.

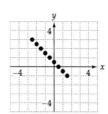

37. a. Find all combinations of 5 points taken 2 at a time.
$$C(5, 2) = \frac{5!}{3!2!} = 10$$

b. All lines not containing points A or B are parallel to $\overline{AB}$. There are 3 such lines.

EXCURSION EXERCISES, SECTION 7.8

1. The folding procedure would require you to fold the paper 10 times. This is not humanly possible with a small piece of paper.

3. Using the rule: $v = 7$

$$v = 7$$

$$\frac{v}{4} = \frac{7}{4} = 1 \text{ R } 3$$

A remainder of 3 corresponds with a left turn.

5. Using the rule:

$$v = 64$$

$$\frac{v}{2} = 32$$

$$\frac{32}{2} = 16$$

$$\frac{16}{2} = 8$$

$$\frac{8}{2} = 4$$

There was a right turn at vertex 4, so there is a right turn at vertex 64.

EXERCISE SET 7.8

1.

Stage 2 — — — —

Stage 3 - - - - - - - -

3.

Stage 2

5.
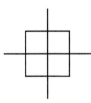
Stage 2

7.

Stage 2

9.
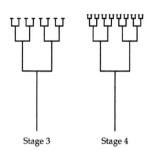
Stage 3   Stage 4

11. Replacement Ratio = 2
Scale Ratio = 3

Similarity Dimension $= \dfrac{\log 2}{\log 3} = 0.631$

13. Replacement Ratio = 5
Scale Ratio = 3

Similarity Dimension $= \dfrac{\log 5}{\log 3} = 1.465$

15. Replacement Ratio = 4
Scale Ratio = 2

Similarity Dimension $= \dfrac{\log 4}{\log 2} = 2.000$

17. Replacement Ratio = 4
Scale Ratio = 2

Similarity Dimension $= \dfrac{\log 4}{\log 2} = 2.000$

19. Replacement Ratio = 18
Scale Ratio = 6

Similarity Dimension $= \dfrac{\log 18}{\log 6} = 1.613$

21. a. Sierpinski carpet (1.893)
Variation 2 (1.771)
Variation 1 (1.465)

    b. The Sierpinski carpet

23. The binary tree fractal is not a strictly self-similar fractal

CHAPTER 7 REVIEW EXERCISES

1. $27 \text{ in.} = 27 \text{ in.} \times \dfrac{1 \text{ ft}}{12 \text{ in.}} = 2\dfrac{1}{4} \text{ ft}$

2. $15 \text{ c} = 15 \text{ c} \times \dfrac{1 \text{ pt}}{2 \text{ c}} = 7\dfrac{1}{2} \text{ pt}$

3. $37 \text{ mm} = 3.7 \text{ cm}$

4. $0.678 \text{ g} = 678 \text{ mg}$

5. $1273 \text{ ml} - 1.273 \text{ L}$

6. $\dfrac{\$3.56}{\text{L}} \approx \dfrac{\$3.56}{\text{L}} \times \dfrac{1 \text{ L}}{1.06 \text{ qt}} \approx \$3.36 / \text{qt}$

7. Calculating:
    $m\angle a = 74° \qquad m\angle b = 52°$
    $m\angle a = m\angle b + m\angle x$
    $\quad 74 = 52 + m\angle x$
    $\quad 22° = m\angle x$

    $180° = m\angle x + m\angle y$
    $180° = 22 + m\angle y$
    $158° = m\angle y$

8. $\dfrac{AC}{DF} = \dfrac{BC}{EF}$
    $\dfrac{AC}{12} = \dfrac{6}{9}$
    $\quad 72 = 9AC$
    Perimeter $ABC = 10 + 6 + 8 = 24 \text{ in.}$

9. Break figure into 2 rectangular prisms.
    $\quad V_1 = (3)(3)(8) = 72$
    $\quad V_2 = (3)(7)(8) = 168$
    $V_{\text{total}} = V_1 + V_2$
    $V_{\text{total}} = 72 + 168 = 240 \text{ in}^3$

10. $x = 180° - 112° = 68°$

11. Calculating:
    $S = 2lw + 2lh + 2wh$
    $S = 2(5)(10) + 2(5)(4) + 2(10)(4)$
    $S = 220 \text{ ft}^2$

12. $r = \dfrac{1}{2}(4) = 2 \qquad h = 8$
    $S = 2\pi r^2 + 2\pi rh$
    $S = 2\pi(2)^2 + 2\pi(2)(8)$
    $S = 8\pi + 32\pi = 40\pi \text{ m}^2$

13. $AB = 3BC \qquad BC = 11$
    $AC = AB + BC$
    $AC = 3(11) + 11 = 44 \text{ cm}$

14. Vertical angles are equal.
    $\angle x$ and $\angle y$ are vertical angles.
    $m\angle x = 150°$
    $m\angle x + m\angle y = 180°$
    $150° + m\angle y = 180°$
    $m\angle y = 30°$
    $m\angle w = 30°$

15. $A = bh$
    $A = 6(4.5) = 27 \text{ in}^2$

16. $s = 6 \qquad h = 8$
    $V = \dfrac{1}{3}s^2 h$
    $V = \dfrac{1}{3}(6)^2(8) = 96 \text{ cm}^3$

17. $C = 2\pi r = \pi d$
    $C = \pi(4.5) \approx 14.1 \text{ m}$

18. $\angle a$ is an alternate interior angle to the given
    angle. $\angle a$ and $\angle b$ are a linear pair.
    $m\angle a = 138° \qquad m\angle b = 42°$

19. $180° - 32° = 148°$
    A $148°$ angle

20. $V = lwh = (6.5)(2)(3) = 39 \text{ ft}^3$

21. Adding:
    $m\angle a + m\angle b + m\angle c = 180°$
    $\quad 37° + 48° m\angle c = 180°$
    $\qquad\qquad m\angle c = 180°$

22. $A = \dfrac{1}{2}bh \qquad A = 28 \qquad h = 7$
    $28 = \dfrac{1}{2}b(7) = \dfrac{7}{2}b$
    $b = 8 \text{ cm}$

23. $V = \dfrac{4}{3}\pi r^3 \qquad r = \dfrac{1}{2}(12) = 6$
    $V = \dfrac{4}{3}\pi(6)^3 = 288\pi \text{ mm}^3$

24. $P = 4s = 86$
    $s = \dfrac{86}{4} = 21.5 \text{ cm}$

25. $r = 6h = 15$

$$S = 2\pi r^2 + 2\pi rh$$

$$S = 2\pi(6)^2 + 2\pi(6)(15)$$

$$S = 252\pi \approx 791.68 \text{ ft}^2$$

$791.68 \div 200 = 3.9584$

4 cans of paint are needed

26. $P = 2l + 2w$

$P = 2(56) + 2(48) = 208$ yd

27. $A = s^2 = (9.5)^2 = 90.25 \text{ m}^2$

28. $A_{\text{walk}} = A_{\text{total}} - A_{\text{grass}}$

$A_{\text{walk}} = 44(29) + 40(25)$

$A_{\text{walk}} = 1276 - 1000 = 276 \text{ m}^2$

29. Yes, by the SAS Theorem.

30. Using the Pythagorean Theorem:

$$a^2 + b^2 = c^2$$

$$a^2 + 7^2 = 12^2$$

$$a^2 = 144 - 49 = 95$$

$$a = \sqrt{95} \approx 9.7 \text{ ft}$$

31. Using the Pythagorean Theorem:

$$a^2 + b^2 = c^2$$

$$5^2 + 8^2 = c^2$$

$$c^2 = 89$$

$$c = \sqrt{89}$$

$$\sin\theta = \frac{5}{\sqrt{89}} = \frac{5\sqrt{89}}{89},$$

$$\cos\theta = \frac{8}{\sqrt{89}} = \frac{8\sqrt{89}}{89}, \quad \tan\theta = \frac{5}{8}$$

32. Using the Pythagorean Theorem:

$$a^2 + b^2 = c^2$$

$$a^2 + 10^2 = 20^2$$

$$a^2 = 400 - 100 = 300$$

$$a = \sqrt{300} = 10\sqrt{3}$$

$$\sin\theta = \frac{10\sqrt{3}}{20} = \frac{\sqrt{3}}{2},$$

$$\cos\theta = \frac{1}{2}, \quad \tan\theta = \frac{10\sqrt{3}}{10} = \sqrt{3}$$

33. $25.7°$

34. $29.2°$

35. $53.8°$

36. $1.9°$

37. $\tan 50° = \dfrac{d}{84}$

$d = 84\tan 50° \approx 100.1$ ft

38. $\cos 40° = \dfrac{d}{200}$

$d = 200\cos 40° \approx 153.2$ mi

39.

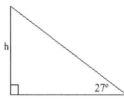

$$\tan 27° = \frac{h}{110}$$

$$h = 110\tan 27° \approx 56.0 \text{ ft}$$

40. Spherical geometry or elliptical geometry

41. Hyperbolic geometry

42. Lobachevskian or hyperbolic geometry

43. Riemannian or spherical geometry

44. $m\angle A = 90°$, $m\angle B = 150°$,
     $m\angle C = 90°$, $r = 12$

$$S = (m\angle A + m\angle B + m\angle C - 180°)\left(\frac{\pi}{180°}\right)r^2$$

$$S = (90° + 150° + 90° - 180°)\left(\frac{\pi}{180°}\right)(12)^2$$

$$S = \frac{5}{6}\pi(144) = 120\pi \text{ in}^2$$

45. $m\angle A = 90°$, $m\angle B = 60°$,
     $m\angle C = 90°$, $r = 5$

$$S = (m\angle A + m\angle B + m\angle C - 180°)\left(\frac{\pi}{180°}\right)r^2$$

$$S = (90° + 60° + 90° - 180°)\left(\frac{\pi}{180°}\right)(5)^2$$

$$S = \frac{\pi}{3}(25) = \frac{25\pi}{3} \text{ ft}^2$$

46. Using the formulas:

$$d_E(P,Q) = \sqrt{(x_2 - x_1)^2 + (y_2 - y_1)^2}$$

$$= \sqrt{(3 - (-1))^2 + (4 - 1)^2}$$

$$= \sqrt{16 + 9} = \sqrt{25} = 5 \text{ blocks}$$

$$d_C(P,Q) = |x_2 - x_1| + |y_2 - y_1|$$

$$= |3 - (-1)| + |4 - 1|$$

$$= |4| + |3| = 7 \text{ blocks}$$

47. Using the formulas:
$$d_E(P,Q) = \sqrt{(x_2 - x_1)^2 + (y_2 - y_1)^2}$$
$$= \sqrt{(2-(-5))^2 + (6-(-2))^2}$$
$$= \sqrt{49 + 64} = \sqrt{113} \approx 10.6 \text{ blocks}$$
$$d_C(P,Q) = |x_2 - x_1| + |y_2 - y_1|$$
$$= |3-(-2)| + |-1-4|$$
$$= |5| + |-5| = 10 \text{ blocks}$$

48. Using the formulas:
$$d_E(P,Q) = \sqrt{(x_2 - x_1)^2 + (y_2 - y_1)^2}$$
$$= \sqrt{(3-2)^2 + (2-8)^2}$$
$$= \sqrt{1 + 36} = \sqrt{37} \approx 6.1 \text{ blocks}$$
$$d_C(P,Q) = |x_2 - x_1| + |y_2 - y_1|$$
$$= |3-2| + |2-8|$$
$$= |1| + |-6| = 7 \text{ blocks}$$

49. Using the formulas:
$$d_E(P,Q) = \sqrt{(x_2 - x_1)^2 + (y_2 - y_1)^2}$$
$$= \sqrt{(5-(-3))^2 + (-2-3)^2}$$
$$= \sqrt{64 + 25} = \sqrt{89} \approx 9.4 \text{ blocks}$$
$$d_C(P,Q) = |x_2 - x_1| + |y_2 - y_1|$$
$$= |5-(-3)| + |-2-3|$$
$$= 8 + |-5| = 13 \text{ blocks}$$

50. a. Using the formulas:
$$d_E(P,Q) = \sqrt{(x_2 - x_1)^2 + (y_2 - y_1)^2}$$
$$= \sqrt{(4-1)^2 + (5-1)^2}$$
$$= \sqrt{9 + 16} = \sqrt{25} = 5 \text{ blocks}$$
$$d_E(P,R) = \sqrt{(x_2 - x_1)^2 + (y_2 - y_1)^2}$$
$$= \sqrt{(-4-1)^2 + (2-1)^2}$$
$$= \sqrt{25 + 1} = \sqrt{26} \approx 5.1 \text{ blocks}$$
$$d_E(R,Q) = \sqrt{(x_2 - x_1)^2 + (y_2 - y_1)^2}$$
$$= \sqrt{(4-(-4))^2 + (5-2)^2}$$
$$= \sqrt{64 + 9} = \sqrt{73} \approx 8.5 \text{ blocks}$$
P and Q are closest.

    b. Using the formulas:
$$d_C(P,Q) = |x_2 - x_1| + |y_2 - y_1|$$
$$= |4-1| + |5-1|$$
$$= |3| + |4| = 7 \text{ blocks}$$
$$d_C(P,R) = |x_2 - x_1| + |y_2 - y_1|$$
$$= |-4-1| + |2-1|$$
$$= |-5| + |1| = 6 \text{ blocks}$$
$$d_C(R,Q) = |x_2 - x_1| + |y_2 - y_1|$$
$$= |4-(-4)| + |5-2|$$
$$= |8| + |3| = 11 \text{ blocks}$$
P and R are closest.

51. The Koch curve is a strictly self-similar fractal, because any portion of the Koch curve replicates the entire fractal.

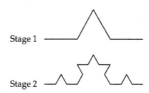

Stage 0

Stage 1

Stage 2

52.

Stage 2

53. a. Replacement Ratio = 2

    b. Scaling Ratio = 2

    c. Similarity Dimension $= \dfrac{\log 2}{\log 2} = 1$

54. $\dfrac{\log 5}{\log 4} \approx 1.161$

CHAPTER 7 TEST

1. Solving:
$$V = \pi r^2 h$$
$$V = \pi(3)^2(6) = 54\pi \approx 169.6 \text{ m}^3$$

2. Solving:
$$P = 2l + 2w$$
$$P = 2(2) + 2(1.4) = 6.8 \text{ m}$$

3. $90 - 32 = 58°$

4. $A = \pi r^2 = \pi(1)^2 = \pi \approx 3.1 \text{ m}^2$

5. Solving:
$$m\angle x + m\angle z = 180°$$
$$30° + m\angle z = 180°$$
$$m\angle z = 150°$$
$$m\angle z = m\angle y \quad \text{(corresponding angles)}$$
$$m\angle y = 150°$$

6. Solving:
$$m\angle x + m\angle b = 180°$$
$$45° + m\angle b = 180°$$
$$m\angle b = 135°$$
$\angle x$ and $\angle a$ are alternate exterior angles, so
$$m\angle x = m\angle a \quad \text{(corresponding angles)}$$
$$m\angle a = 45°$$

7. 1.2 m = 1200 cm

8. Solving:
$$V_{\text{total}} = V_{\text{large}} - V_{\text{small}}$$
$$V_{\text{large}} = \pi(6)^2(14) = 504\pi$$
$$V_{\text{small}} = \pi(2)^2(14) = 56\pi$$
$$V_{\text{total}} = 504\pi - 56\pi = 448\pi \text{ cm}^3$$

9. Solving:
$$\frac{BC}{4} = \frac{\frac{3}{4}}{2\frac{1}{2}}$$
$$4\left(\frac{3}{4}\right) = \left(2\frac{1}{2}\right)BC$$
$$3 = \left(2\frac{1}{2}\right)BC$$
$$BC = 1.2 \text{ ft or } 1\frac{1}{5} \text{ ft}$$

10. By definition, one angle is 90°, while the third will be 90 − 40 = 50°

11. $m\angle x = 38 + 87 = 125°$

12. Solving:
$$A = bh$$
$$A = (8)(4) = 32 \text{ m}^2$$

13. Solving:
$$\frac{5}{x} = \frac{12}{60}$$
$$5(60) = 12x$$
$$12x = 300$$
$$x = 25 \text{ ft}$$

14. Solving:
$$A_{\text{small}} = \pi(8)^2 \quad A_{\text{large}} = \pi(10)^2$$
Difference = $100\pi - 64\pi = 36\pi$
There is about 113.1 in² more.

15. Yes, by the SAS Theorem.

16. Using the Pythagorean Theorem:
$$a^2 + b^2 = c^2$$
$$a^2 + 8^2 = 11^2$$
$$a^2 = 11^2 - 8^2 = 57$$
$$a = \sqrt{57} \approx 7.54$$
Side BC is 7.5 cm long.

17. Using the Pythagorean Theorem:
$$a^2 + b^2 = c^2$$
$$a^2 + 8^2 = 10^2$$
$$a^2 = 100 - 64 = 36$$
$$a = \sqrt{36} = 6$$
$$\sin\theta = \frac{8}{10} = \frac{4}{5}$$
$$\cos\theta = \frac{6}{10} = \frac{3}{5}$$
$$\tan\theta = \frac{8}{6} = \frac{4}{3}$$

18.

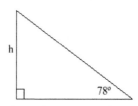

27
$$\tan 78° = \frac{h}{27}$$
$$h = 27\tan 78° \approx 127 \text{ ft}$$

19. Solving:
$$d = 11 \text{ ft } 6 \text{ in.} = 11.5 \text{ ft}$$
$$r = \frac{1}{2}(11.5) = 5.75$$
$$A = \pi r^2$$
$$A = \pi(5.75)^2 \approx 103.9 \text{ ft}^2$$

20. $l = 1 \text{ ft } 1 \text{ in.} = 13 \text{ in.} \quad w = 8 \text{ in.} \quad h = 7.5 \text{ in.}$
$$V = lwh$$
$$V = 13(8)(7.5) = 780 \text{ in}^3$$

21. $\dfrac{88 \text{ ft}}{\text{sec}} \approx \dfrac{88 \text{ ft}}{\text{sec}} \times \dfrac{1 \text{ m}}{3.28 \text{ ft}} \approx 26.82 \text{ m} / \text{sec}$

22. Through a given point not on a given line, exactly one line can be drawn parallel to the given lin.

23. A great circle of a sphere is a circle on the surface of a sphere whose center is at the center of the sphere.
The Spherical Triangle Area Formula is:
$$S = (m\angle A + m\angle B + m\angle C - 180°)\left(\frac{\pi}{180°}\right)r^2$$

where S is the area of spherical triangle ABC on a sphere with radius r.

24. Solving:
$m\angle A = 90°, \ m\angle B = 100°,$
$m\angle C = 90°, \ r = 12$

$$S = (m\angle A + m\angle B + m\angle C - 180°)\left(\frac{\pi}{180°}\right)r^2$$

$$S = (90° + 100° + 90° - 180°)\left(\frac{\pi}{180°}\right)(12)^2$$

$$S = \frac{5}{9}\pi(144) = 80\pi \ \text{ft}^2 \approx 251.3 \ \text{ft}^2$$

25. Using the formulas:

$$d_E(P,Q) = \sqrt{(x_2 - x_1)^2 + (y_2 - y_1)^2}$$
$$= \sqrt{(5 - (-4))^2 + (1 - 2)^2}$$
$$= \sqrt{81 + 1} = \sqrt{82} \approx 9.1 \ \text{blocks}$$

$$d_C(P,Q) = |x_2 - x_1| + |y_2 - y_1|$$
$$= |5 - (-4)| + |1 - 2|$$
$$= |9| + |-1| = 10 \ \text{blocks}$$

26. Using the formula: $4n = 4(4) = 16$ points.

27.

Stage 2

28.

Stage 2

29. Replacement Ratio = 2
Scale Ratio = 2

$$\text{Similarity Dimension} = \frac{\log 2}{\log 2} = 1.000$$

30. Replacement Ratio = 3
Scale Ratio = 2

$$\text{Similarity Dimension} = \frac{\log 3}{\log 2} \approx 1.585$$

Chapter 8: Mathematical Systems

EXCURSION EXERCISES, SECTION 8.1

1. Answers will vary.

3. For January 1, 2050, we have $c = 20$, $y = (50 - 1)$, $d = 1$, $m = 11$. Using these values,

$$x \equiv \left(\left[\frac{13(11) - 1}{5} \right] + \left[\frac{49}{4} \right] + \left[\frac{20}{4} \right] + 1 + 49 - 2(20) \right) \bmod 7$$
$$\equiv (28 + 12 + 5 + 1 + 49 - 40) \bmod 7$$
$$\equiv 55 \bmod 7 \equiv 6 \text{ or Saturday.}$$

EXERCISE SET 8.1

1. 8

3. 12

5. 2

7. 4

9. 4

11. 7

13. 11

15. 7

17. 0300

19. 0400

21. 2000

23. 2100

25. 3

27. 6

29. True. $5 \div 3 = 1$ remainder 2,
   $8 \div 3 = 2$ remainder 2

31. False. $5 \div 4 = 1$ remainder 1,
   $20 \div 4 = 5$ remainder 0.

33. True. $21 \div 6 = 3$ remainder 3,
   $45 \div 6 = 7$ remainder 3.

35. False. $88 \div 9 = 9$ remainder 7,
   $5 \div 9 = 0$ remainder 5.

37. True. $100 \div 8 = 12$ remainder 4,
   $20 \div 8 = 2$ remainder 4.

39. Answers will vary. Adding a multiple of the modulus maintains the congruency. So, possible answers include
   $8 + -1(6) = 2$
   $8 + 0(6) = 8$
   $8 + 1(6) = 14$
   $8 + 2(6) = 20$
   $8 + 3(6) = 26$
   $8 + 4(6) = 32$
   $8 + 5(6) = 38$
   and so on.

41. $(9 + 15) \bmod 7 = 24 \bmod 7$
   $24 \div 7 = 3$ remainder 3
   So $(9 + 15) \bmod 7 \equiv 3$

43. $(5 + 22) \bmod 8 = 27 \bmod 8$
   $27 \div 8 = 3$ remainder 3
   $(5 + 22) \bmod 8 \equiv 3$

45. $(42 + 35) \bmod 3 = 77 \bmod 3$
   $77 \div 3 = 25$ remainder 2
   So $(42 + 35) \bmod 3 \equiv 2$

47. $(37 + 45) \bmod 12 = 82 \bmod 12$
   $82 \div 12 = 6$ remainder 10
   So $(37 + 45) \bmod 12 \equiv 10$

49. $(19 - 6) \bmod 5 = 13 \bmod 5$
   $13 \div 5 = 2$ remainder 3
   So $(19 - 6) \bmod 5 \equiv 3$

51. $(48 - 21) \bmod 6 = 27 \bmod 6$
   $27 \div 6 = 4$ remainder 3
   So $(48 - 21) \bmod 6 \equiv 3$

52. $(60 - 32) \bmod 9 = 28 \bmod 9$
   $28 \div 9 = 3$ remainder 1
   So $(60 - 32) \bmod 9 \equiv 1$

53. $(8 - 15) \bmod 12 = -7 \bmod 12$
   $-7 + 12 = 5$
   So $(8 - 15) \bmod 12 \equiv 5$

55. $(15 - 32) \bmod 7 = -17 \bmod 7$
$-17 + 7 = -10$
$-10 + 7 = -3$
$-3 + 7 = 4$
So $(15 - 32) \bmod 7 \equiv 4$

57. $(6 \cdot 8) \bmod 9 = 48 \bmod 9$
$48 \div 9 = 5$ remainder 3
So $(6 \cdot 8) \bmod 9 \equiv 3$

59. $(9 \cdot 15) \bmod 8 = 135 \bmod 8$
$135 \div 8 = 16$ remainder 7
So $(9 \cdot 15) \bmod 8 \equiv 7$

61. $(14 \cdot 18) \bmod 5 = 252 \bmod 5$
$252 \div 5 = 50$ remainder 2
So $(14 \cdot 18) \bmod 5 \equiv 2$

63. a. $7 + 59 = 66$, which is positive. $66 \div 12 = 5$
remainder 6
It will be 6:00.

   b. $7 - 62 = -55$, which is negative.
$-55 + 12 = -43$
$-43 + 12 = -31$
$-31 + 12 = -19$
$-19 + 12 = -7$
$-7 + 12 = 5$
It was 5:00.

65. a.   The number 5 is associated with Friday, and
$5 + 25 = 30$, which is positive.
$30 \div 7 = 4$ remainder 2
It will be Tuesday 25 days from today.

   b.   The number 5 is associated with Friday, and
$5 - 32 = -27$, which is negative.
$-27 + 7 = -20$
$-20 + 7 = -13$
$-13 + 7 = -6$
$-6 + 7 = 1$
It was Monday 32 days ago.

67. There are 10 years between the two dates, three
of which (2016, 2020, and 2024) are leap years.
So the total number of days between the two
dates is
$10(365) + 3 = 3653$ days
$3653 \div 7 = 521$ remainder 6
So, in 2025, Halloween will occur 6 days after
Saturday, which is Friday.

69. There are 15 years between the two dates, three
of which (2020, 2024 and 2030) are leap years.
So the total number of days between the two
dates is
$15(365) + 3 = 5478$ days
$5478 \div 7 = 782$ remainder 4
So, in 2032, Valentine's Day will occur 4 days
after Tuesday, which is Saturday.

71. Beginning with zero, substitute each whole
number less than 3 into the congruence equation.
$x = 0$ $0 \not\equiv 10 \bmod 3$ NO
$x = 1$ $1 \equiv 10 \bmod 3$ YES
$x = 2$ $2 \not\equiv 10 \bmod 3$ NO
The solutions are 1, 4, 7, 10, 13, 16, . . .

73. Beginning with zero, substitute each whole
number less than 5 into the congruence equation.
$x = 0$ $2(0) \not\equiv 12 \bmod 5$ NO
$x = 1$ $2(1) \equiv 12 \bmod 5$ YES
$x = 2$ $2(2) \not\equiv 12 \bmod 5$ NO
$x = 3$ $2(3) \not\equiv 12 \bmod 5$ NO
$x = 4$ $2(4) \not\equiv 12 \bmod 5$ NO
The solutions are 1, 6, 11, 16, 21, 26, . . .

75. Beginning with zero, substitute each whole
number less than 4 into the congruence.
$x = 0$ $2(0) + 1 \equiv 5 \bmod 4$ YES
$x = 1$ $2(1) + 1 \not\equiv 5 \bmod 4$ NO
$x = 2$ $2(2) + 1 \equiv 5 \bmod 4$ YES
$x = 3$ $2(3) + 1 \not\equiv 5 \bmod 4$ NO
The solutions are 0, 2, 4, 6, 8, 10, 12, . . .

77. Beginning with zero, substitute each whole
number less than 12 into the congruence.
$x = 0$ $2(0) + 3 \not\equiv 8 \bmod 12$ NO
$x = 1$ $2(1) + 3 \not\equiv 8 \bmod 12$ NO
$x = 2$ $2(2) + 3 \not\equiv 8 \bmod 12$ NO
$x = 3$ $2(3) + 3 \not\equiv 8 \bmod 12$ NO
$x = 4$ $2(4) + 3 \not\equiv 8 \bmod 12$ NO
$x = 5$ $2(5) + 3 \not\equiv 8 \bmod 12$ NO
$x = 6$ $2(6) + 3 \not\equiv 8 \bmod 12$ NO
$x = 7$ $2(7) + 3 \not\equiv 8 \bmod 12$ NO
$x = 8$ $2(8) + 3 \not\equiv 8 \bmod 12$ NO
$x = 9$ $2(9) + 3 \not\equiv 8 \bmod 12$ NO
$x = 10$ $2(10) + 3 \not\equiv 8 \bmod 12$ NO
$x = 11$ $2(11) + 3 \not\equiv 8 \bmod 12$ NO
The congruent equation has no solutions.

79. Beginning with zero, substitute each whole
number less than 4 into the congruence.
$x = 0$ $2(0) + 2 \equiv 6 \bmod 4$ YES
$x = 1$ $2(1) + 2 \not\equiv 6 \bmod 4$ NO
$x = 2$ $2(2) + 2 \equiv 6 \bmod 4$ YES
$x = 3$ $2(3) + 2 \not\equiv 6 \bmod 4$ NO
The solutions are 0, 2, 4, 6, 8, 10, 12, ...

81. Beginning with zero, substitute each whole number less than 8 into the congruence.

$x = 0$	$4(0) + 6 \not\equiv 5 \bmod 8$	NO
$x = 1$	$4(1) + 6 \not\equiv 5 \bmod 8$	NO
$x = 2$	$4(2) + 6 \not\equiv 5 \bmod 8$	NO
$x = 3$	$4(3) + 6 \not\equiv 5 \bmod 8$	NO
$x = 4$	$4(4) + 6 \not\equiv 5 \bmod 8$	NO
$x = 5$	$4(5) + 6 \not\equiv 5 \bmod 8$	NO
$x = 6$	$4(6) + 6 \not\equiv 5 \bmod 8$	NO
$x = 7$	$4(7) + 6 \not\equiv 5 \bmod 8$	NO

    The congruence equation has no solutions.

83. $4 + 5 = 9$, so the additive inverse is 5. Substitute whole numbers less than 9 into the congruence $4x \equiv 1 \bmod 9$.

$x = 0$	$4(0) \not\equiv 1 \bmod 9$	NO
$x = 1$	$4(1) \not\equiv 1 \bmod 9$	NO
$x = 2$	$4(2) \not\equiv 1 \bmod 9$	NO
$x = 3$	$4(3) \not\equiv 1 \bmod 9$	NO
$x = 4$	$4(4) \not\equiv 1 \bmod 9$	NO
$x = 5$	$4(5) \not\equiv 1 \bmod 9$	NO
$x = 6$	$4(6) \not\equiv 1 \bmod 9$	NO
$x = 7$	$4(7) \equiv 1 \bmod 9$	YES

    The multiplicative inverse is 7.

85. $7 + 3 = 10$, so the additive inverse is 3. Substitute whole numbers less than 10 into the congruence $7x \equiv 1 \bmod 10$

$x = 0$	$7(0) \not\equiv 1 \bmod 10$	NO
$x = 1$	$7(1) \not\equiv 1 \bmod 10$	NO
$x = 2$	$7(2) \not\equiv 1 \bmod 10$	NO
$x = 3$	$7(3) = 1 \bmod 10$	YES

    The multiplicative inverse is 3.

87. $3 + 5 = 8$, so the additive inverse is 5. Substitute whole numbers less than 8 into the congruence $3x \equiv 1 \bmod 8$

$x = 0$	$3(0) \not\equiv 1 \bmod 8$	NO
$x = 1$	$3(1) \not\equiv 1 \bmod 8$	NO
$x = 2$	$3(2) \not\equiv 1 \bmod 8$	NO
$x = 3$	$3(3) \equiv 1 \bmod 8$	YES

    The multiplicative inverse is 3.

89. Assume $x \equiv (2 \div 7) \bmod 8$; then $7x \equiv 2 \bmod 8$.

$x = 0$	$7(0) \not\equiv 2 \bmod 8$	NO
$x = 1$	$7(1) \not\equiv 2 \bmod 8$	NO
$x = 2$	$7(2) \not\equiv 2 \bmod 8$	NO
$x = 3$	$7(3) \not\equiv 2 \bmod 8$	NO
$x = 4$	$7(4) \not\equiv 2 \bmod 8$	NO
$x = 5$	$7(5) \not\equiv 2 \bmod 8$	NO
$x = 6$	$7(6) \equiv 2 \bmod 8$	YES

    The quotient is 6.

91. Assume $x \equiv (6 \div 4) \bmod 9$; then $4x \equiv 6 \bmod 9$.

$x = 0$	$4(0) \not\equiv 6 \bmod 9$	NO
$x = 1$	$4(1) \not\equiv 6 \bmod 9$	NO
$x = 2$	$4(2) \not\equiv 6 \bmod 9$	NO
$x = 3$	$4(3) \not\equiv 6 \bmod 9$	NO
$x = 4$	$4(4) \not\equiv 6 \bmod 9$	NO
$x = 5$	$4(5) \not\equiv 6 \bmod 9$	NO
$x = 6$	$4(6) \equiv 6 \bmod 9$	YES

93. Assume $x \equiv (5 \div 6) \bmod 7$; then $6x \equiv 5 \bmod 7$.

$x = 0$	$6(0) \not\equiv 5 \bmod 7$	NO
$x = 1$	$6(1) \not\equiv 5 \bmod 7$	NO
$x = 2$	$6(2) \equiv 5 \bmod 7$	YES

    The quotient is 2.

95. Assume $x = (5 \div 8) \bmod 8$; then $8x \equiv 5 \bmod 8$. This congruence has no solution for x because $8x \equiv 0 \bmod 8$ for all whole number values of x. Thus $5 \div 8$ has no solution in modulo 8 arithmetic.

97. Beginning with 0, substitute each whole number less than 11 into the congruence.

$x = 0$	$0^2 + 3(0) + 7 \not\equiv 2 \bmod 11$	NO
$x = 1$	$1^2 + 3(1) + 7 \not\equiv 2 \bmod 11$	NO
$x = 2$	$2^2 + 3(2) + 7 \not\equiv 2 \bmod 11$	NO
$x = 3$	$3^2 + 3(3) + 7 \not\equiv 2 \bmod 11$	NO
$x = 4$	$4^2 + 3(4) + 7 \equiv 2 \bmod 11$	YES

    The solution is 4.

99. a. $x_{n+1} = (2x_n + 13) \bmod 11$

    $x_0 = 3$, so
    $$x_1 = (2 \cdot 3 + 13) \bmod 11$$
    $$= 19 \bmod 11 = 8$$
    $$x_2 = (2 \cdot 8 + 13) \bmod 11$$
    $$= 29 \bmod 11 = 7$$
    $$x_3 = (2 \cdot 7 + 13) \bmod 11$$
    $$= 27 \bmod 11 = 5$$
    $$x_4 = (2 \cdot 5 + 13) \bmod 11$$
    $$= 23 \bmod 11 = 1$$
    $$x_5 = (2 \cdot 1 + 13) \bmod 11$$
    $$= 15 \bmod 11 = 4$$
    $$x_6 = (2 \cdot 4 + 13) \bmod 11$$
    $$= 21 \bmod 11 = 8$$
    $$x_7 = (2 \cdot 10 + 13) \bmod 11$$
    $$= 33 \bmod 11 = 0$$
    $$x_8 = (2 \cdot 0 + 13) \bmod 11$$
    $$= 13 \bmod 11 = 2$$
    $$x_9 = (2 \cdot 2 + 13) \bmod 11$$
    $$= 17 \bmod 11 = 6$$

    b. $x_{10} = (2 \cdot 6 + 13) \bmod 11$
    $$= 25 \bmod 11 = 3$$
    x_{10} is equal to x_0.

c. The sequence of digits repeats after 10 pushes of the garage door opener button. A thief would need only to detect the repeating number and know the next number to which the receiver will respond.

EXCURSION EXERCISES, SECTION 8.2

1. Using the RSA program on the calculator, and the numbers of (13 25 14 15) for the word CODE to get the answer of 4870, 607, 141, 3532.

3. The calculations of (33 25 28 22 14) are deciphered from the message received. Decoded this word is WORLD.

5. Answers will vary.

EXERCISE SET 8.2

1. $d_{13} \equiv 10 - (d_1 + 3d_2 + d_3 + 3d_4 + d_5 + 3d_6 + d_7 + 3d_8 + d_9 + 3d_{10} + d_{11} + 3d_{12}) \bmod 10$
    $d_{13} \equiv 10 - [9 + 3(7) + 8 + 3(0) + 2 + 3(8) + 1 + 3(4) + 4 + 3(2) + 6 + 3(8)] \bmod 10$
    $\equiv 10 - 117 \bmod 10$
    $\equiv 10 - 7 \equiv 3$
    The check digit is 3 and not 5 as it should be. Therefore, ISBN is not valid.

3. $d_{13} \equiv 10 - (d_1 + 3d_2 + d_3 + 3d_4 + d_5 + 3d_6 + d_7 + 3d_8 + d_9 + 3d_{10} + d_{11} + 3d_{12}) \bmod 10$
    $d_{13} \equiv 10 - [9 + 3(7) + 8 + 3(0) + 6 + 3(7) + 1 + 3(5) + 1 + 3(9) + 8 + 3(3)] \bmod 10$
    $\equiv 10 - 126 \bmod 10$
    $\equiv 10 - 6 \equiv 4$
    The check digit is 4. Therefore, ISBN is valid.

5. $d_{13} \equiv 10 - (d_1 + 3d_2 + d_3 + 3d_4 + d_5 + 3d_6 + d_7 + 3d_8 + d_9 + 3d_{10} + d_{11} + 3d_{12}) \bmod 10$
    $d_{13} \equiv 10 - [9 + 3(7) + 8 + 3(0) + 1 + 3(4) + 3 + 3(0) + 3 + 3(9) + 4 + 3(3)] \bmod 10$
    $\equiv 10 - 97 \bmod 10$
    $\equiv 10 - 7 \equiv 3$
    The check digit is 3. Therefore, ISBN is valid.

7. $d_{13} \equiv 10 - (d_1 + 3d_2 + d_3 + 3d_4 + d_5 + 3d_6 + d_7 + 3d_8 + d_9 + 3d_{10} + d_{11} + 3d_{12}) \bmod 10$
    $d_{13} \equiv 10 - [9 + 3(7) + 8 + 3(0) + 4 + 3(3) + 9 + 3(0) + 2 + 3(3) + 5 + 3(2)] \bmod 10$
    $\equiv 10 - 82 \bmod 10$
    $\equiv 10 - 2 \equiv 8$
    The check digit is 8.

9. $d_{13} \equiv 10 - (d_1 + 3d_2 + d_3 + 3d_4 + d_5 + 3d_6 + d_7 + 3d_8 + d_9 + 3d_{10} + d_{11} + 3d_{12}) \bmod 10$
    $d_{13} \equiv 10 - [9 + 3(7) + 8 + 3(0) + 8 + 3(5) + 7 + 3(5) + 2 + 3(2) + 3 + 3(2)] \bmod 10$
    $\equiv 10 - 100 \bmod 10$
    $\equiv 10 - 0 \equiv 10$
    The check digit is 0.

11. $d_{13} \equiv 10 - (d_1 + 3d_2 + d_3 + 3d_4 + d_5 + 3d_6 + d_7 + 3d_8 + d_9 + 3d_{10} + d_{11} + 3d_{12}) \bmod 10$
    $d_{13} \equiv 10 - [9 + 3(7) + 8 + 3(0) + 3 + 3(1) + 6 + 3(2) + 0 + 3(4) + 3 + 3(7)] \bmod 10$
    $\equiv 10 - 92 \bmod 10$
    $\equiv 10 - 2 \equiv 8$
    The check digit is 8.

13. $d_{13} \equiv 10 - (d_1 + 3d_2 + d_3 + 3d_4 + d_5 + 3d_6 + d_7 + 3d_8 + d_9 + 3d_{10} + d_{11} + 3d_{12}) \bmod 10$
    $d_{13} \equiv 10 - [9 + 3(7) + 8 + 3(0) + 5 + 3(1) + 7 + 3(8) + 8 + 3(4) + 4 + 3(1)] \bmod 10$
    $\equiv 10 - 104 \bmod 10$
    $\equiv 10 - 4 \equiv 6$
    The check digit is 6.

15. $d_{12} \equiv 10 - (3d_1 + d_2 + 3d_3 + d_4 + 3d_5 + d_6 + 3d_7 + d_8 + 3d_9 + d_{10} + 3d_{11}) \bmod 10$
    $d_{12} \equiv 10 - [3(0) + 7 + 3(9) + 8 + 3(9) + 3 + 3(4) + 6 + 3(5) + 0 + 3(0)] \bmod 10$
    $\equiv 10 - 105 \bmod 10$
    $\equiv 10 - 5 \equiv 5$
    The check digit is 5.

17. $d_{12} \equiv 10 - (3d_1 + d_2 + 3d_3 + d_4 + 3d_5 + d_6 + 3d_7 + d_8 + 3d_9 + d_{10} + 3d_{11}) \bmod 10$
    $d_{12} \equiv 10 - [3(7) + 1 + 3(4) + 0 + 3(4) + 3 + 3(0) + 1 + 3(1) + 2 + 3(6)] \bmod 10$
    $\equiv 10 - 73 \bmod 10$
    $\equiv 10 - 3 \equiv 7$
    The check digit is 7.

19. $d_{12} \equiv 10 - (3d_1 + d_2 + 3d_3 + d_4 + 3d_5 + d_6 + 3d_7 + d_8 + 3d_9 + d_{10} + 3d_{11}) \bmod 10$
    $d_{12} \equiv 10 - [3(8) + 8 + 3(8) + 4 + 3(6) + 2 + 3(5) + 2 + 3(1) + 4 + 3(8)] \bmod 10$
    $\equiv 10 - 128 \bmod 10$
    $\equiv 10 - 8 \equiv 2$
    The check digit is 2.

21. $d_{12} \equiv 10 - (3d_1 + d_2 + 3d_3 + d_4 + 3d_5 + d_6 + 3d_7 + d_8 + 3d_9 + d_{10} + 3d_{11}) \bmod 10$
    $d_{12} \equiv 10 - [3(0) + 4 + 3(1) + 7 + 3(9) + 0 + 3(2) + 2 + 3(1) + 0 + 3(6)] \bmod 10$
    $\equiv 10 - 70 \bmod 10$
    $\equiv 10 - 0 \equiv 10$
    The check digit is 0.

23. $(0 + 3 + 1 + 6 + 6 + 1 + 5 + 4 + 9 + 8) \bmod 9 \equiv 43 \bmod 9 \equiv 7$

25. $(1 + 3 + 3 + 1 + 4 + 9 + 7 + 5 + 3 + 3) \bmod 9 \equiv 39 \bmod 9 \equiv 3$

27. $(1 + 1 + 8 + 2 + 6 + 4 + 9 + 7 + 5 + 8) \bmod 7 \equiv 51 \bmod 7 \equiv 2$; valid.

29. $(2 + 0 + 2 + 6 + 1 + 7 + 8 + 9 + 1 + 4) \bmod 7 \equiv 40 \bmod 7 \equiv 5$; valid.

31. $\underline{4}\,4\,\underline{1}\,7\,\underline{5}\,4\,\underline{8}\,6\,\underline{1}\,7\,\underline{8}\,5\,\underline{6}\,4\,\underline{1}\,1$ - Double each underlined digit.
    8 4 2 7 10 4 16 6 2 7 16 5 12 4 2 1
    $8 + 4 + 2 + 7 + (1 + 0) + 4 + (1 + 6) + 6 + 2 + 7 + (1 + 6) + 5 + (1 + 2) + 4 + 2 + 1 = 70$
    $70 \equiv 0 \bmod 10$. This credit card number is valid.

33. $\underline{5}\,5\,\underline{9}\,1\,\underline{4}\,9\,\underline{1}\,2\,\underline{7}\,6\,\underline{4}\,4\,\underline{1}\,1\,\underline{0}\,5$ - Double each underlined digit.
    10 5 18 1 8 9 2 2 14 6 8 4 2 1 0 5
    $(1 + 0) + 5 + (1 + 8) + 1 + 8 + 9 + 2 + 2 + (1 + 4) + 6 + 8 + 4 + 2 + 1 + 0 + 5 = 68$
    $68 \neq 0 \bmod 10$. This credit card number is invalid.

35. $\underline{6}\,0\,\underline{1}\,1\,\underline{0}\,4\,\underline{0}\,8\,\underline{4}\,9\,\underline{7}\,7\,\underline{3}\,1\,\underline{5}\,8$ - Double each underlined digit.
    12 0 2 1 0 4 0 8 8 9 14 7 6 1 10 8
    $(1 + 2) + 0 + 2 + 1 + 0 + 4 + 0 + 8 + 8 + 9 + (1 + 4) + 7 + 6 + 1 + (1 + 0) + 8 = 63$
    $63 \neq 0 \bmod 10$. This credit card number is invalid.

37. $3\,\underline{7}\,1\,\underline{5}\,5\,\underline{4}\,8\,\underline{7}\,3\,\underline{1}\,8\,\underline{4}\,4\,\underline{6}\,6$ - Double each underlined digit.
    3 14 1 10 5 8 8 14 3 2 8 8 4 12 6
    $3 + (1 + 4) + 1 + (1 + 0) + 5 + 8 + 8 + (1 + 4) + 3 + 2 + 8 + 8 + 4 + (1 + 2) + 6 = 70$
    $70 \equiv 0 \bmod 10$. This credit card number is valid.

39. T $c \equiv (20 + 8) \bmod 26 \equiv 28 \bmod 26 \equiv 2$ Code T as B.
    H $c \equiv (8 + 8) \bmod 26 \equiv 16 \bmod 26 \equiv 16$ Code H as P.
    R $c \equiv (18 + 8) \bmod 26 \equiv 26 \bmod 26 \equiv 0$ Code R as Z.
    E $c \equiv (5 + 8) \bmod 26 \equiv 13 \bmod 26 \equiv 13$ Code E as M.
    E $c \equiv (5 + 8) \bmod 26 \equiv 13 \bmod 26 \equiv 13$ Code E as M.
    M $c \equiv (13 + 8) \bmod 26 \equiv 21 \bmod 26 \equiv 21$ Code M as U.
    U $c \equiv (21 + 8) \bmod 26 \equiv 29 \bmod 26 \equiv 3$ Code U as C.
    S $c \equiv (19 + 8) \bmod 26 \equiv 27 \bmod 26 \equiv 1$ Code S as A.
    K $c \equiv (11 + 8) \bmod 26 \equiv 19 \bmod 26 \equiv 19$ Code K as S.
    E $c \equiv (5 + 8) \bmod 26 \equiv 13 \bmod 26 \equiv 13$ Code E as M.
    T $c \equiv (20 + 8) \bmod 26 \equiv 28 \bmod 26 \equiv 2$ Code T as B.
    E $c \equiv (5 + 8) \bmod 26 \equiv 13 \bmod 26 \equiv 13$ Code E as M.
    E $c \equiv (5 + 8) \bmod 26 \equiv 13 \bmod 26 \equiv 13$ Code E as M.
    R $c \equiv (18 + 8) \bmod 26 \equiv 26 \bmod 26 \equiv 0$ Code R as Z.
    S $c \equiv (19 + 8) \bmod 26 \equiv 27 \bmod 26 \equiv 1$ Code S as A.
    The plaintext is coded as BPZMM UCASMBMMZA.

41. I $c \equiv (9 + 12) \bmod 26 \equiv 21 \bmod 26 \equiv 21$ Code I as U.
    T $c \equiv (20 + 12) \bmod 26 \equiv 32 \bmod 26 \equiv 6$ Code T as F.
    S $c \equiv (19 + 12) \bmod 26 \equiv 31 \bmod 26 \equiv 5$ Code S as E.
    A $c \equiv (1 + 12) \bmod 26 \equiv 13 \bmod 26 \equiv 13$ Code A as M.
    G $c \equiv (7 + 12) \bmod 26 \equiv 19 \bmod 26 \equiv 19$ Code G as S.
    I $c \equiv (9 + 12) \bmod 26 \equiv 21 \bmod 26 \equiv 21$ Code I as U.
    R $c \equiv (18 + 12) \bmod 26 \equiv 30 \bmod 26 \equiv 4$ Code R as D.
    L $c \equiv (12 + 12) \bmod 26 \equiv 24 \bmod 26 \equiv 24$ Code L as X.
    The plaintext is coded as UF'E M SUDX.

43. S $c \equiv (19 + 3) \bmod 26 \equiv 22 \bmod 26 \equiv 22$ Code S as V.
    T $c \equiv (20 + 3) \bmod 26 \equiv 23 \bmod 26 \equiv 23$ Code T as W.
    I $c \equiv (9 + 3) \bmod 26 \equiv 12 \bmod 26 \equiv 12$ Code I as L.
    C $c \equiv (3 + 3) \bmod 26 \equiv 6 \bmod 26 \equiv 6$ Code C as F.
    K $c \equiv (11 + 3) \bmod 26 \equiv 14 \bmod 26 \equiv 14$ Code K as N.
    S $c \equiv (19 + 3) \bmod 26 \equiv 22 \bmod 26 \equiv 22$ Code S as V.
    A $c \equiv (1 + 3) \bmod 26 \equiv 4 \bmod 26 \equiv 4$ Code A as D.
    N $c \equiv (14 + 3) \bmod 26 \equiv 17 \bmod 26 \equiv 17$ Code N as Q.
    D $c \equiv (4 + 3) \bmod 26 \equiv 7 \bmod 26 \equiv 7$ Code D as G.
    S $c \equiv (19 + 3) \bmod 26 \equiv 22 \bmod 26 \equiv 22$ Code S as V.
    T $c \equiv (20 + 3) \bmod 26 \equiv 23 \bmod 26 \equiv 23$ Code T as W.
    O $c \equiv (15 + 3) \bmod 26 \equiv 18 \bmod 26 \equiv 18$ Code O as R.
    N $c \equiv (14 + 3) \bmod 26 \equiv 17 \bmod 26 \equiv 17$ Code N as Q.
    E $c \equiv (5 + 3) \bmod 26 \equiv 8 \bmod 26 \equiv 8$ Code E as H.
    S $c \equiv (19 + 3) \bmod 26 \equiv 22 \bmod 26 \equiv 22$ Code S as V.
    The plaintext is coded as VWLFNV DQG VWRQHV.

45. $26 - 18 = 8$
    Decode using the congruence $p \equiv (c + 8) \bmod 26$.
    S $p \equiv (19 + 8) \bmod 26 \equiv 27 \bmod 26 \equiv 1$ Decode S as A.
    Y $p \equiv (25 + 8) \bmod 26 \equiv 33 \bmod 26 \equiv 7$ Decode Y as G.
    W $p \equiv (23 + 8) \bmod 26 \equiv 31 \bmod 26 \equiv 5$ Decode W as E.
    G $p \equiv (7 + 8) \bmod 26 \equiv 15 \bmod 26 \equiv 15$ Decode G as O.
    X $p \equiv (24 + 8) \bmod 26 \equiv 32 \bmod 26 \equiv 6$ Decode X as F.
    W $p \equiv (23 + 8) \bmod 26 \equiv 31 \bmod 26 \equiv 5$ Decode W as E.
    F $p \equiv (6 + 8) \bmod 26 \equiv 14 \bmod 26 \equiv 14$ Decode F as N.
    D $p \equiv (4 + 8) \bmod 26 \equiv 12 \bmod 26 \equiv 12$ Decode D as L.
    A $p \equiv (1 + 8) \bmod 26 \equiv 9 \bmod 26 \equiv 9$ Decode A as I.
    Y $p \equiv (25 + 8) \bmod 26 \equiv 33 \bmod 26 \equiv 7$ Decode Y as G.
    Z $p \equiv (0 + 8) \bmod 26 \equiv 8 \bmod 26 \equiv 8$ Decode Z as H.
    L $p \equiv (12 + 8) \bmod 26 \equiv 20 \bmod 26 \equiv 20$ Decode L as T.
    W $p \equiv (23 + 8) \bmod 26 \equiv 31 \bmod 26 \equiv 5$ Decode W as E.
    F $p \equiv (6 + 8) \bmod 26 \equiv 14 \bmod 26 \equiv 14$ Decode F as N.

E $p \equiv (5 + 8) \bmod 26 \equiv 13 \bmod 26 \equiv 13$ Decode E as M.
W $p \equiv (23 + 8) \bmod 26 \equiv 31 \bmod 26 \equiv 5$ Decode W as E.
F $p \equiv (6 + 8) \bmod 26 \equiv 14 \bmod 26 \equiv 14$ Decode F as N.
L $p \equiv (12 + 8) \bmod 26 \equiv 20 \bmod 26 \equiv 20$ Decode L as T.
The ciphertext is decoded as AGE OF ENLIGHTENMENT.

47. $26 - 15 = 11$

Decode using the congruence $p \equiv (c + 11) \bmod 26$.

U $p \equiv (21 + 11) \bmod 26 \equiv 32 \bmod 26 \equiv 6$ Decode U as F.
G $p \equiv (7 + 11) \bmod 26 \equiv 18 \bmod 26 \equiv 18$ Decode G as R.
X $p \equiv (24 + 11) \bmod 26 \equiv 35 \bmod 26 \equiv 9$ Decode X as I.
T $p \equiv (20 + 11) \bmod 26 \equiv 31 \bmod 26 \equiv 5$ Decode T as E.
C $p \equiv (3 + 11) \bmod 26 \equiv 14 \bmod 26 \equiv 14$ Decode C as N.
S $p \equiv (19 + 11) \bmod 26 \equiv 30 \bmod 26 \equiv 4$ Decode S as D.
X $p \equiv (24 + 11) \bmod 26 \equiv 35 \bmod 26 \equiv 9$ Decode X as I.
C $p \equiv (3 + 11) \bmod 26 \equiv 14 \bmod 26 \equiv 14$ Decode C as N.
C $p \equiv (3 + 11) \bmod 26 \equiv 14 \bmod 26 \equiv 14$ Decode C as N.
T $p \equiv (20 + 11) \bmod 26 \equiv 31 \bmod 26 \equiv 5$ Decode T as E.
T $p \equiv (20 + 11) \bmod 26 \equiv 31 \bmod 26 \equiv 5$ Decode T as E.
S $p \equiv (19 + 11) \bmod 26 \equiv 30 \bmod 26 \equiv 4$ Decode S as D.
The ciphertext is decoded as FRIEND IN NEED.

49. Because the encoded message uses a cyclical alphabetic encrypting code, it is possible to rotate through the 26 possible codes until the ciphertext is decoded. Using just the first word:
YVIBZM → ZWJCAN → AXKDBO → BYLECP → CZMFDQ → DANGER
It took 5 transformations to find the plaintext, so the decoding congruence is $p \equiv (c + 5) \bmod 26$. Decode the rest of the ciphertext in the usual way.

R $p \equiv (18 + 5) \bmod 26 \equiv 23 \bmod 26 \equiv 23$ Decode R as W.
D $p \equiv (4 + 5) \bmod 26 \equiv 9 \bmod 26 \equiv 9$ Decode D as I.
G $p \equiv (7 + 5) \bmod 26 \equiv 12 \bmod 26 \equiv 12$ Decode G as L.
G $p \equiv (7 + 5) \bmod 26 \equiv 12 \bmod 26 \equiv 12$ Decode G as L.
M $p \equiv (13 + 5) \bmod 26 \equiv 18 \bmod 26 \equiv 18$ Decode M as R.
J $p \equiv (10 + 5) \bmod 26 \equiv 15 \bmod 26 \equiv 15$ Decode J as O.
W $p \equiv (23 + 5) \bmod 26 \equiv 28 \bmod 26 \equiv 2$ Decode W as B.
D $p \equiv (4 + 5) \bmod 26 \equiv 9 \bmod 26 \equiv 9$ Decode D as I.
I $p \equiv (9 + 5) \bmod 26 \equiv 14 \bmod 26 \equiv 14$ Decode I as N.
N $p \equiv (14 + 5) \bmod 26 \equiv 19 \bmod 26 \equiv 19$ Decode N as S.
J $p \equiv (10 + 5) \bmod 26 \equiv 15 \bmod 26 \equiv 15$ Decode J as O.
I $p \equiv (9 + 5) \bmod 26 \equiv 14 \bmod 26 \equiv 14$ Decode I as N.
The ciphertext is decoded as DANGER WILL ROBINSON.

51. Because the encoded message uses a cyclical alphabetic encrypting code, it is possible to rotate through the 26 possible codes until the ciphertext is decoded. Using just the first word:
UDGIJCT → VEHJKDU → WFIKLEV → XGJLMFW → YHKMNGX → ZILNOHY → AJMOPIZ
→ BKNPQJA → CLOQRKB → DMPRSLC → ENQSTMD → FORTUNE
It took 11 transformations to find the plaintext, so the decoding congruence is $p \equiv (c + 11) \bmod 26$. Decode the rest of the ciphertext in the usual way.

R $p \equiv (18 + 11) \bmod 26 \equiv 29 \bmod 26 \equiv 3$ Decode R as C.
D $p \equiv (4 + 11) \bmod 26 \equiv 15 \bmod 26 \equiv 15$ Decode D as O.
D $p \equiv (4 + 11) \bmod 26 \equiv 15 \bmod 26 \equiv 15$ Decode D as O.
Z $p \equiv (0 + 11) \bmod 26 \equiv 11 \bmod 26 \equiv 11$ Decode Z as K.
X $p \equiv (24 + 11) \bmod 26 \equiv 35 \bmod 26 \equiv 9$ Decode X as I.
T $p \equiv (20 + 11) \bmod 26 \equiv 31 \bmod 26 \equiv 5$ Decode T as E.
The ciphertext is decoded as FORTUNE COOKIE.

53. M $c \equiv (13 + 3) \bmod 26 \equiv 16 \bmod 26 \equiv 16$ Code M as P.
     E $c \equiv (5 + 3) \bmod 26 \equiv 8 \bmod 26 \equiv 8$ Code E as H.
     N $c \equiv (14 + 3) \bmod 26 \equiv 17 \bmod 26 \equiv 17$ Code N as Q.
     W $c \equiv (23 + 3) \bmod 26 \equiv 26 \bmod 26 \equiv 0$ Code W as Z.
     I $c \equiv (9 + 3) \bmod 26 \equiv 12 \bmod 26 \equiv 12$ Code I as L.
     L $c \equiv (12 + 3) \bmod 26 \equiv 15 \bmod 26 \equiv 15$ Code L as O.
     L $c \equiv (12 + 3) \bmod 26 \equiv 15 \bmod 26 \equiv 15$ Code L as O.
     Continuing, the plaintext is coded as PHQ ZLOOLQJRB EHOLHYH ZKDW WKHB ZLVK.

55. T $c \equiv (3 \cdot 20 + 2) \bmod 26 \equiv 62 \bmod 26 \equiv 10$ Code T as J.
     O $c \equiv (3 \cdot 15 + 2) \bmod 26 \equiv 47 \bmod 26 \equiv 21$ Code O as U.
     W $c \equiv (3 \cdot 23 + 2) \bmod 26 \equiv 71 \bmod 26 \equiv 19$ Code W as S.
     E $c \equiv (3 \cdot 5 + 2) \bmod 26 \equiv 17 \bmod 26 \equiv 17$ Code E as Q.
     R $c \equiv (3 \cdot 18 + 2) \bmod 26 \equiv 56 \bmod 26 \equiv 4$ Code R as D.
     O $c \equiv (3 \cdot 15 + 2) \bmod 26 \equiv 47 \bmod 26 \equiv 21$ Code O as U.
     F $c \equiv (3 \cdot 6 + 2) \bmod 26 \equiv 20 \bmod 26 \equiv 20$ Code F as T.
     Continuing, the plaintext is coded as JUSQD UT LURNUR.

57. P $c \equiv (7 \cdot 16 + 8) \bmod 26 \equiv 120 \bmod 26 \equiv 16$ Code P as P.
     A $c \equiv (7 \cdot 1 + 8) \bmod 26 \equiv 15 \bmod 26 \equiv 15$ Code A as O.
     R $c \equiv (7 \cdot 18 + 8) \bmod 26 \equiv 134 \bmod 26 \equiv 4$ Code R as D.
     A $c \equiv (7 \cdot 1 + 8) \bmod 26 \equiv 15 \bmod 26 \equiv 15$ Code A as O.
     L $c \equiv (7 \cdot 12 + 8) \bmod 26 \equiv 92 \bmod 26 \equiv 14$ Code L as N.
     L $c \equiv (7 \cdot 12 + 8) \bmod 26 \equiv 92 \bmod 26 \equiv 14$ Code L as N.
     E $c \equiv (7 \cdot 5 + 8) \bmod 26 \equiv 43 \bmod 26 \equiv 17$ Code E as Q.
     L $c \equiv (7 \cdot 12 + 8) \bmod 26 \equiv 92 \bmod 26 \equiv 14$ Code L as N.
     Continuing, the plaintext is coded as PODONNQN NSBQK.

59. Solve the congruence for p.
     $c = 3p + 4$
     $c - 4 = 3p$
     The multiplicative inverse of 3 is 9.
     $9(c - 4) = 9(3p)$
     $[9(c - 4)] \bmod 26 \equiv p$
     Decode using this congruence.
     L $[9(12 - 4)] \bmod 26 \equiv 72 \bmod 26 \equiv 20$ Decode L as T.
     O $[9(15 - 4)] \bmod 26 \equiv 99 \bmod 26 \equiv 21$ Decode O as U.
     F $[9(6 - 4)] \bmod 26 \equiv 18 \bmod 26 \equiv 18$ Decode F as R.
     T $[9(20 - 4)] \bmod 26 \equiv 144 \bmod 26 \equiv 14$ Decode T as N.
     J $[9(10 - 4)] \bmod 26 \equiv 54 \bmod 26 \equiv 2$ Decode J as B.
     G $[9(7 - 4)] \bmod 26 \equiv 27 \bmod 26 \equiv 1$ Decode G as A.
     M $[9(13 - 4)] \bmod 26 \equiv 81 \bmod 26 \equiv 3$ Decode M as C.
     K $[9(11 - 4)] \bmod 26 \equiv 63 \bmod 26 \equiv 11$ Decode K as K.
     Continuing, the ciphertext is decoded as TURN BACK THE CLOCK.

61. Solve the congruence for p.
     $c = 5p + 9$
     $c - 9 = 5p$
     The multiplicative inverse of 5 is 21.
     $21(c - 9) = 21(5p)$
     $[21(c - 9)] \bmod 26 \equiv p$
     Decode using this congruence.
     S $[21(19 - 9)] \bmod 26 \equiv 210 \bmod 26 \equiv 2$ Decode S as B.
     N $[21(14 - 9)] \bmod 26 \equiv 105 \bmod 26 \equiv 1$ Decode N as A.
     U $[21(21 - 9)] \bmod 26 \equiv 252 \bmod 26 \equiv 18$ Decode U as R.
     U $[21(21 - 9)] \bmod 26 \equiv 252 \bmod 26 \equiv 18$ Decode U as R.
     H $[21(8 - 9)] \bmod 26 \equiv -21 \bmod 26 \equiv 5$ Decode H as E.
     Q $[21(17 - 9)] \bmod 26 \equiv 168 \bmod 26 \equiv 12$ Decode Q as L.
     F $[21(6 - 9)] \bmod 26 \equiv -63 \bmod 26 \equiv 15$ Decode F as O.
     M $[21(13 - 9)] \bmod 26 \equiv 84 \bmod 26 \equiv 6$ Decode M as F.
     Continuing, the ciphertext is decoded as BARREL OF MONKEYS.

63. Because the check digit is simply the sum of the first 10 digits mod 9, the same digits in a different order will give the same sum and hence the same check digit.

65. a. $7(3) + 2(7) + 8(1) + 5(3) + 9(7) + 3(1) + 7(3) + 2(7) + x(1) \equiv 0 \bmod 10.$
    $159 + x \equiv 0 \bmod 10$
    The next multiple of 10 is 160.
    $x = 160 - 159 = 1.$

    b. $5(3) + 8(7) + 4(1) + 9(3) + 2(7) + 6(1) + 1(3) + 0(7) + 5(1) = 130 \equiv 0 \bmod 10.$
    The routing number is valid.

    c. $5(3) + 8(7) + 4(1) + 9(3) + 6(7) + 2(1) + 1(3) + 0(7) + 5(1) = 154 \neq 0 \bmod 10.$
    The typed number is not valid, and the computer will catch the error.

    d.   When determining the check digit for the correct routing number, the digits 6 and 1 are multiplied by 1 and 3 respectively. $(6 \cdot 1 + 1 \cdot 3) \bmod 10 \equiv (1 \cdot 1 + 6 \cdot 3) \bmod 10 \equiv 9 \bmod 10$, so if those digits were transposed, the computer would not catch the error.

EXCURSION EXERCISES, SECTION 8.3

1. Any translation followed by another is itself a translation, so the set of translations is closed. Translations are associative, the identity element is *I*, and any translation can be reversed, so each element has an inverse.

3. a. – c.

5. a.

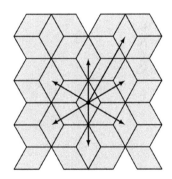

   b.

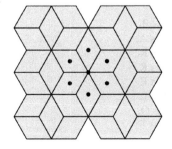

c.

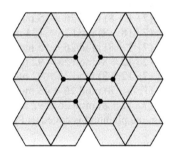

d.

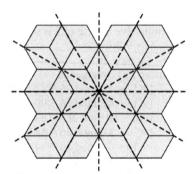

EXERCISE SET 8.3

1. a.   Yes. All possible multiplications within the set yield either −1 or 1.

   b.   No. −1 + 1 = 0, which is not in the set.

3. a.   Yes. The product of two odd integers is always an odd integer. For instance, $3 \cdot 7 = 21$ and $9 \cdot 15 = 135$.

   b.   No. The sum of two odd integers is always an even integer. For instance, $3 + 7 = 10$ and $9 + 15 = 24$.

5. 1. The sum of two even integers is always an integer. Therefore, the closure property holds true.

    2. The associative property of addition holds true for the even integers.

    3. The identity element is 0, which is an even integer. Therefore, the even integers have an identity element for addition.

    4. Each element has an inverse. If a is an even integer, then $-a$ is the inverse of a.

    All of the four properties are satisfied, so the even integers form a group with respect to addition.

7. 1. The sum of two real numbers is always a real number, so the closure property holds true.

    2. The associative property of addition holds true for the real numbers.

    3. The identity element is 0, which is a real number. Therefore the real numbers have an identity element for addition.

    4. Each element has an inverse. If a is a real number, then $-a$ is the inverse of a.

    All of the four properties are satisfied, so the real numbers form a group with respect to addition.

9. 1. The product of two real numbers is always a real number. Therefore, the closure property holds true.

    2. The associative property of multiplication holds true for the real numbers.

    3. The identity element is 1, which is a real number. Therefore, the real numbers have an identity element for multiplication.

    4. Not every element has an inverse. For example, there is no inverse for 0. Therefore, the inverse property fails.

    Property 4 fails, so the real numbers do not form a group with respect to multiplication.

11. 1. The sum of any two rational numbers is always a rational number. Therefore, the closure property holds true.

    2. The associative property of addition holds true for the rational numbers.

    3. The identity element is 0, which is a rational number. Therefore the rational numbers have an identity element for addition.

    4. Every element has an inverse.

    If $\frac{a}{b}$ is rational number, then $\frac{-a}{b}$ is the inverse of $\frac{a}{b}$.

    All of the four properties are satisfied, so the rational numbers form a group with respect to addition.

13. 1. The sum modulo 4 of any two elements in the set is always a member of the set. Therefore, the closure property holds true.

    2. The associative property of addition modulo 4 holds true for the members of the set.

    3. There exists an identity element (0) in the set for addition modulo 4.

    4. Every element has an inverse. If a is an element in the set, then $(4 - a)$ mod 4 is the inverse of a.

    All of the four properties are satisfied, so the set forms a group with respect to addition modulo 4.

15. 1. The product modulo 4 of any two elements in the set is always a member of the set. Therefore, the closure property holds true.

    2. The associative property of multiplication modulo 4 holds true for the members of the set.

    3. There exists an identity element (1) in the set for multiplication modulo 4.

    4. 0 does not have an inverse. Therefore the inverse property fails.

    Property 4 fails, so the set does not form a group with respect to multiplication modulo 4.

17. 1. The product of any two elements in the set is always a member of the set. Therefore, the closure property holds true.

    2. The associative property of multiplication holds true for the members of the set.

    3. There exists an identity element (1) in the set for multiplication.

    4. Both elements have themselves as their inverse. Therefore, the inverse property holds true for the set.

    All of the four properties are satisfied, so the set forms a group with respect to multiplication.

19.

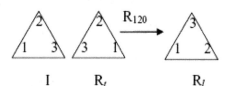

From the diagram, $R_t \Delta R_{120} = R_l$.

21.

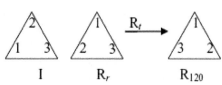

From the diagram, $R_t \Delta R_t = R_{120}$.

23. $R_{240} \Delta R_{120} = I$, so the inverse of R_{240} is R_{120}.

25.

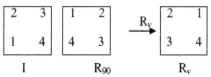

From the diagram, $R_{90} \Delta R_v = R_r$.

27.

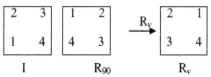

From the diagram, $R_r \Delta R_{180} = R_l$.

29. Perform all possible transformations on the identity element to determine all the elements of the group.

$I = \begin{pmatrix} 1 & 2 & 3 & 4 \\ 1 & 2 & 3 & 4 \end{pmatrix}$,

$R_{90} = \begin{pmatrix} 1 & 2 & 3 & 4 \\ 2 & 3 & 4 & 1 \end{pmatrix}$,

$R_{180} = \begin{pmatrix} 1 & 2 & 3 & 4 \\ 3 & 4 & 1 & 2 \end{pmatrix}$,

$R_{270} = \begin{pmatrix} 1 & 2 & 3 & 4 \\ 4 & 1 & 2 & 3 \end{pmatrix}$,

$R_v = \begin{pmatrix} 1 & 2 & 3 & 4 \\ 4 & 3 & 2 & 1 \end{pmatrix}, R_h = \begin{pmatrix} 1 & 2 & 3 & 4 \\ 2 & 1 & 4 & 3 \end{pmatrix}$,

$R_r = \begin{pmatrix} 1 & 2 & 3 & 4 \\ 3 & 2 & 1 & 4 \end{pmatrix}, R_l = \begin{pmatrix} 1 & 2 & 3 & 4 \\ 1 & 4 & 3 & 2 \end{pmatrix}$

31. Evaluating:

$\begin{pmatrix} 1 & 2 & 3 & 4 \\ 2 & 1 & 4 & 3 \end{pmatrix} \Delta \begin{pmatrix} 1 & 2 & 3 & 4 \\ 2 & 3 & 4 & 1 \end{pmatrix}$
$= \begin{pmatrix} 1 & 2 & 3 & 4 \\ 3 & 2 & 1 & 4 \end{pmatrix} = R_r$

33. Evaluating:

$\begin{pmatrix} 1 & 2 & 3 & 4 \\ 4 & 1 & 2 & 3 \end{pmatrix}^{-1} = \begin{pmatrix} 1 & 2 & 3 & 4 \\ 2 & 3 & 4 & 1 \end{pmatrix} = R_{90}$

35. Evaluating:

$\begin{pmatrix} 1 & 2 & 3 \\ 2 & 3 & 1 \end{pmatrix} \Delta \begin{pmatrix} 1 & 2 & 3 \\ 3 & 1 & 2 \end{pmatrix} = \begin{pmatrix} 1 & 2 & 3 \\ 1 & 2 & 3 \end{pmatrix} = I$

37. Evaluating:

$\begin{pmatrix} 1 & 2 & 3 \\ 3 & 1 & 2 \end{pmatrix} \Delta \begin{pmatrix} 1 & 2 & 3 \\ 2 & 1 & 3 \end{pmatrix} = \begin{pmatrix} 1 & 2 & 3 \\ 3 & 2 & 1 \end{pmatrix} = D$

39. Evaluating:

$\begin{pmatrix} 1 & 2 & 3 \\ 1 & 3 & 2 \end{pmatrix} \Delta \begin{pmatrix} 1 & 2 & 3 \\ 2 & 3 & 1 \end{pmatrix} = \begin{pmatrix} 1 & 2 & 3 \\ 2 & 1 & 3 \end{pmatrix} = E$

41. Evaluating:

$\begin{pmatrix} 1 & 2 & 3 \\ 2 & 3 & 1 \end{pmatrix}^{-1} = \begin{pmatrix} 1 & 2 & 3 \\ 3 & 1 & 2 \end{pmatrix} = B$

43. The list follows:

$\begin{pmatrix} 1 & 2 & 3 & 4 \\ 1 & 2 & 3 & 4 \end{pmatrix}, \begin{pmatrix} 1 & 2 & 3 & 4 \\ 1 & 2 & 4 & 3 \end{pmatrix}, \begin{pmatrix} 1 & 2 & 3 & 4 \\ 1 & 3 & 2 & 4 \end{pmatrix},$

$\begin{pmatrix} 1 & 2 & 3 & 4 \\ 1 & 3 & 4 & 2 \end{pmatrix}, \begin{pmatrix} 1 & 2 & 3 & 4 \\ 1 & 4 & 2 & 3 \end{pmatrix}, \begin{pmatrix} 1 & 2 & 3 & 4 \\ 1 & 4 & 3 & 2 \end{pmatrix},$

$\begin{pmatrix} 1 & 2 & 3 & 4 \\ 2 & 1 & 3 & 4 \end{pmatrix}, \begin{pmatrix} 1 & 2 & 3 & 4 \\ 2 & 1 & 4 & 3 \end{pmatrix}, \begin{pmatrix} 1 & 2 & 3 & 4 \\ 2 & 3 & 1 & 4 \end{pmatrix},$

$\begin{pmatrix} 1 & 2 & 3 & 4 \\ 2 & 3 & 4 & 1 \end{pmatrix}, \begin{pmatrix} 1 & 2 & 3 & 4 \\ 2 & 4 & 1 & 3 \end{pmatrix}, \begin{pmatrix} 1 & 2 & 3 & 4 \\ 2 & 4 & 3 & 1 \end{pmatrix},$

$\begin{pmatrix} 1 & 2 & 3 & 4 \\ 3 & 1 & 2 & 4 \end{pmatrix}, \begin{pmatrix} 1 & 2 & 3 & 4 \\ 3 & 1 & 4 & 2 \end{pmatrix}, \begin{pmatrix} 1 & 2 & 3 & 4 \\ 3 & 2 & 1 & 4 \end{pmatrix},$

$\begin{pmatrix} 1 & 2 & 3 & 4 \\ 3 & 2 & 4 & 1 \end{pmatrix}, \begin{pmatrix} 1 & 2 & 3 & 4 \\ 3 & 4 & 1 & 2 \end{pmatrix}, \begin{pmatrix} 1 & 2 & 3 & 4 \\ 3 & 4 & 2 & 1 \end{pmatrix},$

$\begin{pmatrix} 1 & 2 & 3 & 4 \\ 4 & 1 & 2 & 3 \end{pmatrix}, \begin{pmatrix} 1 & 2 & 3 & 4 \\ 4 & 1 & 3 & 2 \end{pmatrix}, \begin{pmatrix} 1 & 2 & 3 & 4 \\ 4 & 2 & 1 & 3 \end{pmatrix},$

$\begin{pmatrix} 1 & 2 & 3 & 4 \\ 4 & 2 & 3 & 1 \end{pmatrix}, \begin{pmatrix} 1 & 2 & 3 & 4 \\ 4 & 3 & 1 & 2 \end{pmatrix}, \begin{pmatrix} 1 & 2 & 3 & 4 \\ 4 & 3 & 2 & 1 \end{pmatrix},$

45. Evaluating:

$\begin{pmatrix} 1 & 2 & 3 & 4 \\ 1 & 4 & 2 & 3 \end{pmatrix} \Delta \begin{pmatrix} 1 & 2 & 3 & 4 \\ 2 & 3 & 4 & 1 \end{pmatrix}$
$= \begin{pmatrix} 1 & 2 & 3 & 4 \\ 2 & 1 & 3 & 4 \end{pmatrix}$

47. Evaluating:

$$\begin{pmatrix} 1 & 2 & 3 & 4 \\ 2 & 4 & 1 & 3 \end{pmatrix}^{-1} = \begin{pmatrix} 1 & 2 & 3 & 4 \\ 3 & 1 & 4 & 2 \end{pmatrix}$$

49. Evaluating:

$$\begin{pmatrix} 1 & 2 & 3 & 4 \\ 4 & 3 & 2 & 1 \end{pmatrix}^{-1} = \begin{pmatrix} 1 & 2 & 3 & 4 \\ 4 & 3 & 2 & 1 \end{pmatrix}$$

51. From the table, $a\nabla a = d$.

53. From the table, $c\nabla b = c$.

55. The set satisfies the closure property because the table contains only members of the set.
From the question, the set satisfies the associative property.
From the table, note that the identity element is b, because for any element x, $x\nabla b = x$.
Therefore, the set satisfies the identity element property. Also from the table, note that every element has an inverse. The inverse of a is c, the inverse if b is b, the inverse of c is a, and the inverse of d is d. Therefore the set satisfies the inverse property.
Consequently, the set is a group with respect to the operation ∇.

57. The table is symmetrical about the diagonal from top left to bottom right, which indicates that for any two elements m and n in the set, $m\nabla n = n\nabla m$. Therefore, the group is commutative.

59. a. The product modulo 7 of any two members of the set is always within the set. Therefore the set satisfies the closure property.
Multiplication modulo 7 is associative. Therefore the set satisfies the associative property.
$1x \bmod 7 \equiv x \bmod 7$. Therefore, 1 is the identity element, and the set satisfies the identity property.
Every element of the set has an inverse with respect to multiplication modulo 7. Therefore, the set satisfies the inverse property.
The set is a group because it satisfies the four properties of a group.

   b. The set does not satisfy the closure property. For example, $2 \cdot 3 \equiv 6 \bmod 6 \equiv 0$, which is not in the set.

   c. If n is a composite number, then it can be factored into two elements x and y such that $xy = n$ and x and y are elements of the set. Then $xy \equiv n \bmod n \equiv 0$, which is not in the set. Therefore, n cannot be a composite number if the set is to be closed under multiplication. The set is a group only for prime values of n.

61. a. The table is not symmetric about the diagonal from top left to bottom right, which indicates that there are some elements a and b such that $a\nabla b \neq a\nabla b$. Therefore, the set is not commutative.

   b. From the table, note that any element $x\nabla e = e$. Therefore, e is the identity element.

   c. $e\nabla e = e$, so e is the inverse of e.
$r\nabla v = e$, so v is the inverse of r.
$s\nabla w = e$, so w is the inverse of s.
$t\nabla x = e$, so x is the inverse of t.
$u\nabla u = e$, so u is the inverse of u.
$v\nabla r = e$, so r is the inverse of v.
$w\nabla s = e$, so s is the inverse of w.
$x\nabla t = e$, so t is the inverse of x.

   d. $r\nabla r\nabla r\nabla r = (r\nabla r)\nabla(r\nabla r) = u\nabla u = e$

   e. From the table, note that the operation ∇ on each pair of elements of the subset yields a member of the subset. Therefore, the subset satisfies the closure property.
Given that the operation ∇ is associative on Q, and that the subset in question is closed with respect to ∇, the subset must satisfy the associative property with respect to ∇.
Note that the identity element of Q is present in the subset, so the subset satisfies the identity property.
e and u are their own inverses, and r and v are inverses of each other, so the subset satisfies the inverse property.
The subset is a subgroup because it satisfies the four properties.

   f. One element must be e, the identity element of Q. The only two-element subgroup is $\{e, u\}$.

CHAPTER 8 REVIEW EXERCISES

1. 2   2. 2

3. 5   4. 6

5. 9   6. 4

7. 11   8. 7

9. 3   10. 4

11. True. $17 \div 5 = 3$ remainder 2.

12. False. $14 \div 4 = 3$ remainder 2, $24 \div 4 = 6$ remainder 0.

13. False. $35 \div 10 = 3$ remainder 5, $53 \div 10 = 5$ remainder 3.

14. True. $12 \div 8 = 1$ remainder 4,
    $36 \div 8 = 4$ remainder 4.

15. $(8 + 12) \bmod 3 = 20 \bmod 3$
    $20 \div 3 = 6$ remainder 2
    So $(8 + 12) \bmod 3 \equiv 2$.

16. $(15 + 7) \bmod 6 = 22 \bmod 6$
    $22 \div 6 = 3$ remainder 4
    So $(15 + 7) \bmod 6 \equiv 4$.

17. $(42 - 10) \bmod 8 = 32 \bmod 8$
    $32 \div 8 = 4$ remainder 0
    So $(42 - 10) \bmod 8 \equiv 0$.

18. $(19 - 8) \bmod 4 = 11 \bmod 4$
    $11 \div 4 = 2$ remainder 3
    So $(19 - 8) \bmod 4 \equiv 3$.

19. $(7 \cdot 5) \bmod 9 = 35 \bmod 9$
    $35 \div 9 = 3$ remainder 8
    So $(7 \cdot 5) \bmod 9 \equiv 8$.

20. $(12 \cdot 9) \bmod 5 = 108 \bmod 5$
    $108 \div 5 = 21$ remainder 3
    So $(12 \cdot 9) \bmod 5 \equiv 3$.

21. $(15 \cdot 10) \bmod 11 = 150 \bmod 11$
    $150 \div 11 = 13$ remainder 7
    So $(15 \cdot 10) \bmod 11 \equiv 7$.

22. $(41 \cdot 13) \bmod 8 = 533 \bmod 8$
    $533 \div 8 = 66$ remainder 5
    So $(41 \cdot 13) \bmod 8 \equiv 5$

23. a. $(5 + 45) \bmod 12 = 50 \bmod 12$
        $50 \div 12 = 4$ remainder 2
        It will be 2:00.

    b. $(5 - 71) \bmod 12 = -66 \bmod 12$
        $-66 + 12 = -54$
        $-54 + 12 = -42$
        $-42 + 12 = -30$
        $-30 + 12 = -18$
        $-18 + 12 = -6$
        $-6 + 12 = 6$
        It was 6:00.

24. There are 10 years between the two dates, three of which (2016, 2020, and 2024) are leap years. The number of days between the two dates is $10(365) + 3 = 3653$.
    $3653 \div 7 = 521$ remainder 6
    April 15, 2025 will fall 6 days after Wednesday, which is Tuesday.

25. Starting with zero, substitute whole numbers less than 4 for x.

$x = 0$	$0 \not\equiv 7 \bmod 4$	NO
$x = 1$	$1 \not\equiv 7 \bmod 4$	NO
$x = 2$	$2 \not\equiv 7 \bmod 4$	NO
$x = 3$	$3 \equiv 7 \bmod 4$	YES

Adding multiples of the modulus yields other solutions.
The solutions are 3, 7, 11, 15, 19, 23, ...

26. Starting with zero, substitute whole numbers less than 9 for x.

$x = 0$	$2(0) \not\equiv 5 \bmod 9$	NO
$x = 1$	$2(1) \not\equiv 5 \bmod 9$	NO
$x = 2$	$2(2) \not\equiv 5 \bmod 9$	NO
$x = 3$	$2(3) \not\equiv 5 \bmod 9$	NO
$x = 4$	$2(4) \not\equiv 5 \bmod 9$	NO
$x = 5$	$2(5) \not\equiv 5 \bmod 9$	NO
$x = 6$	$2(6) \not\equiv 5 \bmod 9$	NO
$x = 7$	$2(7) \equiv 5 \bmod 9$	YES
$x = 8$	$2(8) \not\equiv 5 \bmod 9$	NO

Adding multiples of the modulus yields other solutions.
The solutions are 7, 16, 25, 34, 43, 52, . . .

27. Starting with zero, substitute whole numbers less than 5 for x.

$x = 0$	$2(0) + 1 \equiv 6 \bmod 5$	YES
$x = 1$	$2(1) + 1 \not\equiv 6 \bmod 5$	NO
$x = 2$	$2(2) + 1 \not\equiv 6 \bmod 5$	NO
$x = 3$	$2(3) + 1 \not\equiv 6 \bmod 5$	NO
$x = 4$	$2(4) + 1 \not\equiv 6 \bmod 5$	NO

Adding multiples of the modulus yields other solutions.
The solution are 0, 5, 10, 15, 20, 25, 30, . . .

28. Starting with zero, substitute whole numbers less than 11 for x.

$x = 0$	$3(0) + 4 \not\equiv 5 \bmod 11$	NO
$x = 1$	$3(1) + 4 \not\equiv 5 \bmod 11$	NO
$x = 2$	$3(2) + 4 \not\equiv 5 \bmod 11$	NO
$x = 3$	$3(3) + 4 \not\equiv 5 \bmod 11$	NO
$x = 4$	$3(4) + 4 \equiv 5 \bmod 11$	YES
$x = 5$	$3(5) + 4 \not\equiv 5 \bmod 11$	NO
$x = 6$	$3(6) + 4 \not\equiv 5 \bmod 11$	NO
$x = 7$	$3(7) + 4 \not\equiv 5 \bmod 11$	NO
$x = 8$	$3(8) + 4 \not\equiv 5 \bmod 11$	NO
$x = 9$	$3(9) + 4 \not\equiv 5 \bmod 11$	NO
$x = 10$	$3(10) + 4 \not\equiv 5 \bmod 11$	NO

Adding multiples of the modulus yields other solutions.
The solution are 4, 15, 26, 37, 48, 59, 70, . . .

29. $5 + 2 = 7$, so the additive inverse is 2. Starting with zero, substitute whole numbers less than 7 for x in the congruence $5x \equiv 1 \bmod 7$.

    $x = 0$ $\qquad$ $5(0) \not\equiv 1 \bmod 7$ $\qquad$ NO

    $x = 1$ $\qquad$ $5(1) \not\equiv 1 \bmod 7$ $\qquad$ NO

    $x = 2$ $\qquad$ $5(2) \not\equiv 1 \bmod 7$ $\qquad$ NO

    $x = 3$ $\qquad$ $5(3) \equiv 1 \bmod 7$ $\qquad$ YES

    $x = 4$ $\qquad$ $5(4) \not\equiv 1 \bmod 7$ $\qquad$ NO

    $x = 5$ $\qquad$ $5(5) \not\equiv 1 \bmod 7$ $\qquad$ NO

    $x = 6$ $\qquad$ $5(6) \not\equiv 1 \bmod 7$ $\qquad$ NO

    The multiplicative inverse is 3.

30. $7 + 5 = 12$, so the additive inverse is 5. Starting with zero, substitute whole numbers less than 12 for x in the congruence $7x \equiv 1 \bmod 12$.

    $x = 0$ $\qquad$ $7(0) \not\equiv 1 \bmod 12$ $\qquad$ NO

    $x = 1$ $\qquad$ $7(1) \not\equiv 1 \bmod 12$ $\qquad$ NO

    $x = 2$ $\qquad$ $7(2) \not\equiv 1 \bmod 12$ $\qquad$ NO

    $x = 3$ $\qquad$ $7(3) \not\equiv 1 \bmod 12$ $\qquad$ NO

    $x = 4$ $\qquad$ $7(4) \not\equiv 1 \bmod 12$ $\qquad$ NO

    $x = 5$ $\qquad$ $7(5) \not\equiv 1 \bmod 12$ $\qquad$ NO

    $x = 6$ $\qquad$ $7(6) \not\equiv 1 \bmod 12$ $\qquad$ NO

    $x = 7$ $\qquad$ $7(7) \equiv 1 \bmod 12$ $\qquad$ YES

    $x = 8$ $\qquad$ $7(8) \not\equiv 1 \bmod 12$ $\qquad$ NO

    $x = 9$ $\qquad$ $7(9) \not\equiv 1 \bmod 12$ $\qquad$ NO

    $x = 10$ $\qquad$ $7(10) \not\equiv 1 \bmod 12$ $\qquad$ NO

    $x = 11$ $\qquad$ $7(11) \not\equiv 1 \bmod 12$ $\qquad$ NO

    The multiplicative inverse is 7.

31. Assume x to be the divisor. Then $5x = 2 \bmod 7$. Starting with zero, substitute whole numbers less than 7 for x into the congruence.

    $x = 0$ $\qquad$ $5(0) \not\equiv 2 \bmod 7$ $\quad$ NO

    $x = 1$ $\qquad$ $5(1) \not\equiv 2 \bmod 7$ $\quad$ NO

    $x = 2$ $\qquad$ $5(2) \not\equiv 2 \bmod 7$ $\quad$ NO

    $x = 3$ $\qquad$ $5(3) \not\equiv 2 \bmod 7$ $\quad$ NO

    $x = 4$ $\qquad$ $5(4) \not\equiv 2 \bmod 7$ $\quad$ NO

    $x = 5$ $\qquad$ $5(5) \not\equiv 2 \bmod 7$ $\quad$ NO

    $x = 6$ $\qquad$ $5(6) \equiv 2 \bmod 7$ $\quad$ YES

    So $(2 \div 5) \bmod 7 \equiv 6$.

32. Assume x to be the divisor. Then $4x = 3 \bmod 5$. Starting with zero, substitute whole numbers less than 5 for x into the congruence.

    $x = 0$ $\qquad$ $4(0) \not\equiv 3 \bmod 5$ $\quad$ NO

    $x = 1$ $\qquad$ $4(1) \not\equiv 3 \bmod 5$ $\quad$ NO

    $x = 2$ $\qquad$ $4(2) \equiv 3 \bmod 5$ $\quad$ YES

    $x = 3$ $\qquad$ $4(3) \not\equiv 3 \bmod 5$ $\quad$ NO

    $x = 4$ $\qquad$ $4(4) \not\equiv 3 \bmod 5$ $\quad$ NO

    So $(3 \div 4) \bmod 5 \equiv 2$.

33. $d_{13} \equiv 10 - (d_1 + 3d_2 + d_3 + 3d_4 + d_5 + 3d_6 + d_7 + 3d_8 + d_9 + 3d_{10} + d_{11} + 3d_{12}) \bmod 10$

    $d_{13} \equiv 10 - [9 + 3(7) + 8 + 3(1) + 4 + 3(0) + 2 + 3(7) + 5 + 3(4) + 2 + 3(5)] \bmod 10$

    $\equiv 10 - 102 \bmod 10$

    $\equiv 10 - 2 \equiv 8$

    The check digit is 8.

34. $d_{13} \equiv 10 - (d_1 + 3d_2 + d_3 + 3d_4 + d_5 + 3d_6 + d_7 + 3d_8 + d_9 + 3d_{10} + d_{11} + 3d_{12}) \bmod 10$

    $d_{13} \equiv 10 - [9 + 3(7) + 8 + 3(0) + 6 + 3(7) + 9 + 3(8) + 0 + 3(5) + 2 + 3(7)] \bmod 10$

    $\equiv 10 - 136 \bmod 10$

    $\equiv 10 - 6 \equiv 4$

    The check digit is 4.

35. $d_{12} \equiv 10 - (3d_1 + d_2 + 3d_3 + d_4 + 3d_5 + d_6 + 3d_7 + d_8 + 3d_9 + d_{10} + 3d_{11}) \bmod 10$

    $d_{12} \equiv 10 - [3(0) + 2 + 3(9) + 0 + 3(0) + 0 + 3(0) + 7 + 3(0) + 0 + 3(4)] \bmod 10$

    $\equiv 10 - 48 \bmod 10$

    $\equiv 10 - 8 \equiv 2$

    The check digit is 2.

36. $d_{12} \equiv 10 - (3d_1 + d_2 + 3d_3 + d_4 + 3d_5 + d_6 + 3d_7 + d_8 + 3d_9 + d_{10} + 3d_{11}) \bmod 10$

    $d_{12} \equiv 10 - [3(0) + 8 + 3(5) + 3 + 3(9) + 1 + 3(8) + 9 + 3(5) + 1 + 3(2)] \bmod 10$

    $\equiv 10 - 109 \bmod 10$

    $\equiv 10 - 9 \equiv 1$

    The check digit is 1.

37. $\underline{5}$ 1 $\underline{2}$ 6 $\underline{6}$ 9 $\underline{9}$ 3 $\underline{4}$ 2 $\underline{3}$ 1 $\underline{2}$ 9 $\underline{5}$ 6 - Double each underlined digit.
    10 1 4 6 12 9 18 3 8 2 6 1 4 9 10 6
    $(1 + 0) + 1 + 4 + 6 + (1 + 2) + 9 + (1 + 8) + 3 + 8 + 2 + 6 + 1 + 4 + 9 + (1 + 0) + 6 = 73$
    $73 \neq 0 \bmod 10$. This credit card number is invalid.

38. $\underline{5}$ 3 $\underline{8}$ 3 $\underline{0}$ 1 $\underline{1}$ 8 $\underline{3}$ 4 $\underline{1}$ 6 $\underline{5}$ 9 $\underline{3}$ 1 - Double each underlined digit.
    10 3 16 3 0 1 2 8 6 4 2 6 10 9 6 1
    $(1 + 0) + 3 + (1 + 6) + 3 + 0 + 1 + 2 + 8 + 6 + 4 + 2 + 6 + (1 + 0) + 9 + 6 + 1 = 60$
    $60 \equiv 0 \bmod 10$. This credit card number is valid.

39. $\underline{3}$ 4 $\underline{1}$ 2 $\underline{4}$ 0 $\underline{8}$ 4 $\underline{3}$ 9 $\underline{8}$ 2 $\underline{5}$ 9 $\underline{4}$ - Double each underlined digit.
    3 8 1 4 4 0 8 8 3 18 8 4 5 18 4
    $3 + 8 + 1 + 4 + 4 + 0 + 8 + 8 + 3 + (1 + 8) + 8 + 4 + 5 + (1 + 8) + 4 = 78$
    $78 \neq 0 \bmod 10$. This credit card number is invalid.

40. $\underline{6}$ 0 $\underline{1}$ 1 $\underline{5}$ 1 $\underline{8}$ 5 $\underline{8}$ 2 $\underline{9}$ 5 $\underline{8}$ 3 $\underline{2}$ 8 - Double each underlined digit.
    12 0 2 1 10 1 16 5 16 2 18 5 16 3 4 8
    $(1 + 2) + 0 + 2 + 1 + (1 + 0) + 1 + (1 + 6) + 5 + (1 + 6) + 2 + (1 + 8) + 5 + (1 + 6) + 3 + 4 + 8 = 65$
    $65 \neq 0 \bmod 10$. This credit card number is invalid.

41.

M	$c \equiv (13 + 7) \bmod 26 \equiv 20 \bmod 26 \equiv 20$	Code M as T.
A	$c \equiv (1 + 7) \bmod 26 \equiv 8 \bmod 26 \equiv 8$	Code A as H.
Y	$c \equiv (25 + 7) \bmod 26 \equiv 32 \bmod 26 \equiv 6$	Code Y as F.
T	$c \equiv (20 + 7) \bmod 26 \equiv 27 \bmod 26 \equiv 1$	Code T as A.
H	$c \equiv (8 + 7) \bmod 26 \equiv 15 \bmod 26 \equiv 15$	Code H as O.
E	$c \equiv (5 + 7) \bmod 26 \equiv 12 \bmod 26 \equiv 12$	Code E as L.
F	$c \equiv (6 + 7) \bmod 26 \equiv 13 \bmod 26 \equiv 13$	Code F as M.
O	$c \equiv (15 + 7) \bmod 26 \equiv 22 \bmod 26 \equiv 22$	Code O as V.
R	$c \equiv (18 + 7) \bmod 26 \equiv 25 \bmod 26 \equiv 25$	Code R as Y.
C	$c \equiv (3 + 7) \bmod 26 \equiv 10 \bmod 26 \equiv 10$	Code C as J.
E	$c \equiv (5 + 7) \bmod 26 \equiv 12 \bmod 26 \equiv 12$	Code E as L.
B	$c \equiv (2 + 7) \bmod 26 \equiv 9 \bmod 26 \equiv 9$	Code B as I.
E	$c \equiv (5 + 7) \bmod 26 \equiv 12 \bmod 26 \equiv 12$	Code E as L.
W	$c \equiv (23 + 7) \bmod 26 \equiv 30 \bmod 26 \equiv 4$	Code W as D.
I	$c \equiv (9 + 7) \bmod 26 \equiv 16 \bmod 26 \equiv 16$	Code I as P.
T	$c \equiv (20 + 7) \bmod 26 \equiv 27 \bmod 26 \equiv 1$	Code T as A.
H	$c \equiv (8 + 7) \bmod 26 \equiv 15 \bmod 26 \equiv 15$	Code H as O.
Y	$c \equiv (25 + 7) \bmod 26 \equiv 32 \bmod 26 \equiv 6$	Code Y as F.
O	$c \equiv (15 + 7) \bmod 26 \equiv 22 \bmod 26 \equiv 22$	Code O as V.
U	$c \equiv (21 + 7) \bmod 26 \equiv 28 \bmod 26 \equiv 2$	Code U as B.

The plaintext is coded as THF AOL MVYJL IL DPAO FVB.

42.

C	$c \equiv (3 + 11) \bmod 26 \equiv 14 \bmod 26 \equiv 14$	Code C as N.
A	$c \equiv (1 + 11) \bmod 26 \equiv 12 \bmod 26 \equiv 12$	Code A as L.
N	$c \equiv (14 + 11) \bmod 26 \equiv 25 \bmod 26 \equiv 25$	Code N as Y.
C	$c \equiv (3 + 11) \bmod 26 \equiv 14 \bmod 26 \equiv 14$	Code C as N.
E	$c \equiv (5 + 11) \bmod 26 \equiv 16 \bmod 26 \equiv 16$	Code E as P.
L	$c \equiv (12 + 11) \bmod 26 \equiv 23 \bmod 26 \equiv 23$	Code L as W.
A	$c \equiv (1 + 11) \bmod 26 \equiv 12 \bmod 26 \equiv 12$	Code A as L.
L	$c \equiv (12 + 11) \bmod 26 \equiv 23 \bmod 26 \equiv 23$	Code L as W.
L	$c \equiv (12 + 11) \bmod 26 \equiv 23 \bmod 26 \equiv 23$	Code L as W.
P	$c \equiv (16 + 11) \bmod 26 \equiv 27 \bmod 26 \equiv 1$	Code P as A.
L	$c \equiv (12 + 11) \bmod 26 \equiv 23 \bmod 26 \equiv 23$	Code L as W.
A	$c \equiv (1 + 11) \bmod 26 \equiv 12 \bmod 26 \equiv 12$	Code A as L.
N	$c \equiv (14 + 11) \bmod 26 \equiv 25 \bmod 26 \equiv 25$	Code N as Y.
S	$c \equiv (19 + 11) \bmod 26 \equiv 30 \bmod 26 \equiv 4$	Code S as D.

The plaintext is coded as NLYNPW LWW AWLYD.

43. Because the encoded message uses a cyclical alphabetic encrypting code, it is possible to rotate through the 26 possible codes until the ciphertext is decoded. Using just the first word:

    PXXM → QYYN → RZZO → SAAP → TBBQ → UCCR → VDDS → WEET → XFFU →YGGV → ZHHW → AIIX → BJJY → CKKZ → DLLA → EMMB → FNNC → GOOD

    It took 17 transformations to find the plaintext, so the decoding congruence is $p \equiv (c + 17)$ mod 26. Decode the rest of the ciphertext normally.

P	$p = (16 + 17)$ mod $26 \equiv 33$ mod $26 \equiv 7$	Decode P as G.
X	$p = (24 + 17)$ mod $26 \equiv 41$ mod $26 \equiv 15$	Decode X as O.
X	$p = (24 + 17)$ mod $26 \equiv 41$ mod $26 \equiv 15$	Decode X as O.
M	$p = (13 + 17)$ mod $26 \equiv 30$ mod $26 \equiv 4$	Decode M as D.
U	$p = (21 + 17)$ mod $26 \equiv 38$ mod $26 \equiv 12$	Decode U as L.
D	$p = (4 + 17)$ mod $26 \equiv 21$ mod $26 \equiv 21$	Decode D as U.
L	$p = (12 + 17)$ mod $26 \equiv 29$ mod $26 \equiv 3$	Decode L as C.
T	$p = (20 + 17)$ mod $26 \equiv 37$ mod $26 \equiv 11$	Decode T as K.
C	$p = (3 + 17)$ mod $26 \equiv 20$ mod $26 \equiv 20$	Decode C as T.
X	$p = (24 + 17)$ mod $26 \equiv 41$ mod $26 \equiv 15$	Decode X as O.
V	$p = (22 + 17)$ mod $26 \equiv 39$ mod $26 \equiv 13$	Decode V as M.
X	$p = (24 + 17)$ mod $26 \equiv 41$ mod $26 \equiv 15$	Decode X as O.
A	$p = (1 + 17)$ mod $26 \equiv 18$ mod $26 \equiv 18$	Decode A as R.
A	$p = (1 + 17)$ mod $26 \equiv 18$ mod $26 \equiv 18$	Decode A as R.
X	$p = (24 + 17)$ mod $26 \equiv 41$ mod $26 \equiv 15$	Decode X as O.
F	$p = (6 + 17)$ mod $26 \equiv 23$ mod $26 \equiv 23$	Decode F as W.

    The ciphertext is decoded as GOOD LUCK TOMORROW.

44. Because the encoded message uses a cyclical alphabetic encrypting code, it is possible to rotate through the 26 possible codes until the ciphertext is decoded. Using just the first word:

    HVS → IWT → JXU → KYV → LZW → MAX → NBY → OCZ → PDA → QEB → RFC → SGD → THE

    It took 12 transformations to find the plaintext, so the decoding congruence is $p \equiv (c + 17)$ mod 26. Decode the rest of the ciphertext normally.

H	$p \equiv (8 + 12)$ mod $26 \equiv 20$ mod $26 \equiv 20$	Decode H as T.
V	$p \equiv (22 + 12)$ mod $26 \equiv 34$ mod $26 \equiv 8$	Decode V as H.
S	$p \equiv (19 + 12)$ mod $26 \equiv 31$ mod $26 \equiv 5$	Decode S as E.
R	$p \equiv (18 + 12)$ mod $26 \equiv 30$ mod $26 \equiv 4$	Decode R as D.
O	$p \equiv (15 + 12)$ mod $26 \equiv 27$ mod $26 \equiv 1$	Decode O as A.
M	$p \equiv (13 + 12)$ mod $26 \equiv 25$ mod $26 \equiv 25$	Decode M as Y.
V	$p \equiv (22 + 12)$ mod $26 \equiv 34$ mod $26 \equiv 8$	Decode V as H.
O	$p \equiv (15 + 12)$ mod $26 \equiv 27$ mod $26 \equiv 1$	Decode O as A.
G	$p \equiv (7 + 12)$ mod $26 \equiv 19$ mod $26 \equiv 19$	Decode G as S.
O	$p \equiv (15 + 12)$ mod $26 \equiv 27$ mod $26 \equiv 1$	Decode O as A.
F	$p \equiv (6 + 12)$ mod $26 \equiv 18$ mod $26 \equiv 18$	Decode F as R.
F	$p \equiv (6 + 12)$ mod $26 \equiv 18$ mod $26 \equiv 18$	Decode F as R.
W	$p \equiv (23 + 12)$ mod $26 \equiv 35$ mod $26 \equiv 9$	Decode W as I.
J	$p \equiv (10 + 12)$ mod $26 \equiv 22$ mod $26 \equiv 22$	Decode J as V.
S	$p \equiv (19 + 12)$ mod $26 \equiv 31$ mod $26 \equiv 5$	Decode S as E.
R	$p \equiv (18 + 12)$ mod $26 \equiv 30$ mod $26 \equiv 4$	Decode R as D.

    The ciphertext is decoded as THE DAY HAS ARRIVED.

45.
E	$c \equiv (3 \cdot 5 + 6)$ mod $26 \equiv 21$ mod $26 \equiv 21$	Code E as U.
N	$c \equiv (3 \cdot 14 + 6)$ mod $26 \equiv 48$ mod $26 \equiv 22$	Code N as V.
D	$c \equiv (3 \cdot 4 + 6)$ mod $26 \equiv 18$ mod $26 \equiv 18$	Code D as R.
O	$c \equiv (3 \cdot 15 + 6)$ mod $26 \equiv 51$ mod $26 \equiv 25$	Code O as Y.
F	$c \equiv (3 \cdot 6 + 6)$ mod $26 \equiv 24$ mod $26 \equiv 24$	Code F as X.
T	$c \equiv (3 \cdot 20 + 6)$ mod $26 \equiv 66$ mod $26 \equiv 14$	Code T as N.
H	$c \equiv (3 \cdot 8 + 6)$ mod $26 \equiv 30$ mod $26 \equiv 4$	Code H as D.
E	$c \equiv (3 \cdot 5 + 6)$ mod $26 \equiv 21$ mod $26 \equiv 21$	Code E as U.
L	$c \equiv (3 \cdot 12 + 6)$ mod $26 \equiv 42$ mod $26 \equiv 16$	Code L as P.
I	$c \equiv (3 \cdot 9 + 6)$ mod $26 \equiv 33$ mod $26 \equiv 7$	Code I as G.
N	$c \equiv (3 \cdot 14 + 6)$ mod $26 \equiv 48$ mod $26 \equiv 22$	Code N as V.
E	$c \equiv (3 \cdot 5 + 6)$ mod $26 \equiv 21$ mod $26 \equiv 21$	Code E as U.

    The plaintext is coded as UVR YX NDU PGVU.

46. Solve the congruence for p.

$$c = 7p + 4$$
$$c - 4 = 7p$$

The multiplicative inverse of 7 is 15.

$$15(c - 4) = 15(7)p$$
$$[15(c - 4)] \bmod 26 \equiv p$$

Decode using this congruence.

W $[15(23 - 4)] \bmod 26 \equiv 285 \bmod 26 \equiv 25$ Decode W as Y.
E $[15(5 - 4)] \bmod 26 \equiv 15 \bmod 26 \equiv 15$ Decode E as O.
U $[15(21 - 4)] \bmod 26 \equiv 255 \bmod 26 \equiv 21$ Decode U as U.
L $[15(12 - 4)] \bmod 26 \equiv 120 \bmod 26 \equiv 16$ Decode L as P.
K $[15(11 - 4)] \bmod 26 \equiv 105 \bmod 26 \equiv 1$ Decode K as A.
G $[15(7 - 4)] \bmod 26 \equiv 45 \bmod 26 \equiv 19$ Decode G as S.
G $[15(7 - 4)] \bmod 26 \equiv 45 \bmod 26 \equiv 19$ Decode G as S.
M $[15(13 - 4)] \bmod 26 \equiv 135 \bmod 26 \equiv 5$ Decode M as E.
F $[15(6 - 4)] \bmod 26 \equiv 30 \bmod 26 \equiv 4$ Decode F as D.
N $[15(14 - 4)] \bmod 26 \equiv 150 \bmod 26 \equiv 20$ Decode N as T.
H $[15(8 - 4)] \bmod 26 \equiv 60 \bmod 26 \equiv 8$ Decode H as H.
M $[15(13 - 4)] \bmod 26 \equiv 135 \bmod 26 \equiv 5$ Decode M as E.
N $[15(14 - 4)] \bmod 26 \equiv 150 \bmod 26 \equiv 20$ Decode N as T.
M $[15(13 - 4)] \bmod 26 \equiv 135 \bmod 26 \equiv 5$ Decode M as E.
G $[15(7 - 4)] \bmod 26 \equiv 45 \bmod 26 \equiv 19$ Decode G as S.
N $[15(14 - 4)] \bmod 26 \equiv 150 \bmod 26 \equiv 20$ Decode N as T.

The ciphertext is decoded as YOU PASSED THE TEST.

47. 1. The product of two rational numbers except 0 is always another non-zero rational number. Therefore, the closure property holds true.
   2. The associative property of multiplication holds true for rational numbers except 0.
   3. There exists an identity element (1) for multiplication in the set of rational numbers except 0.
   4. If $\frac{a}{b}$ rational number with $a \neq 0$, then $\frac{b}{a}$ is its inverse. The inverse property is true.

Because all of the four properties of groups are satisfied, rational numbers except 0 do form a group with respect to multiplication.

48. 1. The sum of two multiples of 3 is always another multiple of 3. Therefore, the closure property holds true.
   2. The associative property of addition holds true for all multiples of 3.
   3. There exists an identity element (0) in the set of multiples of 3 for addition.
   4. If a is a multiple of 3, then $-a$ is the inverse of a. Therefore, the inverse property holds true.

Because all of the four properties of groups are satisfied, multiples of 3 do form a group with respect to addition.

49. 1. The product of two negative integers is always a positive integer. Therefore, the closure property fails.
   2. The associative property of multiplication holds true for all negative integers.
   3. Negative integers do not have an identity element in the set for multiplication. Therefore, the identity property fails.
   4. Negative integers do not have inverses in the set for multiplication. Therefore, the inverse property fails.

Properties 1, 3, and 4 fail, so the negative integers do not form a group with respect to multiplication.

50. 1. The product modulo 11 of two elements of the set is always another member of the set. Therefore, the closure property holds true.
   2. The associative property of multiplication modulo 11 holds true for all members of the set.
   3. There exists an identity element (1) in the set for multiplication modulo 11. Therefore, the identity property holds true.
   4. If a is an element in the set, then $11 - a$ is the inverse of a. Therefore, the inverse property holds true.

Because all of the four properties of groups are satisfied, the set forms a group with respect to multiplication modulo 11.

51.

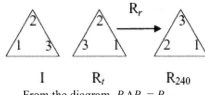

From the diagram, $R_t \Delta R_r = R_{240}$.

52.

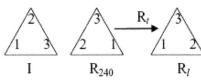

From the diagram, $R_{240} \Delta R_t = R_l$.

53. $R_{240}\Delta R_l = \begin{pmatrix} 1 & 2 & 3 \\ 3 & 1 & 2 \end{pmatrix} \Delta \begin{pmatrix} 1 & 2 & 3 \\ 1 & 3 & 2 \end{pmatrix}$

$= \begin{pmatrix} 1 & 2 & 3 \\ 2 & 1 & 3 \end{pmatrix} = R_r$

54. $R_t\Delta R_r = \begin{pmatrix} 1 & 2 & 3 \\ 3 & 2 & 1 \end{pmatrix} \Delta \begin{pmatrix} 1 & 2 & 3 \\ 2 & 1 & 3 \end{pmatrix}$

$= \begin{pmatrix} 1 & 2 & 3 \\ 3 & 1 & 2 \end{pmatrix} = R_{240}$

55. $C\Delta E = \begin{pmatrix} 1 & 2 & 3 \\ 1 & 3 & 2 \end{pmatrix} \Delta \begin{pmatrix} 1 & 2 & 3 \\ 2 & 1 & 3 \end{pmatrix}$

$= \begin{pmatrix} 1 & 2 & 3 \\ 2 & 3 & 1 \end{pmatrix} = A$

56. $B\Delta A = \begin{pmatrix} 1 & 2 & 3 \\ 3 & 1 & 2 \end{pmatrix} \Delta \begin{pmatrix} 1 & 2 & 3 \\ 2 & 3 & 1 \end{pmatrix}$

$= \begin{pmatrix} 1 & 2 & 3 \\ 1 & 2 & 3 \end{pmatrix} = 1$

57. Inverse of $D = \begin{pmatrix} 1 & 2 & 3 \\ 3 & 2 & 1 \end{pmatrix} = D$

58. Inverse of $A = \begin{pmatrix} 1 & 2 & 3 \\ 3 & 1 & 2 \end{pmatrix} = B$

CHAPTER 8 TEST

1. a. 3
   b. 5

2. There are 9 years between the two dates, two of which (2020 and 2024) are leap years. So the total number of days between the two dates is
   $9(365) + 2 = 3287$ days
   $3287 \div 7 = 469$ remainder 4

So, in 2026, January 1st will occur 4 days after Sunday, which is Thursday.

3. a. True. $8 \div 3 = 2$ remainder 2,
      $20 \div 3 = 6$ remainder 2.

   b. False. $61 \div 7 = 8$ remainder 5,
      $38 \div 7 = 5$ remainder 3.

4. $(25 + 9)$ mod 6 = 34 mod 6.
   $34 \div 6 = 5$ remainder 4
   So $(25 + 9)$ mod 6 = 4.

5. $(31 - 11)$ mod 7 = 20 mod 7.
   $20 \div 7 = 2$ remainder 6
   So $(31 - 11)$ mod 7 = 6.

6. $(5 \cdot 16)$ mod 12 = 80 mod 12.
   $80 \div 12 = 6$ remainder 8.
   So $(5 \cdot 16)$ mod 12 = 8.

7. a. $(3 + 27)$ mod 12 = 30 mod 12
      $30 \div 12 = 2$ remainder 6
      It will be 6:00.

   b. $(3 - 58)$ mod 12 = -55 mod 12
      $-55 + 12 = -43$
      $-43 + 12 = -31$
      $-31 + 12 = -19$
      $-19 + 12 = -7$
      $-7 + 12 = 5$
      It was 5:00.

8. Substitute whole numbers zero through 8 for x in the congruency.

$x = 0$	$0 \not\equiv 5$ mod 9	NO
$x = 1$	$1 \not\equiv 5$ mod 9	NO
$x = 2$	$2 \not\equiv 5$ mod 9	NO
$x = 3$	$3 \not\equiv 5$ mod 9	NO
$x = 4$	$4 \not\equiv 5$ mod 9	NO
$x = 5$	$5 \equiv 5$ mod 9	YES
$x = 6$	$6 \not\equiv 5$ mod 9	NO
$x = 7$	$7 \not\equiv 5$ mod 9	NO
$x = 8$	$8 \not\equiv 5$ mod 9	NO

   Adding multiples of the modulus yields other solutions. The solutions are
   $5, 14, 23, 32, 41, 50, \ldots$

9. Substitute whole numbers zero through 3 for x in the congruency.

$x = 0$	$2(0) + 3 \not\equiv 1$ mod 4	NO
$x = 1$	$2(1) + 3 \equiv 1$ mod 4	YES
$x = 2$	$2(2) + 3 \not\equiv 1$ mod 4	NO
$x = 3$	$2(3) + 3 \equiv 1$ mod 4	YES

   Adding multiples of the modulus yields other solutions. The solutions are $1, 3, 5, 7, 9, 11, \ldots$

10. $5 + 4 = 9$, so the arithmetic inverse is 4.

Substitute whole numbers 0 through 8 for x in the congruency $5x = 1 \bmod 9$.

$x = 0$	$5(0) \not\equiv 1 \bmod 9$	NO
$x = 1$	$5(1) \not\equiv 1 \bmod 9$	NO
$x = 2$	$5(2) \not\equiv 1 \bmod 9$	YES
$x = 3$	$5(3) \not\equiv 1 \bmod 9$	NO
$x = 4$	$5(4) \not\equiv 1 \bmod 9$	NO
$x = 5$	$5(5) \not\equiv 1 \bmod 9$	NO
$x = 6$	$5(6) \not\equiv 1 \bmod 9$	NO
$x = 7$	$5(7) \not\equiv 1 \bmod 9$	NO
$x = 8$	$5(8) \not\equiv 1 \bmod 9$	NO

The multiplicative inverse is 2.

11. $d_{13} \equiv 10 - (d_1 + 3d_2 + d_3 + 3d_4 + d_5 + 3d_6 + d_7 + 3d_8 + d_9 + 3d_{10} + d_{11} + 3d_{12}) \bmod 10$

$d_{13} \equiv 10 - [9 + 3(7) + 8 + 3(0) + 7 + 3(3) + 9 + 3(4) + 9 + 3(4) + 2 + 3(4)] \bmod 10$

$\equiv 10 - 210 \bmod 10$

$\equiv 10 - 0 \equiv 10$

The check digit is 0.

12. $d_{12} \equiv 10 - (3d_1 + d_2 + 3d_3 + d_4 + 3d_5 + d_6 + 3d_7 + d_8 + 3d_9 + d_{10} + 3d_{11}) \bmod 10$

$d_{12} \equiv 10 - [3(6) + 7 + 3(3) + 4 + 3(1) + 9 + 3(2) + 3 + 3(2) + 1 + 3(6)] \bmod 10$

$\equiv 10 - 84 \bmod 10$

$\equiv 10 - 4 \equiv 6$

The check digit is 6.

13. $\underline{4}\,2\,\underline{3}\,2\,\underline{8}\,1\,\underline{8}\,0\,\underline{5}\,7\,\underline{3}\,6\,\underline{4}\,8\,\underline{7}\,6$ - Double each underlined digits.

8 2 6 2 16 1 16 0 10 7 6 6 8 8 14 6

$8 + 2 + 6 + 2 + (1 + 6) + 1 + (1 + 6) + 0 + (1 + 0) + 7 + 6 + 6 + 8 + 8 + (1 + 4) + 6 = 80$

$80 \equiv 0 \bmod 10$. This credit card number is valid.

14.

R	$c \equiv (18 + 10) \bmod 26 \equiv 28 \bmod 26 \equiv 2$	Code R as B.
E	$c \equiv (5 + 10) \bmod 26 \equiv 15 \bmod 26 \equiv 15$	Code E as O.
P	$c \equiv (16 + 10) \bmod 26 \equiv 26 \bmod 26 \equiv 0$	Code P as Z.
O	$c \equiv (15 + 10) \bmod 26 \equiv 25 \bmod 26 \equiv 25$	Code O as Y.
R	$c \equiv (18 + 10) \bmod 26 \equiv 28 \bmod 26 \equiv 2$	Code R as B.
T	$c \equiv (20 + 10) \bmod 26 \equiv 30 \bmod 26 \equiv 4$	Code T as D.
B	$c \equiv (2 + 10) \bmod 26 \equiv 12 \bmod 26 \equiv 12$	Code B as L.
A	$c \equiv (1 + 10) \bmod 26 \equiv 11 \bmod 26 \equiv 11$	Code A as K.
C	$c \equiv (3 + 10) \bmod 26 \equiv 13 \bmod 26 \equiv 13$	Code C as M.
K	$c \equiv (11 + 10) \bmod 26 \equiv 21 \bmod 26 \equiv 21$	Code K as U.

The plaintext is coded as BOZYBD LKMU.

15. Solve the congruence for p.

$$c = 3p + 5$$
$$c - 5 = 3p$$

The multiplicative inverse of 3 is 9.

$$9(c - 5) = 9(3)p$$
$$[9(c - 5)] \bmod 26 = p$$

Decode using this congruence.

U	$[9(21 - 5)] \bmod 26 \equiv 144 \bmod 26 \equiv 14$	Decode U as N.
T	$[9(20 - 5)] \bmod 26 \equiv 135 \bmod 26 \equiv 5$	Decode T as E.
S	$[9(19 - 5)] \bmod 26 \equiv 126 \bmod 26 \equiv 22$	Decode S as V.
T	$[9(20 - 5)] \bmod 26 \equiv 135 \bmod 26 \equiv 5$	Decode T as E.
G	$[9(7 - 5)] \bmod 26 \equiv 18 \bmod 26 \equiv 18$	Decode G as R.
D	$[9(4 - 5)] \bmod 26 \equiv -9 \bmod 26 \equiv 17$	Decode D as Q.
P	$[9(16 - 5)] \bmod 26 \equiv 99 \bmod 26 \equiv 21$	Decode P as U.
F	$[9(6 - 5)] \bmod 26 \equiv 9 \bmod 26 \equiv 9$	Decode F as I.
M	$[9(13 - 5)] \bmod 26 \equiv 72 \bmod 26 \equiv 20$	Decode M as T.

The ciphertext is decoded as NEVER QUIT.

16. 1. The product modulo 5 of any two elements in the set is always a member of the set. Therefore, the closure property holds true.
    2. The associative property of multiplication modulo 5 holds true for the members of the set.
    3. There exists an identity element (1) in the set for multiplication modulo 5.
    4. Every element has an inverse. If a is a member of the set, and c satisfies the congruence $c \cdot a \equiv 1 \bmod 5$, then c is the inverse of a.

    All the four properties are satisfied, so the set forms a group with respect to multiplication modulo 5.

17. 1. The product of any two odd integers is always another odd integer. Therefore, the closure property holds true.
    2. From the question, the associative property of multiplication holds true for odd integers.
    3. There exists an identity element (1) in the set of odd integers with respect to multiplication.
    4. Not every element has an inverse in the set. If a is an odd integer, then $\dfrac{1}{a}$ should be the inverse of a.

    However, $\dfrac{1}{a}$ is a non-reducible fraction for all odd numbers except for –1 and 1. Therefore, the inverse property fails.

    Property 4 fails, so the odd integers do not form a group with respect to multiplication.

18. a.

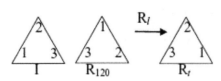

From the diagram, $R_{120}\Delta R_l = R_t$.

b.

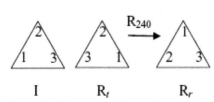

From the diagram, $R_t\Delta R_{240} = R_r$.

19. $\begin{pmatrix} 1 & 2 & 3 \\ 3 & 1 & 2 \end{pmatrix} \Delta \begin{pmatrix} 1 & 2 & 3 \\ 3 & 2 & 1 \end{pmatrix} = \begin{pmatrix} 1 & 2 & 3 \\ 1 & 3 & 2 \end{pmatrix}$

20. Inverse of $\begin{pmatrix} 1 & 2 & 3 \\ 3 & 1 & 2 \end{pmatrix} = \begin{pmatrix} 1 & 2 & 3 \\ 2 & 3 & 1 \end{pmatrix}$

Chapter 9: Applications of Equations

EXCURSION EXERCISES, SECTION 9.1

1. Convert 5'8" to inches. 5'8" = 5(12)" + 8" = 60" + 8" = 68". Substitute 140 for W and 68 for H in the formula.

    $$B = \frac{703(140)}{(68)^2} = \frac{98,420}{4624} \approx 21.3$$

    As 21.3 is less than 25, Amy is at low risk for weight-related disease.

3. Convert 5'9" to inches. 5'9" = 5(12)" + 9" = 69". Substitute 23 for B and 69 for H in the formula.

    $$W = \frac{(23)(69)^2}{703} = \frac{109,503}{703} \approx 156$$

    Bohdan should weigh around 156 pounds; therefore, he needs to lose 185 − 156 = 29 pounds.

5. a. Amy's height = 5'8" = $5\frac{8}{12}$ ft ≈5.67 ft.

    BMI from nomograph ≈ 22 kg/m^2.
    The value calculated and using the nomograph are very close.

    b. Roger's height = 5'11" = $5\frac{11}{12}$ feet ≈ 5.92 feet.
    Weight from nomograph =180 lbs
    Values found using the nomograph should be close to those found using a calculator.

    c. Answers will vary.
    A disadvantage is that the nomograph provides a visual estimate rather than an exact value. An advantage is that the nomograph gives a quick and relatively accurate estimate without having to use a formula.

EXERCISE SET 9.1

1. An equation expresses the equality of two mathematical expressions. An equation contains an equals sign. An expression does not.

3. By substituting the solution for the variable in the original equation. If the resulting equation is a true equation, then the solution is correct.

5. $$9 + b = 21$$
    $$9 - 9 + b = 21 - 9$$
    $$b = 12$$

7. $$b - 11 = 11$$
    $$b - 11 + 11 = 11 + 11$$
    $$b = 22$$

9. $$-48 = 6z$$
    $$\frac{-48}{6} = \frac{6z}{6}$$
    $$z = -8$$

11. $$-\frac{3}{4}x = 15$$
    $$\left(-\frac{4}{3}\right)\left(-\frac{3}{4}\right)x = 15\left(-\frac{4}{3}\right)$$
    $$x = -20$$

13. $$-\frac{x}{4} = -2$$
    $$(-4)\left(-\frac{x}{4}\right) = (-2)(-4)$$
    $$x = 8$$

15. Solving:
    $$4 - 2b = 2 - 4b$$
    $$4 - 4 - 2b + 4b = 2 - 4 - 4b + 4b$$
    $$2b = -2$$
    $$b = -1$$

17. Solving:
    $$5x - 3 = 9x - 7$$
    $$5x - 5x - 3 + 7 = 9x - 5x - 7 + 7$$
    $$4 = 4x$$
    $$x = 1$$

19. Solving:
    $$3m + 5 = 2 - 6m$$
    $$3m + 6m + 5 - 5 = 2 - 5 - 6m + 6m$$
    $$9m = -3$$
    $$m = -\frac{1}{3}$$

21. Solving:
    $$5x + 7 = 8x + 5$$
    $$5x - 5x + 7 - 5 = 8x - 5x + 5 - 5$$
    $$3x = 2$$
    $$x = \frac{2}{3}$$

23. Solving:
    $$4b + 15 = 3 - 2b$$
    $$4b + 2b + 15 - 15 = 3 - 15 - 2b + 2b$$
    $$6b = -12$$
    $$b = -2$$

25. Solving:
    $$9n - 15 = 3(2n - 1)$$
    $$9n - 15 = 6n - 3$$
    $$9n - 6n - 15 = 6n - 6n - 3 + 15$$
    $$3n = 12$$
    $$n = 4$$

27. Solving:
$$5(3-2y)=3-4y$$
$$15-10y=3-4y$$
$$15-3-10y+10y=3-3-4y+10y$$
$$12=6y$$
$$y=2$$

29. Solving:
$$2(3b+5)-1=10b+1$$
$$6b+10-1=10b+1$$
$$6b+9=10b+1$$
$$6b-6b+9-1=10b-6b+1-1$$
$$8=4b$$
$$b=2$$

31. Solving:
$$4a+3=7-(5-8a)$$
$$4a+3=7-5+8a$$
$$4a+3=2+8a$$
$$4a-4a+3-2=2-2+8a-4a$$
$$4a=1$$
$$a=\frac{1}{4}$$

33. Solving:
$$4(3y+1)=2(y-8)$$
$$12y+4=2y-16$$
$$12y-2y+4-4=2y-2y-16-4$$
$$10y=-20$$
$$y=-2$$

35. Solving:
$$3(x-4)=1-(2x+7)$$
$$3x-12=1-2x+7$$
$$3x-12=8-2x$$
$$3x+2x-12+12=8+12-2x+2x$$
$$5x=20$$
$$x=4$$

37. Solving:
$$\frac{x}{8}+2=\frac{3x}{4}-3$$
$$8\left(\frac{x}{8}+2\right)=8\left(\frac{3x}{4}-3\right)$$
$$x-x+16+24=6x-x-24+24$$
$$40=5x$$
$$x=8$$

39. Solving:
$$\frac{2}{3}+\frac{3x+1}{4}=\frac{5}{3}$$
$$12\left(\frac{2}{3}+\frac{3x+1}{4}\right)=12\left(\frac{5}{3}\right)$$
$$8+9x+3=20$$
$$11-11+9x=20-11$$
$$9x=9$$
$$x=1$$

41. Solving:
$$\frac{3}{4}(x-8)=\frac{1}{2}(2x+4)$$
$$4\left(\frac{3}{4}(x-8)\right)=4\left(\frac{1}{2}(2x+4)\right)$$
$$3(x-8)=2(2x+4)$$
$$3x-24=4x+8$$
$$3x-3x-24-8=3x+8-8$$
$$x=-32$$

43. $P=0.018417\,L$
Substitute 350 for P.
$$\frac{350}{0.018417}=L$$
$$19,004.18\approx L$$
The maximum amount is \$19,0004.18.

45. $P=15+\frac{1}{2}D$
Substitute 55 for P.
$$55=15+\frac{1}{2}D$$
$$2(55-15)=D$$
$$80=D$$
The depth is 80 feet.

47. $t=16.11-0.0062y$
Substitute 3.72 for t.
$$3.72=16.11-0.0062y$$
$$-12.39=-0.0062y$$
$$1998\approx y$$
The year 1998.

49. $C=\frac{1}{4}D-45$
Substitute −11 for C.
$$-11=\frac{1}{4}D-45$$
$$4(45-11)=D$$
$$136=D$$
The car will slide 136 feet.

51. $N=7C-30$
Substitute 140 for N.
$$140=7C-30$$
$$170=7C$$
$$24.3\approx C$$
The temperature is approximately 24.3ºC.

53. $H=0.8(200-A)$
Substitute 25 for H.
$$25=0.8(200-H)$$
$$25-160=-0.8H$$
$$\frac{135}{0.8}=H$$
$$168.75=H$$
The bowler's average score is 168.75.

55. a. Let G be the number of children adopted from Guatemala, and E be the number adopted from Ethiopia.
$$G = 3E - 1052$$
$$4123 = 3E - 1052$$
$$5175 = 3E$$
$$1725 = E$$
The number of children adopted from Ethiopia is 1725.

   b. Let C be the number of children adopted from China, and U be the number adopted from Ukraine.
$$C = 8U + 189$$
$$3909 = 8U + 189$$
$$3720 = 8U$$
$$465 = U$$
The number of children adopted from Ukraine is 465.

57. Let H be the number of hours required, and C the cost in dollars.
$$C = 115 + 80H$$
$$355 = 115 + 80H$$
$$240 = 80H$$
$$3 = H$$
3 hours of labor were required.

59. Let r be the number of research assistants. Then there are $3r$ teaching assistants.
$$r + 3r = 600$$
$$4r = 600$$
$$r = 150$$
There are 150 research assistants.

61. Let s be the amount of the smaller deposit. Then the amount of the larger deposit is $2s - 1000$.
$$s + 2s - 1000 = 5000$$
$$3s = 6000$$
$$s = 2000$$
$$2s - 1000 = 3000$$
The amounts are \$2000 and \$3000.

63. Let d be the amount of data used over 6GB, and c the bill for the plan.
$$c = 59.95 + 7.95d$$
$$115.60 = 59.95 + 7.95d$$
$$55.65 = 7.95d$$
$$d = 7$$
$$7 + 6 = 13$$
The amount of data used is 13 GB.

65. $b = P - a - c$

67. $R = \dfrac{E}{I}$

69. $r = \dfrac{I}{Pt}$

71. $C = \dfrac{5}{9}(F - 32)$

73. $t = \dfrac{A - P}{Pr}$

75. $f = \dfrac{T + gm}{m}$

77. $S = C - Rt$

79. $b_2 = \dfrac{2A}{h} - b_1$

81. $h = \dfrac{S}{2\pi r} - r$

83. $y = 2 - \dfrac{4}{3}x$

85. $x = \dfrac{y - y_1}{m} + x_1$

87. Solving:
$$4(x + 5) = 30 - (10 - 4x)$$
$$4x + 20 = 30 - 10 - 4x$$
$$4x + 20 = 20 - 4x$$
$$4x - 4x + 20 - 20 = 20 - 20 - 4x + 4x$$
$$0 = 0$$
Every real number is a solution to this equation.

89. Solving:
$$ax + b = cx + d$$
$$ax + b - b = cx + d - b$$
$$ax - cx = cx - cx + d - b$$
$$x(a - c) = d - b$$
$$x = \dfrac{d - b}{a - c}$$
No, the solution is not valid when $a = c$ and $b \neq d$, as the denominator would equal zero, and x would then be undefined.

91. $T = 750m + 30{,}000$, where T is the total number of miles driven.

93. $C = 29.95d + .50(m - 100)$, where C is the total cost, d is the number of days, and m the number of miles driven.

EXCURSION EXERCISES, SECTION 9.2

1. Using ratios:

$$\frac{4 \text{ earned runs}}{21 \text{ innings}} = \frac{x \text{ earned runs}}{9 \text{ innings}}$$
$$4 \cdot 9 = 21 \cdot x$$
$$\frac{36}{21} = x$$
$$1.71 \approx x$$

Reardon's 1979 ERA was 1.71.

3. Using ratios:

$$\frac{65 \text{ earned runs}}{211.2 \text{ innings}} = \frac{x \text{ earned runs}}{9 \text{ innings}}$$
$$65 \cdot 9 = 211.2 \cdot x$$
$$\frac{585}{211.2} = x$$
$$2.77 \approx x$$

Ryan's 1987 ERA was 2.77.

5. Answers will vary.

EXERCISE SET 9.2

1. Examples will vary.

3. $\frac{582 \text{ miles}}{12 \text{ hours}} = \frac{48.5 \text{ miles}}{1 \text{ hour}} = 48.5$ miles per hour

5. $\frac{544 \text{ words}}{8 \text{ minutes}} = \frac{68 \text{ words}}{1 \text{ minute}} = 68$ words per minute

7. $\frac{\$9100}{350 \text{ shares}} = \frac{\$926}{1 \text{ share}} = \$26$ per share of stock

9. $\frac{\$682.50}{35 \text{ hours}} = \frac{\$19.50}{1 \text{ hour}} = \$19.50$ per hour

11. a. $5.6 \text{ feet/second}\left(\frac{60 \text{ seconds}}{1 \text{ minute}}\right) = 336$ feet/minute

    b. Using ratios:
    $$\frac{5.6 \text{ feet}}{1 \text{ second}} = \frac{500 \text{ feet}}{x \text{ seconds}}$$
    $$5.6x = 500$$
    $$x \approx 89$$
    89 seconds

13. $\frac{\$3.49}{18 \text{ ounces}} \approx \$0.193$ per ounce

    $\frac{\$3.89}{24 \text{ ounces}} \approx \$0.162$ per ounce

    The 24-ounce box is more economical.

15. a. Babe Ruth (1921): $\frac{540}{59} \approx 9.2$

    Babe Ruth (1927): $\frac{540}{60} = 9.0$

    Jimmie Foxx: $\frac{585}{58} \approx 10.1$

    Hank Greenberg: $\frac{556}{58} \approx 9.6$

    Roger Maris: $\frac{590}{61} \approx 9.7$

    Mickey Mantle: $\frac{514}{54} \approx 9.5$

    Willie Mays: $\frac{558}{52} \approx 10.7$

    Mark McGwire: $\frac{540}{59} \approx 9.2$

    Sammy Sosa: $\frac{643}{66} \approx 9.7$

    Barry Bonds: $\frac{476}{73} \approx 6.5$

    Alex Rodriquez: $\frac{624}{57} \approx 10.9$

    Ryan Howard: $\frac{581}{58} \approx 10.0$

    Chris Davis: $\frac{584}{53} \approx 11.0$

    b. Barry Bonds, Mark McGwire

    c. Explanations may vary.

17. a. 2015: $\frac{205,600,000,000}{2,586,000} \approx 79.5$

    2017: $\frac{225,300,000,000}{2,760,000} \approx 81.6$

    2019: $\frac{246,500,000,000}{2,943,000} \approx 83.8$

    b. $\frac{83.8}{79.5} \approx 1.05$ times greater

19. Using ratios:
    $$\frac{67.8697}{\$1} = \frac{x \text{ rupees}}{\$45,000}$$
    $$x = 3,054,136.5$$
    3,054,136.5 Indian rupees

21. Using ratios:
    $$\frac{1.4216}{\$1} = \frac{x \text{ Australian dollars}}{\$29,000}$$
    $$x = 41,226.4$$
    41,226.4 Australian dollars

23. Price-to-rent ratio $= \frac{\text{price of house}}{\text{rent for year}}$

    Rent for year $= \$1150 \cdot 12 = 13,800$

    Price-to-rent ratio $= \frac{155,000}{13,800} \approx 11.2$

    Since the quotient is less than 20, buying is the better option.

25. Calculate each ratio:

    Georgetown: $\dfrac{3244+3982}{1350} \approx 5$

    Syracuse: $\dfrac{6487+8045}{1078} \approx 13$

    Connecticut: $\dfrac{8851+8826}{2007} \approx 9$

    West Virginia: $\dfrac{11,434+9429}{2044} \approx 10$

    Georgetown University has the lowest.

27. Solving:

    $\dfrac{3}{y} = \dfrac{7}{40}$

    $7y = 120$

    $y \approx 17.14$

29. Solving:

    $\dfrac{16}{d} = \dfrac{25}{40}$

    $25d = 640$

    $d = 25.6$

31. Solving:

    $\dfrac{120}{c} = \dfrac{144}{25}$

    $144c = 3000$

    $c \approx 20.83$

33. Solving:

    $\dfrac{4}{a} = \dfrac{9}{5}$

    $9a = 20$

    $a \approx 2.22$

35. Solving:

    $\dfrac{1.2}{2.8} = \dfrac{b}{32}$

    $2.8b = 38.4$

    $b \approx 13.71$

37. Solving:

    $\dfrac{2.5}{0.6} = \dfrac{165}{x}$

    $2.5x = 99$

    $x = 39.6$

39. Solving:

    $\dfrac{5}{4} = \dfrac{x}{52,000}$

    $4x = 260,000$

    $x = 65,000$

    $65,000

41. Solving:

    $\dfrac{70.5}{3} = \dfrac{x}{14}$

    $3x = 987$

    $x = 329$

    329 miles

43. Solving:

    $\dfrac{1.25}{10} = \dfrac{2}{x}$

    $1.25x = 20$

    $x = 16$

    16 miles

45. Solving:

    $\dfrac{1}{54} = \dfrac{1.25}{x}$

    $x = 67.5$

    67.5 inches

47. a. Solving:

    $\dfrac{4}{590} = \dfrac{3}{x}$

    $4x = 1770$

    $x = 442.5$

    $442.50

    b. Solving:

    $\dfrac{4}{590} = \dfrac{5}{x}$

    $4x = 2950$

    $x = 737.5$

    $737.50

49. The amount of fat in a pancake is proportional to the area of the pancake, which is proportional to the square of the pancake's diameter.

    $\dfrac{4^2}{5} = \dfrac{6^2}{x}$

    $16x = 180$

    $x = 11.25$

    11.25 grams

51. The proof is given below:

    $\dfrac{a}{b} = \dfrac{c}{d}$

    $\dfrac{a}{b} + 1 = \dfrac{c}{d} + 1$

    $\dfrac{a}{b} + \dfrac{b}{b} = \dfrac{c}{d} + \dfrac{d}{d}$

    $\dfrac{a+b}{b} = \dfrac{c+d}{d}$

53. a. and b.

    The number of seats per state is AL 7, AK 1, AZ 8, AR 4, CA 53, CO 7, CT 5, DE 1, DC 3, FL 25, GA 13, HI 2, ID 2, IL 19, IN 9, IA 5, KS 4, KY 6, LA 7, ME 2, MD 8, MA 10, MI 15, MN 8, MS 4, MO 9, MT 1, NE 3, NV 3, NH 2, NJ 13, NM 3, NY 29, NC 13, ND 1,

OH 18, OK 5, OR 5, PA 19, RI 2, SC 6, SD 1, TN 9, UT 3, VT 1, VA 11, WA 9, WV 3, WI 8, WY 1. Students will find that the calculated number of representatives per state does not match the actual number of representatives in the following states: AZ, FL, GA, IL, IA, LA, MA, MI, MN, MO, NV, NJ, NY, OH, PA, SC, TX, UT, WA.

EXCURSION EXERCISES, SECTION 9.3

1. A married and filing separate income of $63,850 falls in the range $37,450 to $75,600. The tax is $5,156.25 + 25% of the amount over $37,450.
   Find the amount over $37,450:
   $63,850 – $37,450 = $26,400
   Calculate the tax liability:
   $5,156.25 + 25%($26,400) = $11,756.25
   Joseph's tax liability is $11,756.25.

3. A married and filing jointly income of $58,120 falls in the range $18,450 to $74,900. The tax is $1,845 + 15% of the amount over $18,450. Find the amount over $18,450:
   $58,120 – $18,450 = $39,670
   Calculate the tax liability:
   $1,845 + 15%($39,670) = $7,795.50
   Balance Due: $7,795.50 – $7,124 = $671.50
   Dee's tax liability is $7,795.50, and her balance due is $671.50.

5. No. The taxpayer pays 33% on the amount earned over a given figure.

EXERCISE SET 9.3

1. Answers will vary.

3. $\frac{1}{2} = 0.5 = 50\%$

5. Converting:
   $40\% = 0.4$
   $40\% = 40\left(\frac{1}{100}\right) = \frac{40}{100} = \frac{2}{5}$
   $\frac{2}{5} = 0.4 = 40\%$

7. Converting:
   $0.7 = 70\%$
   $70\% = 70\left(\frac{1}{100}\right) = \frac{70}{100} = \frac{7}{10}$
   $\frac{7}{10} = 0.7 = 70\%$

9. $\frac{11}{20} = 0.55 = 55\%$

11. Converting:
    $15.625\% = 0.15625$
    $15.625\% = 15.625\left(\frac{1}{100}\right) = \frac{15.625}{100}$
    $= \frac{15,625}{100,000} = \frac{5}{32}$
    $\frac{5}{32} = 0.15625 = 15.625\%$

13. 90% is 0.90 Solving:
    $(0.90)(132) = A$
    $119 \approx A$
    119 million returns

15. Solving:
    $(0.15)(358) = A$
    $53.7 = A$
    $53.7 billion

17. Solving:
    $\frac{\text{Percent}}{100} = \frac{\text{amount}}{\text{base}}$
    $\frac{p}{100} = \frac{293}{1236}$
    $1236p = 29,300$
    $p \approx 23.7$
    23.7% were irked most by tailgaters.

19. Solving:
    $\frac{\text{Percent}}{100} = \frac{\text{amount}}{\text{base}}$
    $\frac{16}{100} = \frac{2,240,000}{B}$
    $16B = 224,000,000$
    $B = 14,000,000$
    14,000,000 ounces

21. $\frac{\text{Percent}}{100} = \frac{\text{amount}}{\text{base}}$
    $\frac{22}{100} = \frac{740}{B}$
    $22B = 74,000$
    $B \approx 3364$
    3364 people

23. Solving:
    $\frac{p}{100} = \frac{45}{78}$
    $78p = 4500$
    $p \approx 57.7$
    57.7%

25. a. Using ratios:

$$\frac{x}{100} = \frac{8.1 - 4.0}{4.0}$$

$$4.0x = 410$$

$$x = 102.5$$

102.5% increase

b. Using ratios:

$$\frac{x}{100} = \frac{18.0 - 8.1}{8.1}$$

$$8.1x = 990$$

$$x \approx 122.2$$

122.2% increase

c. Using ratios:

$$\frac{x}{100} = \frac{18.0 - 4.0}{4.0}$$

$$4.0x = 1400$$

$$x = 350$$

350% increase

d. $\frac{18}{4} = 4.5$ times larger. Convert the percent

to a decimal and add 1.

27. Using ratios:

$$\frac{x}{100} = \frac{65.4 - 54.4}{65.4}$$

$$65.4x = 11$$

$$x \approx 16.8$$

16.8% decrease

29. a. Using ratios:

$$\frac{x}{100} = \frac{9,200,000 - 6,300,000}{6,300,000}$$

$$6,300,000x = 2,900,000$$

$$x = 46.0$$

46.0% increase

b. Using ratios:

$$\frac{x}{100} = \frac{9,200,000 - 6,700,000}{9,200,000}$$

$$9,200,000x = 2,500,000$$

$$x = 27.2$$

27.2% decrease

31. $(1.05)(1.06)(1.07) < (1.06)^3$ Your salary is less.

33. a. Solving:

$PB = A$

Since 1% = 1,133,000, then

100% = 113,300,000.

Nielson rating of 10.1 = 10.1%

$0.101(113,300,000) = A$

$11,400,000 \approx A$

11,400,000 TV households

$0.17B = 11,400,000$

$B \approx 67,100,000$

67,100,000 TV households

b. Solving:

$0.056(113,300,000) = A$

$6,300,000 \approx A$

6,300,000 TV households

$0.11B = 6,300,000$

$B \approx 57,300,000$

57,300,000 TV households

c. Solving:

$0.075(113,300,000) = A$

$8,497,500 = A$

$$\frac{19,781,000}{8,497,500} \approx 2.3$$

2.3 people per household

EXCURSION EXERCISES, SECTION 9.4

1. Solving:

$$x^2 - 10 = 3x$$

$$x^2 - 3x - 10 = 3x - 3x$$

$$(x - 5)(x + 2) = 0$$

$$x = -2, \ 5$$

sum: $-2 + 5 = 3 = -\dfrac{(-3)}{1}$

product: $(-2)(5) = -10 = \dfrac{-10}{1}$

3. Solving:

$$3x^2 + 5x = 12$$

$$3x^2 + 5x - 12 = 12 - 12$$

$$(3x - 4)(x + 3) = 0$$

$$x = -3, \ \frac{4}{3}$$

sum: $-3 + \dfrac{4}{3} = -\dfrac{5}{3}$

product: $(-3)\left(\dfrac{4}{3}\right) = -4 = \dfrac{-12}{3}$

5. Solving:

$$x^2 = 6x + 3$$

$$x^2 - 6x - 3 = 6x - 6x + 3 - 3$$

$$x^2 - 6x - 3 = 0$$

$$x = \frac{-(-6) \pm \sqrt{(-6)^2 - 4(1)(-3)}}{2(1)}$$

$$x = \frac{6 \pm \sqrt{48}}{2}$$

$$x = 3 - 2\sqrt{3}, \ 3 + 2\sqrt{3}$$

sum: $3 - 2\sqrt{3} + 3 + 2\sqrt{3} = 6 = -\dfrac{(-6)}{1}$

product: $\left(3 - 2\sqrt{3}\right)\left(3 + 2\sqrt{3}\right) = 9 - 12 = \dfrac{-3}{1}$

7. Solving:
$$4x+1=4x^2$$
$$4x-4x+1-1=4x^2-4x-1$$
$$0=4x^2-4x-1$$
$$x=\frac{-(-4)\pm\sqrt{(-4)^2-4(4)(-1)}}{2(4)}$$
$$x=\frac{4\pm\sqrt{32}}{8}$$
$$x=\frac{1-\sqrt{2}}{2},\ \frac{1+\sqrt{2}}{2}$$
sum: $\dfrac{1-\sqrt{2}}{2}+\dfrac{1+\sqrt{2}}{2}=\dfrac{2}{2}=-\dfrac{(-4)}{4}$

product:$\left(\dfrac{1-\sqrt{2}}{2}\right)\left(\dfrac{1+\sqrt{2}}{2}\right)=\dfrac{1}{4}-\dfrac{1}{2}=\dfrac{-1}{4}$

9. Using the formula:
$$x^2-(-1+6)x+(-1\cdot 6)=0$$
$$x^2-5x-6=0$$

11. Using the formula:
$$x^2-\left(3+\frac{1}{2}\right)x+(3)\left(\frac{1}{2}\right)=0$$
$$2x^2-7x+3=0$$

13. Using the formula:
$$x^2-\left(\frac{1}{4}-\frac{3}{2}\right)x+\left(\frac{1}{4}\right)\left(-\frac{3}{2}\right)=0$$
$$8x^2+10x-3=0$$

15. Using the formula:
$$x^2-(2+\sqrt{2}+2-\sqrt{2})x$$
$$+([2+\sqrt{2}][2-\sqrt{2}])=0$$
$$x^2-4x+2=0$$

EXERCISE SET 9.4

1. Solving:
$$r^2-3r=10$$
$$r^2-3r-10=10-10$$
$$(r-5)(r+2)=0$$
$$r-5=0\quad r+2=0$$
$$r=5\qquad r=-2$$

3. Solving:
$$t^2=t+1$$
$$t^2-t-1=t-t+1-1$$
$$t^2-t-1=0$$
$$t=\frac{-(-1)\pm\sqrt{(-1)^2-4(1)(-1)}}{2(1)}$$
$$t=\frac{1-\sqrt{5}}{2},\ \frac{1+\sqrt{5}}{2}$$
$$t\approx-0.618,\ 1.618$$

5. Solving:
$$y^2-6y=4$$
$$y^2-6y-4=4-4$$
$$y^2-6y-4=0$$
$$y=\frac{-(-6)\pm\sqrt{(-6)^2-4(1)(-4)}}{2(1)}$$
$$y=3-\sqrt{13},\ 3+\sqrt{13}$$
$$y=-0.606,\ 6.606$$

7. Solving:
$$9z^2-18z=0$$
$$9z(z-2)=0$$
$$9z=0\quad z-2=0$$
$$z=0\qquad z=2$$

9. Solving:
$$z^2=z+4$$
$$z^2-z-4=z-z+4-4$$
$$z^2-z-4=0$$
$$z=\frac{-(-1)\pm\sqrt{(-1)^2-4(1)(-4)}}{2(1)}$$
$$z=\frac{1-\sqrt{17}}{2},\ \frac{1+\sqrt{17}}{2}$$
$$z\approx-1.562,\ 2.562$$

11. Solving:
$$2s^2=4s+5$$
$$2s^2-4s-5=4s-4s+5-5$$
$$2s^2-4s-5=0$$
$$s=\frac{-(-4)\pm\sqrt{(-4)^2-4(2)(-5)}}{2(2)}$$
$$s=\frac{2-\sqrt{14}}{2},\ \frac{2+\sqrt{14}}{2}$$
$$s\approx-0.871,\ 2.871$$

13. Solving:
$$r^2 = 4r + 7$$
$$r^2 - 4r - 7 = 4r - 4r + 7 - 7$$
$$r^2 - 4r - 7 = 0$$
$$r = \frac{-(-4) \pm \sqrt{(-4)^2 - 4(1)(-7)}}{2(1)}$$
$$r = 2 - \sqrt{11},\ 2 + \sqrt{11}$$
$$r \approx -1.317,\ 5.317$$

15. Solving:
$$2x^2 = 9x + 18$$
$$2x^2 - 9x - 18 = 9x - 9x + 18 - 18$$
$$(2x + 3)(x - 6) = 0$$
$$2x + 3 = 0 \qquad x - 6 = 0$$
$$x = -\frac{3}{2} \qquad x = 6$$

17. Solving:
$$6x - 11 = x^2$$
$$6x - 6x - 11 + 11 = x^2 - 6x + 11$$
$$0 = x^2 - 6x + 11$$
$$x = \frac{-(-6) \pm \sqrt{(-6)^2 - 4(1)(1)}}{2(1)}$$
$$x = \frac{6 \pm \sqrt{-8}}{2}$$

$\sqrt{-8}$ is not a real number. The equation has no real number solutions.

19. Solving:
$$4 - 15u = 4u^2$$
$$4 - 4 - 15u + 15u = 4u^2 + 15u - 4$$
$$0 = (4u - 1)(u + 4)$$
$$4u - 1 = 0 \qquad u + 4 = 0$$
$$u = \frac{1}{4} \qquad u = -4$$

21. Solving:
$$6y^2 - 4 = 5y$$
$$6y^2 - 5y - 4 = 5y - 5y$$
$$(3y - 4)(2y + 1) = 0$$
$$3y - 4 = 0 \qquad 2y + 1 = 0$$
$$y = \frac{4}{3} \qquad y = -\frac{1}{2}$$

23. Solving:
$$y - 2 = y^2 - y - 6$$
$$y - y - 2 + 2 = y^2 - y - y - 6 + 2$$
$$0 = y^2 - 2y - 4$$
$$y = \frac{-(-2) \pm \sqrt{(-2)^2 - 4(1)(-4)}}{2(1)}$$
$$y = 1 - \sqrt{5},\ 1 + \sqrt{5}$$
$$y = -1.236,\ 3.236$$

25. Answers will vary.

27. Solving:
$$36 = -16t^2 + 60t$$
$$36 - 36 = -16t^2 + 60t - 36$$
$$0 = 4t^2 - 15t + 9$$
$$0 = (4t - 3)(t - 3)$$
$$4t - 3 = 0 \qquad t - 3 = 0$$
$$t = \frac{3}{4} \qquad t = 3$$

The height of the ball will be 36 feet at 0.75 second after the ball is hit and 3 seconds after the ball is hit.

29. Verifying:
$$T(1) = 0.5(1)^2 + 0.5(1) = 1$$
$$T(2) = 0.5(2)^2 + 0.5(2) = 3$$
$$T(3) = 0.5(3)^2 + 0.5(3) = 6$$
$$T(4) = 0.5(4)^2 + 0.5(4) = 10$$

$$55 = 0.5n^2 + 0.5n$$
$$55 - 55 = 0.5n^2 + 0.5n - 55$$
$$0 = 0.5n^2 + 0.5n - 55$$
$$0 = n^2 + n - 110$$
$$0 = (n + 11)(n - 10)$$
$$n + 11 = 0 \qquad n - 10 = 0$$
$$n = -11 \qquad n = 10$$

The number of rows is 10.

31. Solving:
$$y = 0.03x^2 + 0.36x + 34.6$$
$$50 = 0.03x^2 + 0.36x + 34.6$$
$$0 = 0.03x^2 + 0.36x - 15.4$$
$$x = \frac{-0.36 \pm \sqrt{(0.36)^2 - 4(0.03)(-15.4)}}{2(0.03)}$$
$$x = \frac{-0.36 + \sqrt{1.9776}}{0.06},\ x = \frac{-0.36 - \sqrt{1.9776}}{0.06}$$
$$x \approx 17$$

Since $x = 0$ corresponds to the year 2000, $x = 17$ corresponds to the year 2017.

33. a. $d = 0.05(60)^2 + (60) = 240$ feet

    b.   Solving:
$$75 = 0.05r^2 + r$$
$$75 - 75 = 0.05r^2 + r - 75$$
$$0 = 0.05r^2 + r - 75$$
$$0 = r^2 + 20r - 1500$$
$$0 = (r + 50)(r - 30)$$
$$r + 50 = 0 \qquad r - 30 = 0$$
$$r = -50 \qquad r = 30$$
The car was going 30 miles per hour.

35. Solving:
$$0 = -16t^2 + 88t + 1$$
$$t = \frac{-88 \pm \sqrt{(88)^2 - 4(-16)(1)}}{2(-16)}$$
$$t = \frac{11 - \sqrt{122}}{4}, \ \frac{11 + \sqrt{122}}{4}$$
$$t \approx -0.01, \ 5.51$$
The hang time of the football is 5.51 seconds.

37. Solving:
$$5 = -0.005x^2 + 1.2x + 10$$
$$5 - 5 = -0.005x^2 + 1.2x + 10 - 5$$
$$0 = -0.005x^2 + 1.2x + 5$$
$$0 = x^2 - 240x - 1000$$
$$x = \frac{-(-240) \pm \sqrt{(-240)^2 - 4(1)(-1000)}}{2(1)}$$
$$x = 120 - 10\sqrt{154}, \ 120 + 10\sqrt{154}$$
$$x \approx -4.10, \ 244.10$$
The water is 244.10 feet away from the tugboat.

39. $-0.002(36)^2 + 0.36(36) \approx 10.37 > 8$
When the ball reaches the goal, it will be too high off the ground to go into the net.

41. When $h = 0$, the water has traveled its maximum horizontal distance.
$$0 = -0.006x^2 + 1.2x + 10$$
$$x = \frac{-1.2 \pm \sqrt{(1.2)^2 - 4(-0.006)(10)}}{2(-0.006)}$$
$$x \approx -8.012 \text{ and } 208.012$$
The solution -8.012 does not make sense in this context. The water travels 208.012 feet horizontally before falling into the river. Since $208.012 < 220$, the people will not get wet.

43. Substituting $x = 130$:
$y = 0.00657(130)^2 - 0.330(130) - 0.0633 \approx 68.1$
A first-class stamp will be 68 cents.

45. If the discriminant is not a perfect square, the radical expression in the quadratic formula will not simplify to a whole number.

47. Solving:
$$x^2 - 8bx + 15b^2 = 0$$
$$(x - 3b)(x - 5b) = 0$$
$$x - 3b = 0 \qquad x - 5b = 0$$
$$x = 3b \qquad x = 5b$$

49. Solving:
$$2x^2 - xy - 3y^2 = 0$$
$$(2x - 3y)(x + y) = 0$$
$$2x - 3y = 0 \qquad x + y = 0$$
$$x = \frac{3y}{2} \qquad x = -y$$

CHAPTER 9 REVIEW EXERCISES

1.   Solving:
$$5x + 3 = 10x - 17$$
$$5x - 5x + 3 + 17 = 10x - 5x - 17 + 17$$
$$20 = 5x$$
$$x = 4$$

2.   Solving:
$$3x + \frac{1}{8} = \frac{1}{2}$$
$$3x + \frac{1}{8} - \frac{1}{8} = \frac{1}{2} - \frac{1}{8}$$
$$3x = \frac{3}{8}$$
$$x = \frac{1}{8}$$

3.   Solving:
$$6x + 3(2x - 1) = -27$$
$$6x + 6x - 3 + 3 = -27 + 3$$
$$12x = -24$$
$$x = -2$$

4.   Solving:
$$\frac{5}{12} = \frac{n}{8}$$
$$12n = 5(8)$$
$$n = \frac{40}{12} = \frac{10}{3}$$

5.   Solving:
$$4y^2 + 9 = 0$$
$$4y^2 + 9 - 9 = -9$$
$$4y^2 = -9$$
$$y^2 = \frac{-9}{4}$$
$$y = \sqrt{\frac{-9}{4}}$$
$\sqrt{\frac{-9}{4}}$ is not a real number. The equation has no real solutions.

6. Solving:

$$x^2 - x = 30$$
$$x^2 - x - 30 = 30 - 30$$
$$(x+5)(x-6) = 0$$
$$x+5 = 0 \qquad x-6 = 0$$
$$x = -5 \qquad x = 6$$

7. Solving:

$$x^2 = 4x - 1$$
$$x^2 - 4x + 1 = 4x - 4x - 1 + 1$$
$$x^2 - 4x + 1 = 0$$
$$x = \frac{-(-4) \pm \sqrt{(-4)^2 - 4(1)(1)}}{2(1)}$$
$$x = 2 + \sqrt{3}, \ 2 - \sqrt{3}$$

8. Solving:

$$x + 3 = x^2$$
$$0 = x^2 - x - 3$$
$$x = \frac{-(-1) \pm \sqrt{(-1)^2 - 4(1)(-3)}}{2(1)}$$
$$x = \frac{1 + \sqrt{13}}{2}, \ \frac{1 - \sqrt{13}}{2}$$

9. Solving:

$$4x + 3y = 12$$
$$4x - 4x + 3y = 12 - 4x$$
$$\frac{1}{3} \cdot 3y = \frac{1}{3}(12 - 4x)$$
$$y = 4 - \frac{4x}{3}$$

10. Solving:

$$f = v + at$$
$$f - v = v - v + at$$
$$\frac{f - v}{a} = \frac{1}{a}(at)$$
$$t = \frac{f - v}{a}$$

11. Solving:

$$101 = -0.005x + 113.25$$
$$101 - 113.25 = -0.005x + 113.25 - 113.25$$
$$0.005x = 12.25$$
$$x = 2450$$

The elevation is 2450 feet.

12. Solving:

$$100 = 4 + 32t$$
$$100 - 4 = 4 - 4 + 32t$$
$$32t = 96$$
$$t = 3$$

It takes 3 seconds for the velocity to increase from 4 feet per second to 100 feet per second.

13. Solving:

$$100(80 - T) = 50(T - 20)$$
$$8000 - 100T = 50T - 1000$$
$$8000 + 1000 - 100T = 50T + 1000 - 1000$$
$$9000 + 100T - 100T = 50T + 100T$$
$$9000 = 150T$$
$$T = 60$$

The final temperature is 60ºC.

14. Let h be the number of hours worked, and c the number of calls.

$$159.75 = 12h + 0.75c$$
$$159.75 = 12(8) + 0.75c$$
$$159.75 = 96 + 0.75c$$
$$63.75 = 0.75c$$
$$85 = c$$

Completed 85 calls.

15. Solving:

$$\frac{326.6}{11.5} = \frac{x}{1}$$
$$326.6(1) = 11.5x$$
$$x = 28.4 \text{ miles per gallon}$$

16. $\dfrac{350,000 - 280,000}{280,000} = \dfrac{70,000}{280,000} = \dfrac{1}{4}$

17. Using ratios:

$$\frac{x}{100} = \frac{400 - 80}{80}$$
$$80x = 32,000$$
$$x = 400$$

400% increase

18. a. New York: $\dfrac{8,400,000}{321.8} \approx 26,103$

   Los Angeles: $\dfrac{3,900,000}{467.4} \approx 8344$

   Chicago: $\dfrac{2,900,000}{228,469} \approx 12,693$

   Houston: $\dfrac{2,300,000}{594.03} \approx 3872$

   Phoenix: $\dfrac{1,600,000}{136} \approx 11,765$

   New York, Chicago, Phoenix, Los Angeles, Houston

   b. $26,103 - 3872 = 22,231$ more people per square mile

19. a. $\dfrac{156 + 215}{56} \approx 7$

   7:1, or 7 to 1. There are 7 students per faculty member at Prescott College.

b. Arizona State: $\dfrac{20,309+15,955}{2018} \approx 18$

Embry-Riddle: $\dfrac{1441+428}{93} \approx 20$

NAU: $\dfrac{8215+10,958}{1055} \approx 18$

Arizona: $\dfrac{14,054+15,475}{2343} \approx 13$

Prescott College has the lowest student-faculty ratio. Embry-Riddle Aeronautical University has the highest.

c. Arizona State University and the Northern Arizona University have the same student-faculty ratio.

20. Using ratios:

$\dfrac{3}{3+7} = \dfrac{a}{350,000}$

$10a = 3(350,000) = 1,050,000$

$a = 105,000$

Department A has \$105,000 allocated, and Department B has 350,000 – 105,000 = \$245,000 allocated.

21. Using ratios:

$\dfrac{3}{4} = \dfrac{t}{10}$

$4t = 3(10) = 30$

$t = 7.5$

7.5 tablespoons of fertilizer are required.

22. a. More.

Health care $= \dfrac{21}{21+20+20+14+19+6}$

$= \dfrac{21}{100}$

$\dfrac{1}{5}$ is less than $\dfrac{21}{100}$.

b. $\dfrac{21+20+14+19+6}{20} = \dfrac{80}{20} = \dfrac{4}{1}$

$4:1$

c. Using ratios:

$\dfrac{80}{80+20} = \dfrac{x}{3.6}$

$100x = 3.6(80)$

$100x = 288$

$x = 2.88$

\$2.88 trillion will be spent on fixed expenditures.

d. Using ratios:

$\dfrac{20}{100} = \dfrac{x}{3.6}$

$100x = 72$

$x = 0.72$

\$0.72 trillion or \$720 billion will be spent on fixed expenditures.

23. a. $\dfrac{170,931}{170,931+164,119} \approx 0.51 = 51.0\%$

51.0% of the projected population in 2025 is female.

b. 2050:

$\dfrac{200,696}{200,696+193,234} \approx 0.509 = 50.9\%$

Less than one percent.

24. Using ratios:

$\dfrac{35}{100} = \dfrac{7}{B}$

$35B = 700$

$B = 20$

20 billion hotdogs

25. Using ratios:

$\dfrac{x}{100} = \dfrac{36,598}{27,167}$

$27,167x = 3,659,800$

$x \approx 97.9$

97.9%

26. a. Using ratios:

$\dfrac{p}{100} = \dfrac{24-4}{24}$

$24p = 20(100) = 2000$

$p = 83.\overline{3}$

The fat content is decreased by $83.\overline{3}\%$.

b. The cholesterol content is decreased by 100%.

c. Using ratios:

$\dfrac{p}{100} = \dfrac{280-140}{280}$

$280p = 140(100) = 14,000$

$p = 50$

The calorie content is decreased by 50%.

27. Using ratios:

$\dfrac{x}{100} = \dfrac{26.6-23.1}{26.6}$

$26.6x = 350$

$x \approx 13.2$

13.2% decrease

28. Using ratios:

$\dfrac{p}{100} = \dfrac{38}{38+38+24+9}$

$\dfrac{p}{100} = \dfrac{38}{109}$

$109p = 3800$

$p \approx 35$

35%

29. Solving:
$$0.11(64,000) = A$$
$$7040 = A$$
$7040

30. Let d be the number of vacation days in United States.
$$3 + 3d = 42$$
$$3d = 39$$
$$d = 13$$
13 days

31. Solving:
$$25 = 30t - 5t^2$$
$$25 - 25 = -25 + 30t - 5t^2$$
$$0 = t^2 - 6t + 5$$
$$0 = (t - 1)(t - 5)$$
$$t - 1 = 0 \quad t - 5 = 0$$
$$t = 1 \qquad t = 5$$
The rocket is 25 meters above the cliff 1 second and at 5 seconds after it is shot.

32. Solving:
$$18 = -16t^2 + 32t + 6$$
$$18 - 18 = -16t^2 + 32t + 6 - 18$$
$$0 = 4t^2 - 8t + 3$$
$$0 = (2t - 3)(2t - 1)$$
$$t = 0.5, \ 1.5$$
The ball is 18 feet off the ground 0.5 second and 1.5 seconds after it is thrown.

CHAPTER 9 TEST

1. Solving:
$$\frac{x}{4} - 3 = \frac{1}{2}$$
$$\frac{x}{4} - 3 + 3 = \frac{1}{2} + 3$$
$$4 \cdot \frac{x}{4} = 4 \cdot \frac{7}{2}$$
$$x = 14$$

2. Solving:
$$x + 5(3x - 20) = 10(x - 4)$$
$$16x - 100 = 10x - 40$$
$$16x - 10x - 100 = 10x - 10x - 40$$
$$6x - 100 + 100 = 100 - 40$$
$$6x = 60$$
$$x = 10$$

3. Solving:
$$\frac{7}{16} = \frac{x}{12}$$
$$16x = 84$$
$$x = \frac{84}{16} = \frac{21}{4}$$

4. Solving:
$$x^2 = 12x - 27$$
$$x^2 - 12x + 27 = 12x - 12x - 27 + 27$$
$$(x - 3)(x - 9) = 0$$
$$x - 3 = 0 \quad x - 9 = 0$$
$$x = 3 \qquad x = 9$$

5. Solving:
$$3x^2 - 4x = 1$$
$$3x^2 - 4x - 1 = 1 - 1$$
$$3x^2 - 4x - 1 = 0$$
$$x = \frac{-(-4) \pm \sqrt{(-4)^2 - 4(3)(-1)}}{2(3)}$$
$$x = \frac{2 + \sqrt{7}}{3}, \ \frac{2 - \sqrt{7}}{3}$$
$$x \approx -0.215, \ 1.549$$

6. Solving:
$$x - 2y = 15$$
$$x - x - 2y = 15 - x$$
$$-\frac{1}{2}(-2y) = -\frac{1}{2}(15 - x)$$
$$y = \frac{x - 15}{2}, \text{ or } y = \frac{1}{2} - \frac{15}{2}$$

7. Solving:
$$C = \frac{5}{9}(F - 32)$$
$$\frac{9}{5}C = F - 32$$
$$F = \frac{9}{5}C + 32$$

8. Solving:
$$63 = 12.4L + 32$$
$$12.4L = 63 - 32 = 31$$
$$L = 2.5$$
The duration of the last eruption was 2.5 minutes.

9. Let n be the number of kilowatt-hours used. $[n > 300]$
$$74.25 = (0.16)(300) + 0.2(n - 300)$$
$$74.25 = 48 + 0.2n - 60$$
$$74.25 = 0.2n - 12$$
$$86.25 = 0.2n$$
$$431.25 = n$$
431.25 kilowatt-hours used.

10. Using ratios:
$$\frac{246.6}{4.5} = \frac{x}{1}$$
$$(246.6)(1) = 4.5x$$
$$x = 54.8 \text{ miles per hour}$$

11. Solving:
$$4218 = 3 + 5c$$
$$5c = 4215$$
$$c = 843$$
Central Park has 843 acres.

12. a.   Ty Cobb: $\dfrac{11,429}{4191} \approx 2.727$

Billy Hamilton: $\dfrac{6284}{2163} \approx 2.905$

Roger Hornsby: $\dfrac{8137}{2930} \approx 2.777$

Joe Jackson: $\dfrac{4981}{1774} \approx 2.808$

Tris Speaker: $\dfrac{10,195}{3514} \approx 2.901$

Ted Williams: $\dfrac{7706}{2654} \approx 2.904$

b.   Ty Cobb, Roger Hornsby, Joe Jackson, Tris Speaker, Ted Williams, Billy Hamilton

13. Using ratios:
$$\frac{80}{100} = \frac{A}{320}$$
$$100A = 25,600$$
$$A = 256$$
256 million active users

14. Using ratios:
$$\frac{5}{15+3} = \frac{x}{360,000}$$
$$8x = (5)(360,000)$$
$$8x = 1,800,000$$
$$x = 225,000$$
The amounts are $225,000 and $360,000 - 225,000 = \$135,000$.

15. Using ratios:
$$\frac{0.5}{50} = \frac{x}{275}$$
$$50x = 137.5$$
$$x = 2.75$$
2.75 pounds of plant food should be used.

16. a.   Detroit, Michigan

b.   Using ratios:
$$\frac{13.34}{1000} = \frac{x}{624,000}$$
$$1000x = 8,324,160$$
$$x \approx 8324$$
There are about 8324 violent crimes committed in Baltimore.

17. $\dfrac{172,481}{1,200,000} \approx 0.144$

14.4% of registrations are Lab retrievers.

18. Using ratios:
$$\frac{x}{100} = \frac{74,779,000 - 48,046,000}{48,046,000}$$
$$48,046,000x = 2,673,300,000$$
$$x \approx 55.7$$
55.7% increase

19. a.   Using ratios:
$$\frac{x}{100} = \frac{598.9 - 125.6}{598.9}$$
$$598.9x = 47,330$$
$$x \approx 79.0$$
79.0% decrease

b. $(1-0.79)(125.6 \text{ million}) \approx \$26.4$ million

20. Solving:
$$10 = -16t^2 + 28t + 6$$
$$10 - 10 = -16t^2 + 28t + 6 - 10$$
$$0 = -16t^2 + 28t - 4$$
$$0 = -4t^2 - 7t + 1$$
$$t = \frac{-(-7) \pm \sqrt{(-7)^2 - 4(4)(1)}}{2(4)}$$
$$t = \frac{7 + \sqrt{33}}{8}, \ \frac{7 - \sqrt{33}}{8}$$
$$t \approx 1.6, \ 0.2$$
The shot put is 10 feet above the ground at 0.2 second and 1.6 seconds.

Chapter 10: Applications of Functions

EXCURSION EXERCISES, SECTION 10.1

1. a.

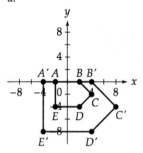

b.

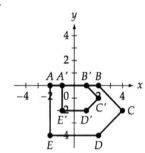

3. Drawings will vary.

5. The center of dilation is the center of the edge of the paper along the width.

EXERCISE SET 10.1

1.

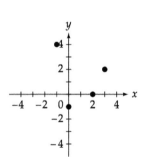

3.

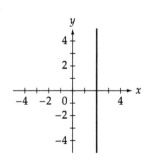

5.

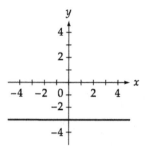

7.

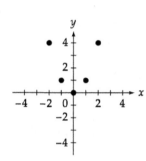

9.

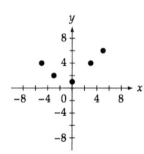

11.

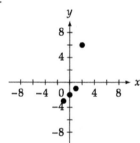

13.

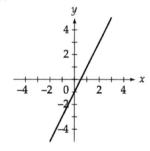

15.

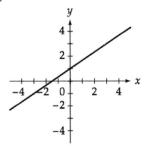

17.

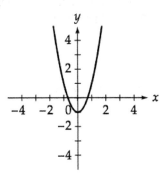

19.

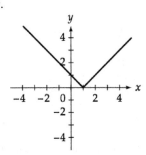

21. $f(x) = 2x + 7$
$f(-2) = 2(-2) + 7 = 3$

23. $f(t) = t^2 - t - 3$
$f(3) = 3^2 - 3 - 3 = 3$

25. $T(p) = \dfrac{p^2}{p-2}$

$T(0) = \dfrac{0^2}{0-2} = 0$

27.   a. $P(s) = 4s$
$P(4) = 4(4) = 16$
The perimeter of the square is 16 meters.

b. $P(s) = 4s$
$P(5) = 4(5) = 20$
The perimeter of the square is 20 feet.

29.   a. $h(t) = -16t^2 + 80t + 4$
$h(2) = -16(2)^2 + 80(2) + 4 = 100$
The height of the ball is 100 feet.

31.   a. $s(t) = \dfrac{1087\sqrt{t+273}}{16.52}$

$s(0) = \dfrac{1087\sqrt{0+273}}{16.52} \approx 1087$

The speed is approximately 1087 feet per second.

b. $s(t) = \dfrac{1087\sqrt{t+273}}{16.52}$

$s(25) = \dfrac{1087\sqrt{25+273}}{16.52} \approx 1136$

The speed is approximately 1136 feet per second.

33.   a. $T(L) = 2\pi\sqrt{\dfrac{L}{32}}$

$T(3) = 2\pi\sqrt{\dfrac{3}{32}} \approx 1.93$

The time is approximately 1.92 seconds.

b.   Since L is measured in feet, change 9 inches to 0.75 feet.

$T(L) = 2\pi\sqrt{\dfrac{L}{32}}$

$T(0.75) = 2\pi\sqrt{\dfrac{0.75}{32}} \approx 1.0$

The time is approximately 1.0 seconds.

35.

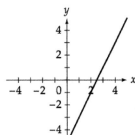

37.

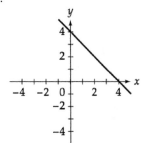

39.

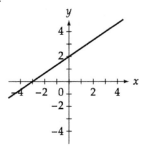

41.

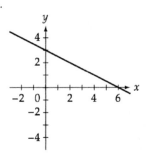

43.

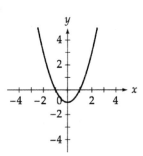

45.

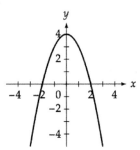

47.

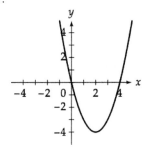

49.

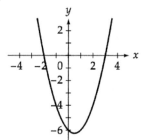

51. No. A function cannot have different elements in the range corresponding to one element in the domain.

53. $f(x) = 2x + 5$
$f(a) = 2a + 5$
Since $f(a) = 9$, $2a + 5 = 9$, so $a = 2$.

55. $f(a, b) = a + b$
$f(2, 5) = 2 + 5 = 7$
$g(a, b) = (a)(b)$
$g(2, 5) = (2)(5) = 10$
$f(2, 5) + g(2, 5) = 7 + 10 = 17$

EXCURSION EXERCISES, SECTION 10.2

1. A car moving to the right has a positive velocity. On a distance-time graph, positive velocity is indicated by positive slope. This graph has a positive slope from 0 to 2 hours, so the car is moving to the right from 0 to 2 hours.

3. A car moving to the left has a negative velocity. On a distance-time graph, negative velocity is indicated by negative slope. This graph has a negative slope from 2 to 6 hours, so the car is moving to the left from 2 to 6 hours.

5. The height of graph at $t = 2$ hours is $d = 120$ miles. The car is 120 miles from its starting position after 2 hours.

7. The velocity of the car during its first 2 hours is equal to the slope of the graph from $t = 0$ to $t = 2$.
$$m = \frac{\text{change in } d}{\text{change in } t} = \frac{120 - 0}{2 - 0} = \frac{120}{2} = 60 \text{ mph}$$

9. The slope of a horizontal line is zero.

11. Where the line is steepest, the absolute value of the slope and hence the absolute value of the velocity is greatest. This occurs on the interval from 2 hours to 3 hours.

EXERCISE SET 10.2

1. Solving:
    $$f(x) = 3x - 6$$
    $$0 = 3x - 6$$
    $$3x = 6$$
    $$x = 2$$
    x-intercept: $(2, 0)$
    $$f(x) = 3x - 6$$
    $$f(0) = 3(0) - 6 = -6$$
    y-intercept: $(0, -6)$

3. Solving:
    $$y = \frac{2}{3}x - 4$$
    $$0 = \frac{2}{3}x - 4$$
    $$-\frac{2}{3}x = -4$$
    $$x = 6$$
    x-intercept: $(6, 0)$
    $$y = \frac{2}{3}x - 4$$
    $$y = \frac{2}{3}(0) - 4 = -4$$
    y-intercept: $(0, -4)$

5. Solving:
    $$y = -x - 4$$
    $$0 = -x - 4$$
    $$x = -4$$
    x-intercept: $(-4, 0)$
    $$y = -x - 4$$
    $$y = -0 - 4 = -4$$
    y-intercept: $(0, -4)$

7. Solving:
    $$3x + 4y = 12$$
    $$3x + 4(0) = 12$$
    $$3x = 12$$
    $$x = 4$$
    x-intercept: $(4, 0)$
    $$3x + 4y = 12$$
    $$3(0) + 4y = 12$$
    $$4y = 12$$
    $$y = 3$$
    y-intercept: $(0, 3)$

9. Solving:
    $$2x - 3y = 9$$
    $$2x - 3(0) = 9$$
    $$2x = 9$$
    $$x = \frac{9}{2}$$
    x-intercept: $\left(\frac{9}{2}, 0\right)$

$$2x - 3y = 9$$
$$2(0) - 3y = 9$$
$$-3y = 9$$
$$y = -3$$
y-intercept: $(0, -3)$

11. Solving:
    $$\frac{x}{2} + \frac{y}{3} = 1$$
    $$\frac{x}{2} + \frac{0}{3} = 1$$
    $$\frac{x}{2} = 1$$
    $$x = 2$$
    x-intercept: $(2, 0)$
    $$\frac{x}{2} + \frac{y}{3} = 1$$
    $$\frac{0}{2} + \frac{y}{3} = 1$$
    $$\frac{y}{3} = 1$$
    $$y = 3$$
    y-intercept: $(0, 3)$

13. Solving:
    $$x - \frac{y}{2} = 1$$
    $$x - \frac{0}{2} = 1$$
    $$x = 1$$
    x-intercept: $(1, 0)$
    $$x - \frac{y}{2} = 1$$
    $$0 - \frac{y}{2} = 1$$
    $$-\frac{y}{2} = 1$$
    $$y = -2$$
    y-intercept: $(0, -2)$

15. Finding the intercept:
    $$f(x) = 7x - 30$$
    $$0 = 7x - 30$$
    $$-7x = -30$$
    $$x = \frac{30}{7}$$
    x-intercept: $\left(\frac{30}{7}, 0\right)$; At $\frac{30}{7}^{\circ}$ C, the cricket stops chirping.

17. Finding the intercepts:
    $$T(x) = 3x - 15$$
    $$0 = 3x - 15$$
    $$-3x = -15$$
    $$x = 5$$

The intercept on the horizontal axis is (5, 0).
This means that it takes 5 minutes for the
temperature of the object to reach 0°F.

$T(x) = 3x - 15$
$T(0) = 3(0) - 15$
$T(0) = -15$

The intercept on the vertical axis is (0, –15).
This means that the temperature of the object is
–15°F before it is removed from the freezer.

19. $m = \dfrac{1-3}{3-1} = \dfrac{-2}{2} = -1$

21. $m = \dfrac{5-4}{2-(-1)} = \dfrac{1}{3}$

23. $m = \dfrac{5-3}{(-4)-(-1)} = \dfrac{2}{-3} = -\dfrac{2}{3}$

25. $m = \dfrac{0-3}{4-0} = \dfrac{-3}{4} = -\dfrac{3}{4}$

27. $m = \dfrac{(-2)-4}{2-2} = \dfrac{-6}{0}$; the slope is undefined.

29. $m = \dfrac{(-2)-5}{(-3)-2} = \dfrac{-7}{-5} = \dfrac{7}{5}$

31. $m = \dfrac{3-3}{(-1)-2} = \dfrac{0}{-3} = 0$

33. $m = \dfrac{5-4}{(-2)-0} = \dfrac{1}{-2} = -\dfrac{1}{2}$

35. $m = \dfrac{240-80}{6-2} = \dfrac{160}{4} = 40$

    The slope is 40, which means the motorist was
    traveling at 40 miles per hour.

37. Finding slope:
    $$m = \dfrac{20{,}000-5000}{10-25} = \dfrac{15{,}000}{-15} = -1000$$
    A slope of –1000 means that the height of the
    plane is decreasing at the rate of 1,000 ft/min.

39. Finding slope:
    $$m = \dfrac{3.7-1.8}{2-5} = \dfrac{1.9}{-3} = -\dfrac{19}{30} = 0.6\overline{3}$$

    A slope of $-\dfrac{19}{30}$ means that each minute the

    water in the lock decreases by $0.6\overline{3}$ million
    gallons.

41.
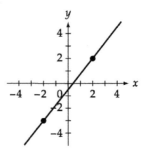

43.
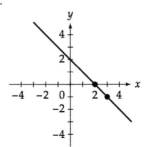

45.
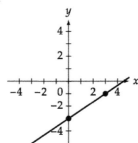

47.
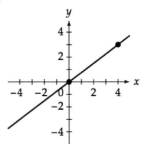

49.
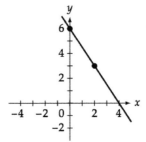

51. a. Line *A* represents the depth of the water in can 1, since after *x* minutes the water will be deeper in the can with the smaller diameter.

    b. When $x = 15$, line *A* has a value of 15 and line *B* has a value of 10. So the difference in the depths of the two cans after 15 seconds is $15 - 10 = 5$ millimeters.

53. Graph the point $(-1, 2)$ and then find another point on the line with slope -3 by moving 3 units down and 1 unit to the right. Then move 3 units down and 1 unit right again. The value when $x = 1$ is $y = -4$

55. It rotates the line clockwise.

57. It lowers the line on the rectangular coordinate system.

EXCURSION EXERCISES, SECTION 10.3

1. The total cost of the building is $7000 + \$6500$ $= \$13,500$. Each cartridge costs \$28. The linear function is $C = 28n + 13,500$.

3. The business will break even when revenue equals cost.
$$R = C$$
$$45n = 28n + 13,500$$
$$17n = 13,500$$
$$n \approx 794$$
The business must restore and sell approximately 794 cartridges annually.

5. The number of cartridges sold for the year is $n = 25(260) = 6500$.
The total revenue is $R = 45(6500) = \$292,500$.
The total cost is
$C = 28(6500) + 13,500 = \$195,500$.
The difference between revenue and cost is the profit: $292,500 - 195,500 = 97,000$
The two entrepreneurs worked 10 hours a day for 260 days, so together they worked $2(260)(10) = 5200$ hours. The hourly wage is:
$\$97,000 \div 5200$ hours $\approx \$18.65$ per hour

7. The \$400 per month spent on advertising will raise costs by $12(400) = \$4800$ per year.
The cost function is $C = 28n + 13,500 + 4800$ or $C = 28n + 18,300$.
The revenue function will not change, since the \$4800 is added to the costs. So, $R = 45n$.

The business breaks even when revenue equals cost.
$$R = C$$
$$45n = 28n + 18,300$$
$$17n = 18,300$$
$$n \approx 1076$$
Approximately 1076 cartridges must be sold to break even.
The additional \$4800 is added to the costs. So the total cost is now
$\$195,500 + \$4800 = \$200,300$.
The revenue is the same, \$292,500.
The difference is
$\$292,500 - \$200,300 = \$92,200$.
The entrepreneurs worked 5200 hours, so the hourly wage will be:
$\$92,200 \div 5200$ hours $\approx \$17.73$ per hour

EXERCISE SET 10.3

1. $y - y_1 = m(x - x_1)$
$y - 5 = 2(x - 0)$
$y - 5 = 2x$
$\quad y = 2x + 5$

3. $y - y_1 = m(x - x_1)$
$y - 7 = -3(x - (-1))$
$y - 7 = -3(x + 1)$
$y - 7 = -3x - 3$
$\quad y = -3x + 4$

5. $y - y_1 = m(x - x_1)$
$y - 5 = -\dfrac{2}{3}(x - 3)$
$y - 5 = -\dfrac{2}{3}x + 2$
$\quad y = -\dfrac{2}{3}x + 7$

7. $y - y_1 = m(x - x_1)$
$y - (-3) = 0(x - (-2))$
$\quad y + 3 = 0$
$\qquad y = -3$

9. Find the slope of the line.
$$m = \frac{y_2 - y_1}{x_2 - x_1} = \frac{5 - 2}{3 - 0} = 1$$
Use the point-slope formula to find the equation of the line.
$y - y_1 = m(x - x_1)$
$y - 2 = 1(x - 0)$
$\quad y = x + 2$

11. Find the slope of the line.

$$m = \frac{y_2 - y_1}{x_2 - x_1} = \frac{0 - 3}{2 - 0} = -\frac{3}{2}$$

Use the point-slope formula to find the equation of the line.

$$y - y_1 = m(x - x_1)$$
$$y - 3 = -\frac{3}{2}(x - 0)$$
$$y = -\frac{3}{2}x + 3$$

13. Find the slope of the line.

$$m = \frac{y_2 - y_1}{x_2 - x_1} = \frac{-1 - 0}{0 - 2} = \frac{1}{2}$$

Use the point-slope formula to find the equation of the line.

$$y - y_1 = m(x - x_1)$$
$$y - 0 = \frac{1}{2}(x - 2)$$
$$y = \frac{1}{2}x - 1$$

15. Find the slope of the line.

$$m = \frac{y_2 - y_1}{x_2 - x_1} = \frac{(-5) - 5}{2 - (-2)} = -\frac{10}{4} = -\frac{5}{2}$$

Use the point-slope formula to find the equation of the line.

$$y - y_1 = m(x - x_1)$$
$$y - 5 = -\frac{5}{2}(x - (-2))$$
$$y = -\frac{5}{2}x$$

17. Let $f(x)$ represent the monthly cost of the phone, and x represent the number of minutes used.

$$f(x) = 0.59x + 4.95$$

Evaluate the function when $x = 13$.

$$f(13) = 0.59(13) + 4.95 = 12.62$$

The cost of using the cellular phone for 13 minutes in 1 month is $12.62.

19. Let y represent the cost of building a house and let x represent the number of square feet of floor space. Then $y = 30,000$ when $x = 0$. So, the y-intercept is $(0, 30,000)$. The cost of the house increases by $85 for each square foot of floor space. Therefore, the slope is 85.

$$y = mx + b$$
$$y = 85x + 30,000$$

Evaluate the function when $x = 1800$.

$$C(x) = 85x + 30,000$$
$$C(1800) = 85(1800) + 30,000 = 183,000$$

The cost to build a house with 1800 square feet is $183,000.

21. Let N represent the number of economy cars that would be sold per month and let x be the price of the truck. Then when $x = 18,000$, $N = 50,000$, and when $x = 17,500$, $N = 55,000$. So the ordered pairs are $(18,000, 50,000)$ and $(17,500, 55,000)$.

Find the slope.

$$m = \frac{N_2 - N_1}{x_2 - x_1} = \frac{55,000 - 50,000}{17,500 - 18,000} = -10$$

Use the point-slope formula to find the equation of the line.

$$N - N_1 = m(x - x_1)$$
$$N - 50,000 = -10(x - 18,000)$$
$$N = -10x + 230,000$$

Evaluate the function when $x = 17,000$ dollars.

$$N(x) = -10x + 230,000$$
$$N(17,000) = -10(17,000) + 230,000$$
$$= 60,000$$

At a price of $ 17,000, there should be 60,000 economy cars sold.

23. a. Find the slope.

$$m = \frac{y_2 - y_1}{x_2 - x_1} = \frac{76 - 64}{10.0 - 8.5} = 8$$

Use the point-slope formula to find the equation of the line.

$$y - y_1 = m(x - x_1)$$
$$y - 86 = 8(x - 10.0)$$
$$y = 8x - 4$$
$$f(x) = 8x - 4$$

b. Evaluate the equation when $x = 10.5$

$$f(x) = 8x - 4$$
$$f(10.5) = 8(10.5) - 4 = 80$$

A student with a reading test grade of 10.5 has a history final exam grade of 80.

25. a. Find the slope.

$$m = \frac{y_2 - y_1}{x_2 - x_1} = \frac{2300 - 600}{60 - 10} = \frac{1700}{50} = 34$$

Use the point-slope formula to find the equation of the line.

$$y - y_1 = m(x - x_1)$$
$$y - 600 = 34(x - 10)$$
$$y = 34x + 260$$
$$f(x) = 34x + 260$$

b. Evaluate the equation when $x = 25$.

$$f(x) = 34x + 260$$
$$f(25) = 34(25) + 260 = 1110$$

After playing a tennis match for 25 minutes, a tennis player's water loss is 1110 milliliters.

27. a. Find the slope.
$$m = \frac{y_2 - y_1}{x_2 - x_1} = \frac{21 - 80}{56 - 22} = \frac{-59}{34} \approx -1.735$$
Use the point-slope formula to find the equation of the line.
$$y - y_1 = m(x - x_1)$$
$$y - 80 = -1.735(x - 22)$$
$$y = -1.735x + 118.17$$
$$f(x) = -1.735x + 118.17$$

 b. Evaluate the equation when $x = 45$.
$$f(x) = -1.735x + 118.17$$
$$f(45) = -1.735(45) + 118.17 = 40.095$$
$$\approx 40$$
At 45°N latitude, the expected maximum temperature in January is 40°F.

29. Answers will vary.
Any points on the line $y = -x + 3$.
For example (0, 3), (1, 2), and (3, 0).

31. Solutions to the same linear equation lie on the same line. The equation of the line through the first two points, (0, 1), and (4, 9), is $y = 2x + 1$. When $x = 3$, $y = 2(3) + 1 = 7$, so $n = 7$.

33. No. The three points do not lie on a straight line.

35. Let y represent the steepness of the hill and let x represent the speed of the car. Then when $x = 77$, $y = 5$, and when $x = 154$, $y = -2$.
Find the slope.
$$m = \frac{5 - (-2)}{77 - 154} = -\frac{7}{77} = -\frac{1}{11}$$
Use the point-slope formula to find the equation of the line.
$$y - y_1 = m(x - x_1)$$
$$y - 5 = -\frac{1}{11}(x - 77)$$
$$y = -\frac{1}{11}x + 12$$
Evaluate the function when $x = 99$.
$$y = -\frac{1}{11}x + 12$$
$$y = -\frac{1}{11}(99) + 12 = 3$$
The car is climbing 3°.

EXCURSION EXERCISES, SECTION 10.4

1. $\frac{1}{4p} = 0.4$
$$1 = 1.6p$$
$$\frac{1}{1.6} = p$$
$$p = 0.625$$
The coordinates of the focus are (0, 0.625).

3. Answers will vary. The filament of the light bulb should be placed at the focus so that all light hitting the paraboloid will be bounced off to form a strong beam of light from the flashlight.

EXERCISE SET 10.4

1. $x = -\frac{b}{2a} = -\frac{0}{2(1)} = 0$
Find y by replacing x by 0 in the equation.
$$y = x^2 - 2$$
$$y = 0^2 - 2$$
$$y = -2$$
Vertex: (0, –2)

3. $x = -\frac{b}{2a} = -\frac{0}{2(-1)} = 0$
Find y by replacing x by 0 in the equation.
$$y = -x^2 - 1$$
$$y = -0^2 - 1$$
$$y = -1$$
Vertex: (0,–1)

5. $x = -\frac{b}{2a} = -\frac{0}{2\left(-\frac{1}{2}\right)} = 0$
Find y by replacing x by 0 in the equation.
$$y = -\frac{1}{2}x^2 + 2$$
$$y = -\frac{1}{2}(0)^2 + 2$$
$$y = 2$$
Vertex: (0, 2)

7. $x = -\frac{b}{2a} = -\frac{0}{2(2)} = 0$
Find y by replacing x by 0 in the equation.
$$y = 2x^2 - 1$$
$$y = 2(0)^2 - 1$$
$$y = -1$$
Vertex: (0,–1)

9. $x = -\frac{b}{2a} = -\frac{(-1)}{2(1)} = \frac{1}{2}$
Find y by replacing x by $\frac{1}{2}$ in the equation.
$$y = x^2 - x + 2$$
$$y = \left(\frac{1}{2}\right)^2 - \left(\frac{1}{2}\right) - 2$$
$$y = -\frac{9}{4}$$
Vertex: $\left(\frac{1}{2}, -\frac{9}{4}\right)$

11. $x = -\dfrac{b}{2a} = -\dfrac{(-1)}{2(2)} = \dfrac{1}{4}$

Find y by replacing x by $\dfrac{1}{4}$ in the equation.

$y = 2x^2 - x - 5$

$y = 2\left(\dfrac{1}{4}\right)^2 - \left(\dfrac{1}{4}\right) - 5$

$y = -\dfrac{41}{8}$

Vertex: $\left(\dfrac{1}{4}, -\dfrac{41}{8}\right)$

13. Solving:

$y = 2x^2 - 4x$

$0 = 2x^2 - 4x$

$0 = 2x(x-2)$

$2x = 0 \quad x - 2 = 0$

$x = 0 \qquad x = 2$

x-intercepts: $(0, 0)$ and $(2, 0)$

15. Solving:

$y = 4x^2 + 11x + 6$

$0 = 4x^2 + 11x + 6$

$0 = (4x+3)(x+2)$

$4x + 3 = 0 \qquad x + 2 = 0$

$x = -\dfrac{3}{4} \qquad x = -2$

x-intercepts: $\left(-\dfrac{3}{4},\ 0\right)$ and $(-2,0)$

17. Solving:

$y = x^2 + 2x - 1$

$0 = x^2 + 2x - 1$

$x = \dfrac{-b \pm \sqrt{b^2 - 4ac}}{2a}$

$x = \dfrac{-2 \pm \sqrt{2^2 - 4(1)(-1)}}{2(1)}$

$x = \dfrac{-2 \pm \sqrt{8}}{2} = \dfrac{-2 \pm 2\sqrt{2}}{2}$

x-intercepts: $\left(-1 + \sqrt{2},\ 0\right)$ and $\left(-1 - \sqrt{2},\ 0\right)$

19. Solving:

$y = -x^2 - 4x - 5$

$0 = -x^2 - 4x - 5$

$x = \dfrac{-b \pm \sqrt{b^2 - 4ac}}{2a}$

$x = \dfrac{-(-4) \pm \sqrt{(-4)^2 - 4(-1)(-5)}}{2(-1)}$

$x = \dfrac{4 \pm \sqrt{16 - 20}}{-2} = \dfrac{4 \pm \sqrt{-4}}{-2}$

Since $\sqrt{-4}$ is not a real number, this graph has no x-intercepts.

21. Solving:

$y = -x^2 + 4x + 1$

$0 = -x^2 + 4x + 1$

$x = \dfrac{-b \pm \sqrt{b^2 - 4ac}}{2a}$

$x = \dfrac{-4 \pm \sqrt{4^2 - 4(-1)(1)}}{2(-1)}$

$x = \dfrac{-4 \pm \sqrt{20}}{-2} = \dfrac{-4 \pm 2\sqrt{5}}{-2} = 2 \pm \sqrt{5}$

x-intercepts: $\left(2 + \sqrt{5},\ 0\right)$ and $\left(2 - \sqrt{5},\ 0\right)$

23. Solving:

$y = 2x^2 - 5x - 3$

$0 = 2x^2 - 5x - 3$

$0 = (2x+1)(x-3)$

$2x + 1 = 0 \qquad x - 3 = 0$

$x = -\dfrac{1}{2} \qquad x = 3$

x-intercepts: $\left(-\dfrac{1}{2},\ 0\right)$ and $(3, 0)$

25. $x = -\dfrac{b}{2a} = -\dfrac{(-2)}{2(1)} = 1$

Find y by replacing x by 1 in the function.

$f(x) = x^2 - 2x + 3$

$f(1) = 1^2 - 2(1) + 3 = 2$

The vertex is $(1, 2)$. The minimum value of the function is 2.

27. $x = -\dfrac{b}{2a} = -\dfrac{4}{2(-2)} = 1$

Find y by replacing x by 1 in the function.

$f(x) = -2x^2 + 4x - 5$

$f(1) = -2(1)^2 + 4(1) - 5 = -3$

The vertex is $(1, -3)$. The maximum value of the function is -3.

29. $x = -\dfrac{b}{2a} = -\dfrac{(-5)}{2(1)} = 2.5$

Find y by replacing x by 2.5 in the function.

$f(x) = x^2 - 5x + 3$

$f(2.5) = (2.5)^2 - 5(2.5) + 3 = -3.25$

The vertex is $(2.5, -3.25)$. The minimum value of the function is -3.25.

31. $x = -\dfrac{b}{2a} = -\dfrac{(-1)}{2(-1)} = -\dfrac{1}{2}$

Find y by replacing x by $-\dfrac{1}{2}$ in the function.

$$f(x) = -x^2 - x + 2$$
$$f\left(-\dfrac{1}{2}\right) = -\left(-\dfrac{1}{2}\right)^2 - \left(-\dfrac{1}{2}\right) + 2 = \dfrac{9}{4}$$

The vertex is $\left(-\dfrac{1}{2}, \dfrac{9}{4}\right)$. The maximum value of the function is $\dfrac{9}{4}$.

33. The vertex of $y = y = x^2 - 2x - 3$ is $(1, -4)$, so its minimum value is -4.

The vertex of $y = x^2 - 10x + 20$ is $(5, -5)$ so its minimum value is -5.

The vertex of $y = 3x^2 - 1$ is $(0, -1)$ so its minimum value is -1.

The largest minimum value is -1, so the answer is **c.**

35. Find the coordinates of the vertex.

$$t = -\dfrac{b}{2a} = -\dfrac{80}{2(-16)} = 2.5$$

$$s(t) = -16t^2 + 80t + 50$$
$$s(2.5) = -16(2.5)^2 + 80(2.5) + 50 = 150$$

The maximum height above the ground that the ball will attain is 150 feet.

37. Find the x-coordinate of the vertex.

$$x = -\dfrac{b}{2a} = -\dfrac{-20}{2(0.1)} = 100$$

The company should produce 100 lenses to minimize the average cost.

39. Find the coordinates of the vertex.

$$x = -\dfrac{b}{2a} = -\dfrac{-0.8}{2(0.25)} = 1.6$$

$$h(x) = 0.25x^2 - 0.8x + 25$$
$$h(1.6) = 0.25(1.6)^2 - 0.8(1.6) + 25 = 24.36$$

The minimum height of the cable above the bridge is 24.36 feet.

41. Find the maximum height of the orange.

$$t = -\dfrac{b}{2a} = -\dfrac{32}{2(-16)} = 1$$

$$h(t) = -16t^2 + 32t + 4$$
$$h(1) = -16(1)^2 + 32(1) + 4 = 20$$

The maximum height the orange will attain is 20 feet. The orange will be at 18 feet on its way up to the maximum height and on its way back down to the ground. So, the answer is yes.

43. a. Evaluate when $x = 6$.

$$h(x) = -\dfrac{1}{9}x^2 + \dfrac{8}{3}x$$
$$h(6) = -\dfrac{1}{9}(6)^2 + \dfrac{8}{3}(6) = 12$$

The height of the water 6 feet to the right of the nozzle is 12 feet.

b. Evaluate when $x = 15$.

$$h(x) = -\dfrac{1}{9}x^2 + \dfrac{8}{3}x$$
$$h(15) = -\dfrac{1}{9}(15)^2 + \dfrac{8}{3}(15) = 15$$

The height of the water 15 feet to the right of the nozzle is 15 feet.

c. Find the coordinates of the vertex.

$$x = -\dfrac{b}{2a} = -\dfrac{\dfrac{8}{3}}{2\left(-\dfrac{1}{9}\right)} = 12$$

$$h(x) = -\dfrac{1}{9}x^2 + \dfrac{8}{3}x$$
$$h(12) = -\dfrac{1}{9}(12)^2 + \dfrac{8}{3}(12) = 16$$

The maximum height the water attains is 16 feet.

45. a. Find the x-coordinate of the vertex.

$$x = -\dfrac{b}{2a} = -\dfrac{1.476}{2(-0.018)} = 41$$

The speed that will yield the maximum fuel efficiency is 41 miles per hour.

b. Find the y-coordinate of the vertex.

$$E(v) = -0.018v^2 + 1.476v + 3.4$$
$$E(41) = -0.018(41)^2 + 1.476(41) + 3.4$$
$$E(41) = 33.658$$

The maximum fuel efficiency is 33.658 miles per gallon.

47. Find t when the altitude is 9000 m or greater.

$$A(t) = -4.9t^2 + 90t + 9000$$
$$9000 = -4.9t^2 + 90t + 9000$$
$$0 = -4.9t^2 + 90t$$
$$0 = t(-4.9t + 90)$$
$$t = 0 \quad -4.9t + 90 = 0$$
$$t \approx 18.4$$
$$t = 18.4 - 0 = 18.4$$

The astronauts experience microgravity for about 18.4 seconds.

EXCURSION EXERCISES, SECTION 10.5

1. a.   1, 2, 4, 8, 16, 32, 64

    b. $1 + 2 = 3$

    c. $3 + 4 = 7$

    d. $7 + 8 = 15$

    e. $15 + 16 = 31$

    f. $31 + 32 = 63$

    g. $63 + 64 = 127$

3. Use your solution to Excursion Exercise 2, and evaluate it when $n = 64$.
    For example,
    $$f(n) = 2(2^{n-1}) - 1$$
    $$f(64) = 2(2^{64-1}) - 1$$
    $$f(64) = 2(2^{63}) - 1$$
    $$f(64) \approx 18,000,000,000,000,000,000 \text{ grains}$$
    of wheat

5. (144 trillion kilograms) × (1 metric ton)/(1000 kg) × (1 year)/(6.5×10^8 metric tons) ≈ 222 years

EXERCISE SET 10.5

1. a. $f(2) = 3^2 = 9$

    b. $f(0) = 3^0 = 1$

    c. $f(-2) = 3^{-2} = \dfrac{1}{9}$

3. a. $g(3) = 2^{3+1} = 2^4 = 16$

    b. $g(1) = 2^{1+1} = 2^2 = 4$

    c. $g(-3) = 2^{-3+1} = 2^{-2} = \dfrac{1}{4}$

5. a. $G(0) = \left(\dfrac{1}{2}\right)^{2(0)} = \left(\dfrac{1}{2}\right)^0 = 1$

    b. $G\left(\dfrac{3}{2}\right) = \left(\dfrac{1}{2}\right)^{2(3/2)} = \left(\dfrac{1}{2}\right)^3 = \dfrac{1}{8}$

    c. $G(-2) = \left(\dfrac{1}{2}\right)^{2(-2)} = \left(\dfrac{1}{2}\right)^{-4} = 16$

7. a. $h(4) = e^{4/2} = e^2 \approx 7.3891$

    b. $h(-2) = e^{-2/2} = e^{-1} \approx 0.3679$

    c. $h\left(\dfrac{1}{2}\right) = e^{(1/2)/2} = e^{1/4} \approx 1.2840$

9. a. $H(-1) = e^{-(-1)+3} = e^4 \approx 54.5982$

    b. $H(3) = e^{-3+3} = e^0 = 1$

    c. $H(5) = e^{-5+3} = e^{-2} \approx 0.1353$

11. a. $F(2) = 2^{2^2} = 2^4 = 16$

    b. $F(-2) = 2^{(-2)^2} = 2^4 = 16$

    c. $F\left(\dfrac{3}{4}\right) = 2^{(3/4)^2} = 2^{9/16} \approx 1.4768$

13. a. $f(-2) = e^{-(-2)^2/2} = e^{-2} \approx 0.1353$

    b. $f(-2) = e^{-(2)^2/2} = e^{-2} \approx 0.1353$

    c. $f(-3) = e^{-(-3)^2/2} = e^{-9/2} \approx 0.0111$

15.

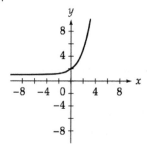

17.

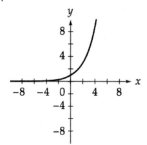

19.

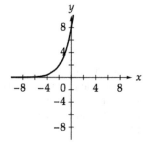

21.

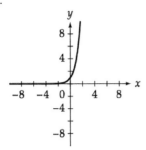

23.

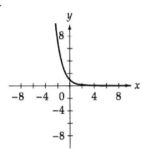

25. $A = P\left(1+\dfrac{r}{n}\right)^{nt}$

$= 2500\left(1+\dfrac{0.075}{365}\right)^{365(20)}$

$\approx 2500(4.480999) \approx 11,202.50$

After 20 years, there is $11,202.50 in the account.

27. $A = Pe^{rt}$

$A = 2500e^{0.05(5)} \approx 3210.06$

The value of the investment is $3210.06.

29. $A = 8\left(\dfrac{1}{2}\right)^{t/8}$

$A = 8\left(\dfrac{1}{2}\right)^{5/8} \approx 5.2$

The amount of iodine-131 present is 5.2 micrograms.

31. Make a table to find the pattern.

Note	Frequency
1 (concert A)	$440 = 440 \times 2^0 =$ $440 \times 2^{0/12}$
12 (next A)	$880 = 440 \times 2^1 =$ $440 \times 2^{12/12}$

The function to represent the frequency f of note n above concert A is $f(n) = 440(2^{n/12})$.

33. a. Since n is measured in months of the lease, change 5 years to 60 months. The annual interest rate is 0.06 as a decimal, so the monthly interest rate is $\dfrac{0.06}{12} = 0.005$.

Substitute 60 for n, 10,000 for A, 6000 for V, and 0.005 for r.

$P = \dfrac{Ar(1+r)^n - Vr}{(1+r)^n - 1}$

$= \dfrac{10,000 \cdot 0.005(1+0.005)^{60} - 6000 \cdot 0.005}{(1+0.005)^{60} - 1}$

≈ 107.33

The monthly lease payment is $107.33.

b. Since n is measured in months of the lease, change 5 years to 60 months. The annual interest rate is 0.06 as a decimal, so the monthly interest rate is $\dfrac{0.06}{12} = 0.005$.

Substitute 60 for n, 10,000 for A, and 0.005 for r.

$P = \dfrac{Ar(1+r)^n - Vr}{(1+r)^n - 1}$

$P = \dfrac{10,000 \cdot 0.005(1+0.005)^{60}}{(1+0.005)^{60} - 1}$

$P \approx 193.33$

The monthly payment is $193.33.

c. There is no residual value on the car when it is purchased.

d. The lease payment in part (a) is $107.33. Substitute 107.33 for P, 10,000 for A, 0.005 for r, and 60 for n.

$C = \dfrac{(P - Ar)[(1+r)^n - 1]}{r}$

$= \dfrac{(107.33 - 10000 \cdot 0.005)[(1+0.005)^{60} - 1]}{0.005}$

≈ 3999.92

The total amount to be repaid in 5 years is $3999.92.
Since $10,000 was the amount borrowed, the amount remaining to be paid is $10,000 − $3999.92 = $6000.08.

e. The monthly payment in part (b) is $193.33. Substitute 193.33 for P, 10,000 for A, 0.005 for r, and 60 for n.

$C = \dfrac{(P - Ar)[(1+r)^n - 1]}{r}$

$= \dfrac{(193.33 - 10000 \cdot 0.005)[(1+0.005)^{60} - 1]}{0.005}$

$\approx 10,000.14$

The total amount to be repaid in 5 years is $10,000.14.
Since $10,000 was the amount borrowed, the amount remaining to be paid is $10,000 - $10,000.14 = -$0.14 or approximately $0.

f. The $6000 remaining on the leased car is its residual value.

EXCURSION EXERCISES, SECTION 10.6

1.

d	$P(d) = \log_{10}(1 + 1/d)$
1	0.301
2	0.176
3	0.125
4	0.097
5	0.079
6	0.067
7	0.058
8	0.051
9	0.046

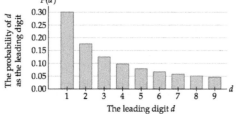

3. $P(1) = 0.301$
    $P(9) = 0.046$

$\dfrac{0.301}{0.046} \approx 6.5$

You are about 6.5 times more likely to make purchases for dollar amounts that start with a 1 than for dollar amounts starting with a 9.

5. There is not a wide range of values.

EXERCISE SET 10.6

1. $\log_7 49 = 2$

3. $\log_5 625 = 4$

5. $\log 0.0001 = -4$

7. $\log_{10} x = y$

9. $3^4 = 81$

11. $5^3 = 125$

13. $4^{-2} = \dfrac{1}{16}$

15. $e^y = x$

17. $\log_3 81 = x$
    $3^x = 81$
    $3^x = 3^4$
    $x = 4$
    $\log_3 81 = 4$

19. $\log 100 = x$
    $10^x = 100$
    $10^x = 10^2$
    $x = 2$
    $\log 100 = 2$

21. $\log_3 \dfrac{1}{9} = x$
    $3^x = \dfrac{1}{9}$
    $3^x = 3^{-2}$
    $x = -2$
    $\log_3 \dfrac{1}{9} = -2$

23. $\log_2 64 = x$
    $2^x = 64$
    $2^x = 2^6$
    $x = 6$
    $\log_2 64 = 6$

25. $\log_3 x = 2$
    $3^2 = x$
    $9 = x$

27. $\log_7 x = -1$
    $7^{-1} = x$
    $\dfrac{1}{7} = x$

29. $\log_3 x = -2$
    $3^{-2} = x$
    $\dfrac{1}{9} = x$

31. $\log_4 x = 0$
    $4^0 = x$
    $1 = x$

33. $\log x = 2.5$
    $10^{2.5} = x$
    $316.23 \approx x$

35. $\ln x = 2$
    $e^2 = x$
    $7.39 \approx x$

36. $\ln x = 4$
$e^4 = x$
$54.60 \approx x$

38. $\log x = 0.127$
$10^{0.127} = x$
$1.34 \approx x$

39. $\ln x = \dfrac{8}{3}$
$e^{8/3} = x$
$14.39 \approx x$

41.

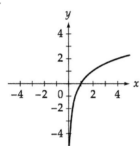

43.

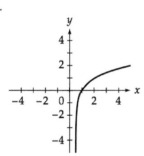

45.

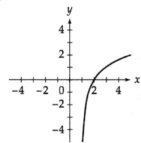

47. $\log P = -kd$
$\log P = -(0.2)(0.5)$
$\log P = -0.1$
$10^{-0.1} = P$
$0.79 \approx P$
The amount of light passing through the material is about 79%.

49. $D = 10(\log I + 16)$
$D = 10[\log(3.2 \times 10^{-10}) + 16] \approx 65$
Normal conversation is about 65 decibels.

51. $\text{pH} = -\log(\text{H}^+)$
$\text{pH} = -\log(0.045) \approx 1.35$
The digestive solution of the stomach has a pH of about 1.35.

53. $M = \log \dfrac{I}{I_0}$
$M = \log\left(\dfrac{630{,}957{,}345}{I_0}\right)$
$M = \log 630{,}957{,}345 = 8.8$
$M = \log 630{,}957{,}345 = 8.8$
The earthquake had a Richter scale magnitude of 8.8.

55. $M = \log \dfrac{I}{I_0}$
$8.9 = \log \dfrac{I}{I_0}$
$\dfrac{I}{I_0} = 10^{8.9} \approx 794{,}328{,}235$
$I = 794{,}328{,}235 I_0$
The earthquake had an intensity of $794{,}328{,}235 I_0$.

57.   Find the intensity of an earthquake of magnitude 8.
$M = \log \dfrac{I}{I_0}$
$8 = \log \dfrac{I}{I_0}$
$\dfrac{I}{I_0} = 10^8 = 100{,}000{,}000$
$I = 100{,}000{,}000 I_0$
Find the intensity of an earthquake of magnitude 6.
$M = \log \dfrac{I}{I_0}$
$6 = \log \dfrac{I}{I_0}$
$\dfrac{I}{I_0} = 10^6 = 1{,}000{,}000$
$I = 1{,}000{,}000 I_0$
Since 100,000,000 is 100 times more than 1,000,000, the earthquake of magnitude 8 is 100 times stronger than an earthquake of magnitude 6.

59. $M = 5\log r - 5$
$-1.11 = 5\log r - 5$
$3.89 = 5\log r$
$\dfrac{3.89}{5} = \log r$
$0.778 = \log r$
$r = 10^{0.778} \approx 6.0$
Earth is approximately 6.0 parsecs from Alpha Centauri.

61. $x = \dfrac{3}{2}$

$\quad = \dfrac{10^{0.47712}}{10^{0.30103}} = 10^{0.47712 - 0.30103} = 10^{0.176609}$

In logarithmic form, the equation is
$\log x = 0.17609.$

$10^{0.17609} \approx 1.5$

63. Answers will vary.

65. $M = \log A + 3\log 8t - 2.92$
$M = \log(26) + 3\log(8 \cdot 26) - 2.92 \approx 4.9$
The magnitude of the earthquake was about 4.9.

CHAPTER 10 REVIEW EXERCISES

1.

2.

3.
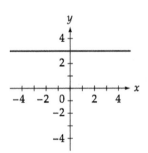

4.
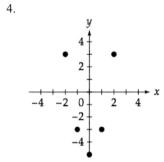

5.
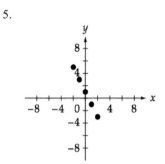

6.
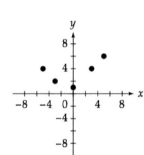

7.
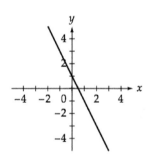

8.
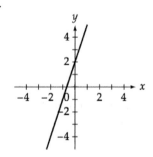

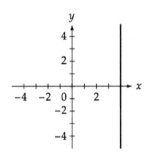
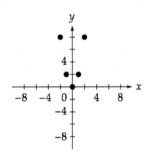

9.

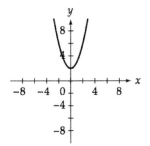

10.

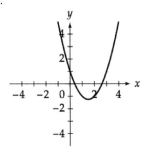

11.

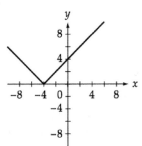

12.

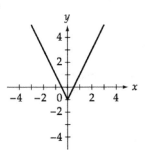

13.

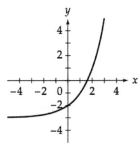

14.

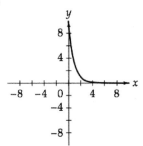

15.

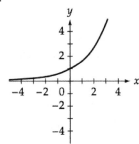

16.

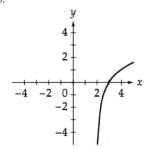

17. $f(x) = 4x - 5$
$f(-2) = 4(-2) - 5 = -13$

18. $g(x) = 2x^2 - x - 2$
$g(3) = 2(3)^2 - 3 - 2 = 13$

19. $s(t) = \dfrac{4}{3t - 5}$
$s(-1) = \dfrac{4}{3(-1) - 5} = \dfrac{4}{-8} = -\dfrac{1}{2}$

20. $R(s) = s^2 - 2s^2 + s - 3$
$R(-2) = (-2)^3 - 2(-2)^2 + (-2) - 3 = -21$

21. $f(x) = 2^{x-3}$
$f(5) = 2^{5-3} = 2^2 = 4$

22. $g(x) = \left(\dfrac{2}{3}\right)^x$
$g(2) = \left(\dfrac{2}{3}\right)^2 = \dfrac{4}{9}$

23. $T(r) = 2e^r + 1$

    $T(2) = 2e^2 + 1 \approx 15.78$

24. a. $V(r) = \dfrac{4\pi r^3}{3}$

    $V(3) = \dfrac{4\pi(3)^3}{3} \approx 113.1$

    The volume is about 113.1 cubic inches.

    b. $V(r) = \dfrac{4\pi r^3}{3}$

    $V(12) = \dfrac{4\pi(12)^3}{3} \approx 7238.2$

    The volume of the sphere with radius 12 inches is about 7238.2 cubic centimeters.

25. a. $h(t) = -16t^2 + 64t + 5$

    $h(2) = -16(2)^2 + 64(2) + 5 = 69$

    The height of the ball is 69 feet.

    b. $h(t) = -16t^2 + 64t + 5$

    $h(4) = -16(4)^2 + 64(4) + 5 = 5$

    The height of the ball is 5 feet.

26. a. $P(x) = \dfrac{100x + 100}{x + 20}$

    $P(0) = \dfrac{100(0) + 100}{0 + 20} = 5$

    The original concentration of sugar is 5%.

    b. $P(x) = \dfrac{100x + 100}{x + 20}$

    $P(10) = \dfrac{100(10) + 100}{10 + 20} \approx 36.7$

    The concentration of sugar after 10 grams of sugar have been added is 36.7%.

27. $f(x) = 2x + 10$

    $0 = 2x + 10$

    $-2x = 10$

    $x = -5$

    x-intercept: $(-5, 0)$

    $f(x) = 2x + 10$

    $f(0) = 2(0) + 10 = 10$

    y-intercept: $(0, 10)$

28. $f(x) = \dfrac{3}{4}x - 9$

    $0 = \dfrac{3}{4}x - 9$

    $9 = \dfrac{3}{4}x$

    $12 = x$

    x-intercept: $(12, 0)$

$f(x) = \dfrac{3}{4}x - 9$

$f(0) = \dfrac{3}{4}(0) - 9 = -9$

y-intercept: $(0, -9)$

29. $3x - 5y = 15$

    $3x - 5(0) = 15$

    $3x = 15$

    $x = 5$

    x-intercept: $(5, 0)$

    $3x - 5y = 15$

    $3(0) - 5y = 15$

    $-5y = 15$

    $y = -3$

    y-intercept: $(0, -3)$

30. $4x + 3y = 24$

    $4x + 3(0) = 24$

    $4x = 24$

    $x = 6$

    x-intercept: $(6, 0)$

    $4x + 3y = 24$

    $4(0) + 3y = 24$

    $3y = 24$

    $y = 8$

    y-intercept: $(0, 8)$

31. $V(t) = 25{,}000 - 5000t$

    $V(0) = 25{,}000 - 5000(0) = 25{,}000$

    The intercept on the vertical axis is $(0, 25{,}000)$. This means that the value of the van was $25,000 when it was new.

    $V(t) = 25{,}000 - 5000t$

    $0 = 25{,}000 - 5000t$

    $5000t = 25{,}000$

    $t = 5$

    The intercept on the horizontal axis is $(5, 0)$. This means that after 5 years the van will be worth $0.

32. $m = \dfrac{-3 - 2}{2 - 3} = \dfrac{-5}{-1} = 5$

33. $m = \dfrac{-1 - 4}{-3 - (-1)} = \dfrac{-5}{-2} = \dfrac{5}{2}$

34. $m = \dfrac{-5 - (-5)}{-4 - 2} = \dfrac{0}{-6} = 0$

35. $m = \dfrac{7 - 2}{5 - 5} = \dfrac{5}{0}$; the slope is undefined.

36. Find the slope.
$$m = \frac{13-6}{40-180} = \frac{7}{-140} = -0.05$$
The slope is –0.05, which means for each mile the car is driven, 0.05 gallon of fuel is used.

37.
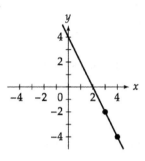

38.
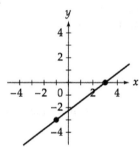

39. $y - y_1 = m(x - x_1)$
$y - 3 = 2(x - (-2))$
$y - 3 = 2x + 4$
$y = 2x + 7$

40. $y - y_1 = m(x - x_1)$
$y - (-4) = 1(x - 1)$
$y + 4 = x - 1$
$y = x - 5$

41. $y - y_1 = m(x - x_1)$
$y - 1 = \frac{2}{3}(x - (-3))$
$y - 1 = \frac{2}{3}x + 2$
$y = \frac{2}{3}x + 3$

42. $y - y_1 = m(x - x_1)$
$y - 1 = \frac{1}{4}(x - 4)$
$y - 1 = \frac{1}{4}x - 1$
$y = \frac{1}{4}x$

43.
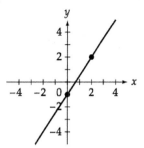

44. a. Let x represent the number of square yards of carpet, and let $f(x)$ represent the cost. The cost of the carpet increases by \$12 for each additional square yard. So, the slope is 12. The initial cost of the carpet is \$100, for installation, so the y-intercept is 100.
$$y = mx + b$$
$$f(x) = 12x + 1000$$

b. $f(x) = 12x + 1000$
$f(288) = 12(288) + 100 = 3556$
The cost to carpet 288 square yards of floor space is \$3556.

45. a. Let A represent the number of gallons sold and let p represent the price relative to Western QuickMart. When the price increases by \$.02, the number of gallons sold decreases by 500 gallons.
$$m = \frac{\text{change in } A}{\text{change in } p} = \frac{-500}{0.02} = -25,000$$
When the prices are the same, the price relative to Western QuickMart is 0, so an ordered pair is (0, 10,000). The y-intercept is 10,000.
$$y = mx + b$$
$$A = mp + b$$
$$A(p) = -25,000p + 10,000$$

b. The price is \$0.03 below Western QuickMart, so $p = -0.03$.
$$A(p) = -25,000p + 10,000$$
$$A(-0.03) = -25,000(-0.03) + 10,000$$
$$= 10,750$$
Valley Gas Mart will sell 10,750 gallons of gasoline.

46. $x = -\frac{b}{2a} = -\frac{2}{2(1)} = -1$
Find y by replacing x by –1 in the equation.
$$y = x^2 + 2x + 4$$
$$y = (-1)^2 + 2(-1) + 4 = 3$$
Vertex: (–1, 3)

47. $x = -\dfrac{b}{2a} = -\dfrac{(-6)}{2(-2)} = -\dfrac{3}{2}$

    Find y by replacing x by $-\dfrac{3}{2}$ in the equation.

    $y = -2x^2 - 6x + 1$

    $y = -2\left(-\dfrac{3}{2}\right)^2 - 6\left(-\dfrac{3}{2}\right) + 1 = \dfrac{11}{2}$

    Vertex: $\left(-\dfrac{3}{2}, \dfrac{11}{2}\right)$

48. $x = -\dfrac{b}{2a} = -\dfrac{6}{2(-3)} = 1$

    Find $f(x)$ by replacing x by 1 in the equation.

    $f(x) = -3x^2 + 6x - 1$

    $f(1) = -3(1)^2 + 6(1) - 1 = 2$

    Vertex: $(1, 2)$

49. $x = -\dfrac{b}{2a} = -\dfrac{5}{2(1)} = -2.5$

    Find $f(x)$ by replacing x by –2.5 in the equation.

    $f(x) = x^2 + 5x - 1$

    $f(-2.5) = (-2.5)^2 + 5(-2.5) - 1 = -7.25$

    Vertex: $(-2.5, -7.25)$

50. $y = x^2 + x - 20$

    $0 = (x - 4)(x + 5)$

    $x - 4 = 0 \quad x + 5 = 0$

    $x = 4 \qquad x = -5$

    x-intercepts: $(4, 0)$ and $(-5, 0)$

51. $y = x^2 + 2x - 1$

    $0 = x^2 + 2x - 1$

    $x = \dfrac{-b \pm \sqrt{b^2 - 4ac}}{2a}$

    $x = \dfrac{-2 \pm \sqrt{2^2 - 4(1)(-1)}}{2(1)}$

    $x = \dfrac{-2 \pm \sqrt{8}}{2} = \dfrac{-2 \pm 2\sqrt{2}}{2} = -1 \pm \sqrt{2}$

    x-intercepts: $(-1 + \sqrt{2}, 0)$ and $(-1 - \sqrt{2}, 0)$

52. $f(x) = 2x^2 + 9x + 4$

    $0 = (2x + 1)(x + 4)$

    $2x + 1 = 0 \qquad x + 4 = 0$

    $x = -\dfrac{1}{2} \qquad x = -4$

    x-intercepts: $\left(-\dfrac{1}{2}, 0\right)$ and $(-4, 0)$

53. $f(x) = x^2 + 4x + 6$

    $0 = x^2 + 4x + 6$

    $x = \dfrac{-b \pm \sqrt{b^2 - 4ac}}{2a}$

    $x = \dfrac{-4 \pm \sqrt{4^2 - 4(1)(6)}}{2(1)}$

    $x = \dfrac{-4 \pm \sqrt{16 - 24}}{2} = \dfrac{-4 \pm \sqrt{-8}}{2}$

    Since $\sqrt{-8}$ is not a real number, the graph has no x-intercepts.

54. $x = -\dfrac{b}{2a} = -\dfrac{4}{2(-1)} = 2$

    Find y by replacing x by 2 in the function.

    $y = -x^2 + 4x + 1$

    $y = -(2)^2 + 4(2) + 1 = 5$

    The vertex is $(2, 5)$. The maximum value of the function is 5.

55. $x = -\dfrac{b}{2a} = -\dfrac{6}{2(2)} = -\dfrac{3}{2}$

    Find y by replacing x by $-\dfrac{3}{2}$ in the function.

    $y = 2x^2 + 6x - 3$

    $y = 2\left(-\dfrac{3}{2}\right)^2 + 6\left(-\dfrac{3}{2}\right) - 3 = -\dfrac{15}{2}$

    The vertex is $\left(-\dfrac{3}{2}, -\dfrac{15}{2}\right)$. The minimum value of the function is $-\dfrac{15}{2}$.

56. $x = -\dfrac{b}{2a} = -\dfrac{(-4)}{2(1)} = 2$

    Find $f(x)$ by replacing x by 2 in the function.

    $f(x) = x^2 - 4x - 1$

    $f(2) = 2^2 - 4(2) - 1 = -5$

    The vertex is $(2, -5)$. The minimum value of the function is –5.

57. $x = -\dfrac{b}{2a} = -\dfrac{3}{2(-2)} = \dfrac{3}{4}$

    Find $f(x)$ by replacing x by $\dfrac{3}{4}$ in the function.

    $f(x) = -2x^2 + 3x - 1$

    $f\left(\dfrac{3}{4}\right) = -2\left(\dfrac{3}{4}\right)^2 + 3\left(\dfrac{3}{4}\right) - 1 = \dfrac{1}{8}$

    The vertex is $\left(\dfrac{3}{4}, \dfrac{1}{8}\right)$. The maximum value of the function is $\dfrac{1}{8}$.

58. Find the coordinates of the vertex.

$$t = -\frac{b}{2a} = -\frac{80}{2(-16)} = \frac{5}{2}$$

$$s(t) = -16t^2 + 80t + 25$$

$$s(2.5) = -16\left(\frac{5}{2}\right)^2 + 80\left(\frac{5}{2}\right) + 25 = 125$$

The maximum height that the rock will attain is 125 feet above the ground.

59. Find the x-coordinate of the vertex.

$$x = -\frac{b}{2a} = -\frac{(-40)}{2(0.01)} = 2000$$

The company should produce 2000 DVDs.

60. $A = P\left(1 + \frac{r}{n}\right)^{nt}$

$$= 5000\left(1 + \frac{0.06}{365}\right)^{365(15)}$$

$$\approx 5000(2.459421) \approx 12,297.11$$

After 15 years, there is $12,297.11 in the account.

61. $A = 10\left(\frac{1}{2}\right)^{t/6}$

$$A(2) = 10\left(\frac{1}{2}\right)^{2/6} \approx 7.94$$

There will be 7.94 milligrams of the isotope present after 2 hours.

62. $A = 8\left(\frac{1}{2}\right)^{t/8}$

$$A(10) = 8\left(\frac{1}{2}\right)^{10/8} \approx 3.36$$

There will be 3.36 micrograms present after 10 days.

63. a. Make a table to find the pattern.

n = Number of bounces	H = Height of ball
0	6
1	6(2/3)= 4
2	4(2/3) $= 6(2/3)^2$
3	$6(2/3)^2 (2/3)$ $= 6(2/3)^3$

$$H(n) = 6\left(\frac{2}{3}\right)^n$$

b. $H(n) = 6\left(\frac{2}{3}\right)^n$

$$H(5) = 6\left(\frac{2}{3}\right)^5 \approx 0.79$$

The height of the ball after the fifth bounce is about 0.79 feet.

64. $\log_3 243 = x$

$$3^x = 243$$
$$3^x = 3^5$$
$$x = 5$$
$$\log_3 243 = x$$

65. $\log_2 \frac{1}{16} = x$

$$2^x = \frac{1}{16}$$
$$2^x = x^{-4}$$
$$x = -4$$
$$\log_2 \frac{1}{16} = -4$$

66. $\log_4 \frac{1}{4} = x$

$$4^x = \frac{1}{4}$$
$$4^x = 4^{-1}$$
$$x = -1$$
$$\log_4 \frac{1}{4} = -1$$

67. $\log_2 64 = x$

$$2^x = 64$$
$$2^x = 2^6$$
$$x = 6$$
$$\log_2 64 = 6$$

68. $\log_4 x = 3$

$$4^3 = x$$
$$64 = x$$

69. $\log_3 x = \frac{1}{3}$

$$3^{1/3} = x$$
$$1.4422 \approx x$$

70. $\ln x = 2.5$

$$e^{2.5} = x$$
$$12.1825 \approx x$$

71. $\log x = 2.4$

$$10^{2.4} = x$$
$$251.1886 \approx x$$

72. $M = 5\log r - 5$

$$3.2 = 5\log 4 - 5$$
$$8.2 = 5\log r$$
$$\frac{8.2}{5} = \log r$$
$$10^{8.2/5} = r$$
$$43.7 \approx r$$

The distance to the star is 43.7 parsecs.

73. $D = 10(\log I + 16)$
    $D = 10[\log(0.01) + 16]$
    $D = 140$
    The pain threshold for most humans is 140 decibels.

CHAPTER **10** TEST

1. $s(t) = -3t^2 + 4t - 1$
   $s(-2) = -3(-2)^2 - 4(-2) - 1 = -21$

2. $f(x) = 3^{x-4}$
   $f(2) = 3^{2-4} = 3^{-2} = \dfrac{1}{9}$

3. $\log_5 125 = x$
   $5^x = 125$
   $5^x = 5^3$
   $x = 3$
   $\log_5 125 = 3$

4. $\log_6 x = 2$
   $6^2 = x$
   $36 = x$

5.
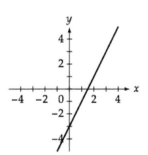

6.
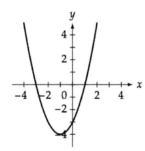

7.
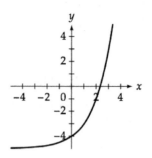

8.
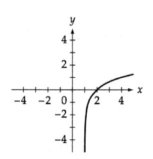

9. $m = \dfrac{(-4)-(-1)}{(-2)-3} = \dfrac{-3}{-5} = \dfrac{3}{5}$

10. $y - y_1 = m(x - x_1)$
    $y - 5 = \dfrac{2}{3}(x - 3)$
    $y - 5 = \dfrac{2}{3}x - 2$
    $y = \dfrac{2}{3}x + 3$

11. $x = -\dfrac{b}{2a} = -\dfrac{6}{2(1)} = -3$
    Find y by replacing x with -3 in the equation.
    $f(x) = x^2 + 6x - 1$
    $f(-3) = (-3)^2 + 6(-3) - 1 = -10$
    Vertex: $(-3, -10)$

12. $y = x^2 + 2x - 8$
    $0 = (x - 2)(x + 4)$
    $x - 2 = 0 \quad x + 4 = 0$
    $x = 2 \qquad x = -4$
    x-intercepts: $(2, 0)$ and $(-4, 0)$

13. $x = -\dfrac{b}{2a} = -\dfrac{-3}{2(-1)} = -\dfrac{3}{2}$
    Find j by replacing x with $-\dfrac{3}{2}$ in the function.
    $y = -x^2 - 3x + 10$
    $y = -\left(-\dfrac{3}{2}\right)^2 - 3\left(-\dfrac{3}{2}\right) + 10 = \dfrac{49}{4}$
    The vertex is $\left(-\dfrac{3}{2}, \dfrac{49}{4}\right)$. The maximum value of the function is $\dfrac{49}{4}$.

14. $d = 250 - 100t$
    $d = 250 - 100(0)$
    $d = 250$
    The vertical intercept is (0, 250). This means that the plane starts 250 miles from its destination.
    $$d = 250 - 100t$$
    $$0 = 250 - 100t$$
    $$100t = 250$$
    $$t = 2.5$$
    The horizontal intercept is (2.5, 0). This means that it takes the plane 2.5 hours to reach its destination.

15. Find the coordinates of the vertex.
    $$t = -\frac{b}{2a} = -\frac{48}{2(-16)} = 1.5$$
    $$h(t) = -16t^2 + 48t + 8$$
    $$h(1.5) = -16(1.5)^2 + 48(1.5) + 8 = 44$$
    The maximum height above the ground that the ball will attain is 44 feet.

16. $$A = 10\left(\frac{1}{2}\right)^{t/5}$$
    $$A(8) = 10\left(\frac{1}{2}\right)^{8/5} \approx 3.30$$
    About 3.30 grams of the isotope will remain after 8 hours.

17. Find the intensity of an earthquake of magnitude 7.3.
    $$M = \log\frac{I}{I_0}$$
    $$7.3 = \log\left(\frac{I}{I_0}\right)$$
    $$\frac{I}{I_0} = 10^{7.3} \approx 19,952,623$$
    $$I \approx 19,952,623 I_0$$

    Find the intensity of an earthquake of magnitude 7.0.
    $$M = \log\frac{I}{I_0}$$
    $$7.0 = \log\left(\frac{I}{I_0}\right)$$
    $$\frac{I}{I_0} = 10^7 = 10,000,000$$
    $$I = 10,000,000 I_0$$
    Since 19,952,623 is about 1.995 times more than 10,000,000, the earthquake of magnitude 7.3 is about 2.0 times stronger than an earthquake of magnitude 7.0.

18. Let x represent the number of trees per acre and let y represent the average yield per tree. Then, when $x = 320$, $y = 260$, and when $x = 330$, $y = 245$. Find the slope.
    $$m = \frac{245 - 260}{330 - 320} = -1.5$$
    Find the equation of the linear model.
    $$y - y_1 = m(x - x_1)$$
    $$y - 260 = -1.5(x - 320)$$
    $$y = -1.5x + 740$$

19. a. Find the slope.
    $$m = \frac{y_2 - y_1}{x_2 - x_1} = \frac{117 - 74}{3915 - 2456} \approx 0.029$$
    Use the point-slope formula to find the equation of the line.
    $$y - y_1 = m(x - x_1)$$
    $$y - 74 = 0.029(x - 2456)$$
    $$y = 0.029x - 71.224$$
    $$f(x) = 0.029x + 2.776$$

    b. Evaluate the equation when $x = 2600$.
    $$f(x) = 0.029x + 2.776$$
    $$f(2600) = 0.029(2600) + 2.776$$
    $$\approx 78.176$$
    For 2600 cars traveling on the road, 78 accidents are expected.

20. $$\log P = -kd$$
    $$\log(0.75) = -0.05d$$
    $$\frac{\log(0.75)}{-0.05} = d$$
    $$2.5 \approx d$$
    At a depth of about 2.5 meters, 75% of the light will pass through the surface.

Chapter 11: The Mathematics of Finance

EXCURSION EXERCISES, SECTION 11.1

1. $P = 9621.97$, $r = \dfrac{0.05}{12}$, $t = 1$

   $I = Prt$

   $= 9621.97\left(\dfrac{0.05}{12}\right)(1)$

   ≈ 40.09

   You owe $40.09 in interest for the third month.

3. $9621.97 - $190.20 = 9431.77

   You owe $9431.77 after making the third month payment.

5. $230.29 - $39.30 = 190.99

   Thus, $190.99 goes toward reducing the principal in the fourth month.

7. Final payment = remaining balance + interest

   $230.29 = P + Prt$

   $230.29 = P + P\left(\dfrac{0.05}{12}\right)(1)$

   $230.29 = P\left(1 + \dfrac{0.05}{12}\right)$

   $\dfrac{230.29}{1 + \dfrac{0.05}{12}} = P$

   $P \approx 229.33$

   The remaining balance is $229.33.

EXERCISE SET 11.1

1. Divide the number of months by 12.

3. I is the interest, P is the principal, r is the interest rate, and t is the time period.

5. $I = Prt = 8000(0.07)(1) = 560$ The simple interest earned is $560.

7. Because the interest rate is an annual rate, the time must be measured in years:

   $t = \dfrac{6 \text{ months}}{1 \text{ year}} = \dfrac{6 \text{ months}}{12 \text{ months}} = \dfrac{6}{12}$

   $I = Prt = 7000(0.065)\left(\dfrac{6}{12}\right) = 227.50$

   The interest earned is $227.50.

9. Because the interest rate is an annual rate, the time must be measured in years:

   $t = \dfrac{4 \text{ months}}{1 \text{ year}} = \dfrac{4 \text{ months}}{12 \text{ months}} = \dfrac{4}{12}$

   $I = Prt = 9000(0.0675)\left(\dfrac{4}{12}\right) = 202.50$

   The interest earned is $202.50.

11. Because the interest rate is an annual rate, the time must be measured in years:

    $t = \dfrac{\text{number of days}}{360} = \dfrac{21}{360}$

    $I = Prt = 3000(0.096)\left(\dfrac{21}{360}\right) = 16.80$

    The interest earned is $16.80.

13. Because the interest rate is an annual rate, the time must be measured in years:

    $t = \dfrac{\text{number of days}}{360} = \dfrac{114}{360}$

    $I = Prt = 7000(0.072)\left(\dfrac{114}{360}\right) = 159.60$

    The interest earned is $159.60.

15. Because the interest rate is per month, the time period of the loan is $t = 5$ months.

    $I = Prt = 2000(0.0125)(5) = 125$

    The interest earned is $125.

17. Use the simple interest formula. Because the interest rate is per month, the time period of the loan is $t = 6$ months.

    $I = Prt = 1600(0.0175)(6) = 168$

    The interest earned is $168.

19. Since the interest rate is per year, the time period of the loan is $t = \dfrac{6}{12}$ year.

    $A = P(1 + rt) = 15,000\left[1 + 0.089\left(\dfrac{6}{12}\right)\right]$

    $= 15,000(1.034) = 16,667.50$

    The maturity value of the loan is $15,667.50.

21. Since the interest rate is per year, the time period of the loan is $t = \dfrac{4}{12}$ year.

    $A = P(1 + rt) = 7200\left[1 + 0.00795\left(\dfrac{4}{12}\right)\right]$

    $= 7200(1.0265) = 7390.80$

    The maturity value of the loan is $7390.80.

23. Since the interest rate is per year, the time period of the loan is $t = \dfrac{3}{12}$ year.

    $A = P(1 + rt) = 2800\left[1 + 0.092\left(\dfrac{3}{12}\right)\right]$

    $= 2800(1.023) = 2864.40$

    The maturity value of the loan is $2864.40

25. Solve the simple interest formula for r.
$$I = Prt$$
$$120 = 1600r(1)$$
$$120 = 1600r$$
$$\frac{120}{1600} = r$$
$$0.075 = r$$
$$r = 7.5\%$$
The simple interest rate is 7.5%.

27. Solve the simple interest formula for r.
$$I = Prt$$
$$80 = 2000r\left(\frac{6}{12}\right)$$
$$80 = 1000r$$
$$\frac{80}{1000} = r$$
$$0.08 = r$$
$$r = 8\%$$
The simple interest rate is 8%.

29. Solve the simple interest formula for r.
$$I = Prt$$
$$37.20 = 1200r\left(\frac{4}{12}\right)$$
$$37.20 = 400r$$
$$\frac{37.20}{400} = r$$
$$0.093 = r$$
$$r = 9.3\%$$
The simple interest rate is 9.3%.

31. $I = Prt = 1500(0.052)\left(\frac{6}{12}\right) = 39$

The simple interest earned is $39.

33. $I = Prt = 1600(0.09)\left(\frac{45}{360}\right) = 18$

The simple interest earned is $18.

35. Using the formula:
$$A = P(1 + rt) = 7000\left[1 + 0.087\left(\frac{8}{12}\right)\right]$$
$$= 7000(1.058) = 7406$$
The maturity value of the loan is $7406.

37. Using the formula:
$$A = P(1 + rt) = 5200\left[1 + 0.051(1)\right]$$
$$= 5200(1.051) = 5465.20$$
The maturity value of the loan is $5465.20.

39. Using the formula:
$$A = P(1 + rt) = 750\left[1 + 0.073(1)\right]$$
$$= 750(1.073) = 804.75$$
The future value of the investment is $804.75.

41. Solve the simple interest formula for r.
$$I = Prt$$
$$918 = 18,000r\left(\frac{6}{12}\right)$$
$$918 = 9000r$$
$$\frac{918}{9000} = r$$
$$0.102 = r$$
$$r = 10.2\%$$
The simple interest rate is 10.2%.

43. First find the amount of interest paid. Subtract the principal from the maturity value.
$I = A - P = 5125 - 5000 = 125$
Find the simple interest rate by solving the simple interest formula for r.
$$I = Prt$$
$$125 = 5000r\left(\frac{3}{12}\right)$$
$$125 = 1250r$$
$$0.10 = r$$
$$r = 10\%$$
The simple interest rate on the loan is 10%.

45. $A = P(1 + rt) = 132\left[1 + 0.09\left(\frac{1}{12}\right)\right] = 132.99$

You will owe $132.99.

47. The exact method uses 365 days a year.
$$A = P(1 + rt) = 10,000\left[1 + 0.08\left(\frac{140}{365}\right)\right]$$
$$\approx 10,306.85$$
The exact method yields a maturity value of $10,306.85.

The ordinary method uses 360 days a year.
$$A = P(1 + rt) = 10,000\left[1 + 0.08\left(\frac{140}{360}\right)\right]$$
$$\approx 10,311.11$$
The ordinary method yields a maturity value of $10,311.11.
The ordinary method yields the greater maturity value. The lender benefits from using the ordinary method.

49. There are fewer days in September than there are in August.

EXCURSION EXERCISES, SECTION 11.2

1. The CPI is the percent written without a percent symbol, so the CPI of 218.058 indicates 218.058% of the base period prices.

3. CPI for 2015 = CPI for 2012 + (percent increase)(CPI for 2012)
$$237.017 = 229.939 + p(229.939)$$
$$7.078 = 229.939p$$
$$0.03078 \approx p$$
$$3.08\% \approx p$$
The increase was approximately 3.08%.

5. Cost in 2015 = (Cost in base years)(CPI for entertainment)
$$9 = x(115.626\%)$$
$$9 = x(1.15616)$$
$$\frac{9}{1.15616} = x$$
$$7.78 \approx x$$
The cost of a movie ticket during the base years would be $7.78.

EXERCISE SET 11.2

1. $P = 1200, r = 7\% = 0.07, n = 2, t = 12$
$$A = P\left(1 + \frac{r}{n}\right)^{nt}$$
$$= 1200\left(1 + \frac{0.07}{2}\right)^{(2 \cdot 12)}$$
$$= 1200(1.035)^{24} \approx 2739.99$$
The compound amount after 12 years is approximately $2739.99.

3. $P = 500, r = 9\% = 0.09, n = 4, t = 6$
$$A = P\left(1 + \frac{r}{n}\right)^{nt}$$
$$= 500\left(1 + \frac{0.09}{2}\right)^{(4 \cdot 6)}$$
$$= 500(1.0225)^{24} \approx 852.88$$
The compound amount after 6 years is approximately $852.88.

5. $P = 8500, r = 9\% = 0.09, n = 12, t = 10$
$$A = P\left(1 + \frac{r}{n}\right)^{nt}$$
$$= 8500\left(1 + \frac{0.09}{2}\right)^{(12 \cdot 10)}$$
$$= 8500(1.075)^{120} \approx 20,836.54$$
The compound amount after 10 years is approximately $20,836.54.

7. Interest is compounded daily. $P = 9600$, $n = 360, r = 9\% = 0.09, t = 3$
$$A = P\left(1 + \frac{r}{n}\right)^{nt}$$
$$= 9600\left(1 + \frac{0.09}{360}\right)^{(360 \cdot 3)}$$
$$= 9600(1.00025)^{1080} \approx 12,575.23$$

The compound amount after 3 years is approximately $12,575.23.

9. Use the financial mode on the calculator. Since interest is compounded quarterly, there are $4(10) = 40$ payments.
The compound amount is $3532.86.

11. Use the financial mode on the calculator. Since interest is compounded monthly, there are $12(5) = 60$ payments.
The compound amount is $3850.08.

13. Use the financial mode on the calculator. Since interest is compounded semiannually, there are $2(3) = 6$ payments.
The compound amount is $2213.84.

15. $P = 7500, r = 6\% = 0.06, n = 12, t = 5$
$$A = P\left(1 + \frac{r}{n}\right)^{nt}$$
$$= 7500\left(1 + \frac{0.06}{12}\right)^{(12 \cdot 5)}$$
$$= 7500(1.005)^{60} \approx 10,116.38$$
The future value is approximately $10,116.38.

17. $P = 4600, r = 5\% = 0.05, n = 2, t = 12$
$$A = P\left(1 + \frac{r}{n}\right)^{nt}$$
$$= 4600\left(1 + \frac{0.05}{2}\right)^{(2 \cdot 12)}$$
$$= 4600(1.025)^{24} \approx 8320.14$$
The future value is approximately $8320.14.

19. $P = 22,000, r = 9\% = 0.09, n = 12, t = 7$
$$A = P\left(1 + \frac{r}{n}\right)^{nt}$$
$$= 22,000\left(1 + \frac{0.09}{12}\right)^{(12 \cdot 7)}$$
$$= 22,000(1.0075)^{84} \approx 41,210.44$$
The future value is approximately $41,210.44.

21. $A = 25,000, r = 10\% = 0.10, n = 4, t = 12$
$$P = \frac{A}{\left(1 + \frac{r}{n}\right)^{nt}}$$
$$= \frac{25,000}{\left(1 + \frac{0.10}{4}\right)^{(4 \cdot 12)}}$$
$$= \frac{25,000}{(1.025)^{48}}$$
$$\approx 7641.78$$
$7641.78 should be invested in the account in order to have $25,000 in 12 years.

23. $A = 40,000, r = 7.5\% = 0.075, n = 1, t = 35$

$$P = \frac{A}{\left(1 + \dfrac{r}{n}\right)^{nt}}$$

$$= \frac{40,000}{\left(1 + \dfrac{0.075}{1}\right)^{(1 \cdot 35)}}$$

$$= \frac{40,000}{(1.075)^{35}}$$

$$\approx 3182.47$$

$3182.47 should be invested in the account in order to have $40,000 in 35 years.

25. $A = 15,000, r = 8\% = 0.08, n = 4, t = 5$

$$P = \frac{A}{\left(1 + \dfrac{r}{n}\right)^{nt}}$$

$$= \frac{15,000}{\left(1 + \dfrac{0.08}{4}\right)^{(4 \cdot 5)}}$$

$$= \frac{15,000}{(1.02)^{20}}$$

$$\approx 10,094.57$$

$10,094.57 should be invested in the account in order to have $15,000 in 5 years.

27. $P = 8000, r = 8\% = 0.08, n = 4, t = 5$

$$A = P\left(1 + \frac{r}{n}\right)^{nt}$$

$$= 8000\left(1 + \frac{0.08}{4}\right)^{(4 \cdot 5)}$$

$$= 8000(1.02)^{20} \approx 11,887.58$$

The compound amount after 5 years is approximately $11,887.58.

29. Interest is compounded daily.
$P = 2500, n = 360, r = 9\% = 0.09, t = 4$

$$A = P\left(1 + \frac{r}{n}\right)^{nt}$$

$$= 2500\left(1 + \frac{0.09}{360}\right)^{(360 \cdot 4)}$$

$$= 2500(1.00025)^{1440} \approx 3583.16$$

The compound amount after 4 years is approximately $3583.16.

31. $P = 4000, r = 12\% = 0.12, n = 12, t = 6$

$$A = P\left(1 + \frac{r}{n}\right)^{nt}$$

$$= 4000\left(1 + \frac{0.06}{12}\right)^{(12 \cdot 6)}$$

$$= 4000(1.005)^{72} \approx 5728.18$$

The future value is approximately $5728.18.

33. Calculate the compound amount.
$P = 2000, r = 6\% = 0.06, n = 4, t = 3$

$$A = P\left(1 + \frac{r}{n}\right)^{nt}$$

$$= 2000\left(1 + \frac{0.06}{4}\right)^{(4 \cdot 3)}$$

$$= 2000(1.015)^{10} \approx 2391.24$$

Calculate the interest earned.
$I = A - P = 2391.24 - 2000 = 391.24$
The amount of interest earned is approximately $391.24.

35. Calculate the compound amount.
$P = 15,000, r = 10\% = 0.10, n = 4, t = 8$

$$A = P\left(1 + \frac{r}{n}\right)^{nt}$$

$$= 15,000\left(1 + \frac{0.10}{4}\right)^{(4 \cdot 8)}$$

$$= 15,000(1.025)^{32} \approx 33,056.35$$

Calculate the interest earned.
$I = A - P = 33,056.35 - 15,000 = 18,056.35$
The amount of interest earned is approximately $18,056.35.

37. $A = 15,000, r = 6\% = 0.06, n = 12, t = 5$

$$P = \frac{A}{\left(1 + \dfrac{r}{n}\right)^{nt}}$$

$$= \frac{15,000}{\left(1 + \dfrac{0.06}{12}\right)^{(12 \cdot 5)}}$$

$$= \frac{15,000}{(1.005)^{60}}$$

$$\approx 11,120.58$$

$11,120.58 should be invested in the account in order to have $15,000 in 5 years.

39. a. $P = 1000, r = 0.09,$ and $t = 5$.
$I = Prt$
$I = 1000(0.09)(5)$
$I = 450$
The simple interest earned is $450.

   b. Calculate the compound amount. Interest is compounded daily.
$P = 1000, r = 9\% = 0.09, t = 5, n = 360$

$$A = P\left(1 + \frac{r}{n}\right)^{nt}$$

$$= 1000\left(1 + \frac{0.09}{360}\right)^{(360 \cdot 5)}$$

$$= 1000(1.00025)^{1800} \approx 1568.22$$

Calculate the interest earned.
$I = A - P = 1568.22 - 1000 = 568.22$
The amount of interest earned is approximately $568.22.

c. $568.22 - 450 = 118.22$
$118.22 more is earned when interest is compounded daily.

41. a. $P = 15,000, r = 8\% = 0.08, n = 2, t = 4$

$$A = P\left(1 + \frac{r}{n}\right)^{nt}$$

$$= 15,000\left(1 + \frac{0.08}{2}\right)^{(2 \cdot 4)}$$

$$= 15,000(1.04)^8 \approx 20,528.54$$

The future value of the investment is $20,528.54.

b. $P = 15,000, r = 8\% = 0.08, n = 2, t = 4$

$$A = P\left(1 + \frac{r}{n}\right)^{nt}$$

$$= 15,000\left(1 + \frac{0.08}{4}\right)^{(4 \cdot 4)}$$

$$= 15,000(1.02)^{16} \approx 20,591.79$$

The future value of the investment is $20,591.79.

c. $20,591.79 - 20,528.54 = 63.25$
The investment is $63.25 greater when interest is compounded quarterly.

43. a.   Calculate the compound amount.
$P = 10,000, r = 8\% = 0.08, n = 2, t = 2$

$$A = P\left(1 + \frac{r}{n}\right)^{nt}$$

$$= 10,000\left(1 + \frac{0.08}{2}\right)^{(2 \cdot 2)}$$

$$= 10,000(1.04)^4 \approx 11,698.59$$

Calculate the interest earned.
$I = A - P = 11,698.59 - 10,000 = 1698.59$
The amount of interest earned is approximately $1698.59.

b.   Calculate the compound amount.
$P = 10,000, r = 8\% = 0.08, n = 4, t = 2$

$$A = P\left(1 + \frac{r}{n}\right)^{nt}$$

$$= 10,000\left(1 + \frac{0.08}{4}\right)^{(4 \cdot 2)}$$

$$= 10,000(1.02)^8 \approx 11,716.59$$

Calculate the interest earned.
$I = A - P$
$I = 11,716.59 - 10,000$
$I = 1716.59$
The amount of interest earned is approximately $1716.59.

c. $1716.59 - 1698.59 = 18.00$
The investment is $18.00 greater when interest is compounded quarterly.

45. a. $P = 7000, r = 8\% = 0.08, n = 4, t = 5$

$$A = P\left(1 + \frac{r}{n}\right)^{nt}$$

$$= 7000\left(1 + \frac{0.08}{4}\right)^{(4 \cdot 5)}$$

$$= 7000(1.02)^{20} \approx 10,401.63$$

The maturity value of the loan is $10,401.63.

b. $10,401.63 - 7000 = 3401.63$
The interest you are paying is $3401.63.

47. $A = 40,000, r = 8\% = 0.08, n = 4, t = 18$

$$P = \frac{A}{\left(1 + \frac{r}{n}\right)^{nt}}$$

$$= \frac{40,000}{\left(1 + \frac{0.08}{4}\right)^{(4 \cdot 18)}}$$

$$= \frac{40,000}{(1.02)^{72}}$$

$$\approx 9612.75$$

$9612.75 should be invested in the account to have $40,000 in 18 years.

49. a. $A = 1,000,000, r = 9\% = 0.09, n = 360,$
$t = 30$

$$P = \frac{A}{\left(1 + \frac{r}{n}\right)^{nt}}$$

$$= \frac{1,000,000}{\left(1 + \frac{0.09}{360}\right)^{(360 \cdot 30)}}$$

$$= \frac{1,000,000}{(1.00025)^{10,800}}$$

$$\approx 67,228.19$$

$67,228.19 should be invested in the account in order to have $1,000,000 in 30 years.

b.   Calculate the compound amount. $t = 1,$
$P = 1,000,000, r = 9\% = 0.09, n = 360$

$$A = P\left(1 + \frac{r}{n}\right)^{nt}$$

$$= 1,000,000\left(1 + \frac{0.09}{360}\right)^{(360 \cdot 1)}$$

$$= 1,000,000(1.00025)^{360} \approx 1,094,161.98$$

Calculate the interest earned.
$I = A - P$
$= 1,094,161.98 - 1,000,000$
$= 94,161.98$
The amount of interest earned each year is approximately $94,161.98.

51. Calculate the compound amount for the first two years. $P = 5000, r = 3.1\% = 0.031, n = 360, t = 2$

$$A = P\left(1 + \frac{r}{n}\right)^{nt}$$

$$= 5000\left(1 + \frac{0.031}{360}\right)^{(360 \cdot 2)}$$

$$\approx 5319.80$$

Calculate the compound amount for the second two years. The principal is now $5319.80. $r = 2.2\% = 0.022, n = 360, t = 2$

$$A = P\left(1 + \frac{r}{n}\right)^{nt}$$

$$= 5000\left(1 + \frac{0.022}{360}\right)^{(360 \cdot 2)}$$

$$\approx 5559.09$$

The compound amount when the second CD matures is $5559.09.

53. $P = 1249, r = 7\% = 0.07, n = 1, t = 15$

$$A = P\left(1 + \frac{r}{n}\right)^{nt}$$

$$= 1249\left(1 + \frac{0.07}{1}\right)^{(1 \cdot 15)}$$

$$= 1249(1.07)^{15} \approx 3346.03$$

Fifteen years from now, the average rent for an apartment will be approximately $3446.03.

55. $P = 40,000, r = 7\% = 0.07, n = 1, t = 5$

$$A = P\left(1 + \frac{r}{n}\right)^{nt}$$

$$= 40,000\left(1 + \frac{0.07}{1}\right)^{(1 \cdot 5)}$$

$$= 40,000(1.07)^5 \approx 56,102.07$$

In 2020, you need a salary of approximately $56,102.07 to have the same purchasing power.

57. $P = 125,000, r = 6\% = 0.06$ and $t = 40$. The inflation rate is an annual rate, so $n = 1$.

$$P = \frac{A}{\left(1 + \frac{r}{n}\right)^{nt}}$$

$$= \frac{125,000}{\left(1 + \frac{0.06}{1}\right)^{(1 \cdot 40)}}$$

$$= \frac{125,000}{(1.06)^{40}}$$

$$\approx 12,152.77$$

The purchasing power of the $125,000 in 2014 will be approximately $12,152.77 in 2054.

59. $A = 3500, r = 7\% = 0.07$ and $t = 5$. The inflation rate is an annual rate, so $n = 1$.

$$P = \frac{A}{\left(1 + \frac{r}{n}\right)^{nt}}$$

$$= \frac{3500}{\left(1 + \frac{0.07}{1}\right)^{(1 \cdot 5)}}$$

$$= \frac{3500}{(1.07)^5}$$

$$\approx 2495.45$$

The purchasing power of the $3500 in will be approximately $2495.45 in 5 years.

61. Use the compound amount formula to find the future value of $100 after 1 year.
$P = 100, r = 7.2\%, n = 4, t = 1$

$$A = P\left(1 + \frac{r}{n}\right)^{nt}$$

$$= 100\left(1 + \frac{0.072}{4}\right)^{(4 \cdot 1)}$$

$$= 100(1.018)^4 \approx 107.40$$

Find the interest earned on the $100.
$I = A - P = 107.40 - 100 = 7.40$
The effective interest rate is 7.40%.

63. Use the compound amount formula to find the future value of $100 after 1 year.
$P = 100, r = 7.5\%, n = 12, t = 1$

$$A = P\left(1 + \frac{r}{n}\right)^{nt}$$

$$= 100\left(1 + \frac{0.075}{1}\right)^{(4 \cdot 1)}$$

$$= 100(1.00625)^4 \approx 107.76$$

Find the interest earned on the $100.
$I = A - P$
$I = 107.76 - 1\,00$
$I = 7.76$
The effective interest rate is 7.76%.

65. Use the compound amount formula to find the future value of $100 after 1 year.
$P = 100, r = 8.1\%, n = 360, t = 1$

$$A = P\left(1 + \frac{r}{n}\right)^{nt}$$

$$= 100\left(1 + \frac{0.081}{360}\right)^{(360 \cdot 1)}$$

$$= 100(1.000225)^{360} \approx 108.44$$

Find the interest earned on the $100.
$I = A - P = 108.44 - 100 = 8.44$
The effective interest rate is 8.44%.

67. Use the compound amount formula to find the future value of $100 after 1 year.
$P = 100, r = 5.94\%, n = 12, t = 1$

$$A = P\left(1 + \frac{r}{n}\right)^{nt}$$

$$= 100\left(1 + \frac{0.0594}{12}\right)^{(12 \cdot 1)}$$

$$= 100(1.00495)^{12} \approx 106.10$$

Find the interest earned on the $100.
$I = A - P = 106.10 - 100 = 6.10$
The effective interest rate is 6.10%.

69. The inflation rate is an annual rate, so $n = 1$.
$P = 3.00, r = 6\% = 0.06$ In 2014, $t = 5$.

$$A = P\left(1 + \frac{r}{n}\right)^{nt}$$

$$= 3.00\left(1 + \frac{0.06}{1}\right)^{(1 \cdot 5)}$$

$$= 3.00(1.06)^5 \approx 4.01$$

Gasoline will cost about $4.01 in 2014.

In 2019, $t = 10$.

$$A = P\left(1 + \frac{r}{n}\right)^{nt}$$

$$= 3.00\left(1 + \frac{0.06}{1}\right)^{(10 \cdot 1)}$$

$$= 3.00(1.06)^{10} \approx 5.37$$

Gasoline will cost about $5.37 in 2019.

In 2029, $t = 20$.

$$A = P\left(1 + \frac{r}{n}\right)^{nt}$$

$$= 3.00\left(1 + \frac{0.06}{1}\right)^{(1 \cdot 20)}$$

$$= 3.00(1.06)^{20} \approx 9.62$$

Gasoline will cost about $9.62 in 2029.

71. The inflation rate is an annual rate, so $n = 1$.
$P = 2.69, r = 6\% = 0.06$ In 2014, $t = 5$.

$$A = P\left(1 + \frac{r}{n}\right)^{nt}$$

$$= 2.69\left(1 + \frac{0.06}{1}\right)^{(1 \cdot 5)}$$

$$= 2.69(1.06)^5 \approx 3.60$$

A loaf of bread will cost approximately $3.60 in 2014.

In 2019, $t = 10$.

$$A = P\left(1 + \frac{r}{n}\right)^{nt}$$

$$= 2.69\left(1 + \frac{0.06}{1}\right)^{(1 \cdot 10)}$$

$$= 2.69(1.06)^{10} \approx 4.82$$

A loaf of bread will cost approximately $4.82 in 2019.

In 2029, $t = 20$.

$$A = P\left(1 + \frac{r}{n}\right)^{nt}$$

$$= 2.69\left(1 + \frac{0.06}{1}\right)^{(1 \cdot 20)}$$

$$= 2.69(1.06)^{20} \approx 8.63$$

A loaf of bread will cost approximately $8.63 in 2029.

73. The inflation rate is an annual rate, so $n = 1$.
$P = 9, r = 6\% = 0.06$ In 2014, $t = 5$.

$$A = P\left(1 + \frac{r}{n}\right)^{nt}$$

$$= 9\left(1 + \frac{0.06}{1}\right)^{(1 \cdot 5)}$$

$$= 9(1.06)^5 \approx 12.04$$

A ticket to a movie will cost approximately $12.04 in 2014.

In 2019, $t = 10$.

$$A = P\left(1 + \frac{r}{n}\right)^{nt}$$

$$= 9\left(1 + \frac{0.06}{1}\right)^{(1 \cdot 10)}$$

$$= 9(1.06)^{10} \approx 16.12$$

A ticket to a movie will cost approximately $16.12 in 2019.

In 2029, $t = 20$.

$$A = P\left(1 + \frac{r}{n}\right)^{nt}$$

$$= 9\left(1 + \frac{0.06}{1}\right)^{(1 \cdot 20)}$$

$$= 9(1.06)^{20} \approx 28.86$$

A ticket to a movie will cost approximately $28.86 in 2029.

75. The inflation rate is an annual rate, so $n = 1$.
$P = 275,000, r = 6\% = 0.06$ In 2014, $t = 5$.

$$A = P\left(1 + \frac{r}{n}\right)^{nt}$$

$$= 275,000\left(1 + \frac{0.06}{1}\right)^{(1 \cdot 5)}$$

$$= 275,000(1.06)^5 \approx 368,012.03$$

A house will cost approximately $368,012.03 in 2014.

In 2019, $t = 10$.

$$A = P\left(1 + \frac{r}{n}\right)^{nt}$$

$$= 275,000\left(1 + \frac{0.06}{1}\right)^{(1 \cdot 10)}$$

$$= 275,000(1.06)^{10} \approx 492,483.12$$

A house will cost approximately $492,483.12 in 2019.

In 2029, $t = 20$.

$$A = P\left(1 + \frac{r}{n}\right)^{nt}$$

$$= 275,000\left(1 + \frac{0.06}{1}\right)^{(1 \cdot 20)}$$

$$= 275,000(1.06)^{20} \approx 881,962.25$$

A house will cost approximately \$881,962.25 in 2029.

77. $A = 50,000$, $r = 7\% = 0.07$ and $t = 10$. The inflation rate is an annual rate, so $n = 1$.

$$P = \frac{A}{\left(1 + \frac{r}{n}\right)^{nt}}$$

$$= \frac{50,000}{\left(1 + \frac{0.07}{1}\right)^{(1 \cdot 10)}}$$

$$= \frac{50,000}{(1.07)^{10}} \approx 25,417.46$$

The purchasing power of the \$50,000 will be approximately \$25,417.46 in 10 years.

79. $A = 100,000$, $r = 7\% = 0.07$ and $t = 20$. The inflation rate is an annual rate, so $n = 1$.

$$P = \frac{A}{\left(1 + \frac{r}{n}\right)^{nt}}$$

$$= \frac{100,000}{\left(1 + \frac{0.07}{1}\right)^{(1 \cdot 20)}}$$

$$= \frac{100,000}{(1.07)^{20}} \approx 25,841.90$$

The purchasing power of the \$100,000 will be approximately \$25,841.90 in 20 years.

81. $A = 75,000$, $r = 7\% = 0.07$ and $t = 5$. The inflation rate is an annual rate, so $n = 1$.

$$P = \frac{A}{\left(1 + \frac{r}{n}\right)^{nt}}$$

$$= \frac{75,000}{\left(1 + \frac{0.07}{1}\right)^{(1 \cdot 5)}}$$

$$= \frac{75,000}{(1.07)^{5}} \approx 53,473.96$$

The purchasing power of the \$75,000 will be approximately \$53,473.96 in 5 years.

83. a. $r = 4\% = 0.04$, $P = 1$, $n = 1$, $and\ t = 1$. Use the compound amount formula,

$A = P\left(1 + \frac{r}{n}\right)^{nt}$ to find the amount after 1

year. Then use $I = A - P$ to find the effective rate. See the table:

Nominal Rate	Effective Rate
4% annual compounding	4.0%
4% semiannual compounding	4.04%
4% quarterly compounding	4.06%
4% monthly compounding	4.07%
4% daily compounding	4.08%

b.   As the number of compounding periods increase, the effective rate increases.

85. Find the future value of \$100 after 1 year.
$P = 100$, $r = 3\% = 0.03$, $n = 12$

$$A = P\left(1 + \frac{r}{n}\right)^{nt}$$

$$= 100\left(1 + \frac{0.03}{12}\right)^{(12 \cdot 1)}$$

$$= 100(1.0025)^{12} \approx 103.04$$

Find the interest earned on the \$100.
$I = A - P = 103.04 - 100 = 3.04$ The effective interest rate is 3.04%.

87. Calculate $\left(1 + \frac{r}{n}\right)^{nt}$ for each investment.

6% compounded quarterly: $n = 4$,
$r = 6\% = 0.06$, $t = 1$

$$\left(1 + \frac{r}{n}\right)^{nt} = \left(1 + \frac{0.06}{4}\right)^{(4 \cdot 1)} \approx 1.061363551$$

6.25% compounded semiannually: $n = 2$,
$r = 6.25\% = 0.0625$, $t = 1$

$$\left(1 + \frac{r}{n}\right)^{nt} = \left(1 + \frac{0.0625}{4}\right)^{(2 \cdot 1)} \approx 1.063476563$$

Compare the two compound amounts.
$1.063476563 > 1.061363551$
6.25% compounded semiannually has a higher yield than 6% compounded quarterly.

89. Calculate $\left(1 + \frac{r}{n}\right)^{nt}$ for each investment. 5.8%

compounded quarterly: $n = 4$, $r = 5.8\% = 0.058$,
$t = 1$

$$\left(1 + \frac{r}{n}\right)^{nt} = \left(1 + \frac{0.058}{4}\right)^{(4 \cdot 1)} \approx 1.05927373$$

5.6% compounded monthly: $n = 12$,
$r = 5.6\% = 0.056$, $t = 1$

$$\left(1 + \frac{r}{n}\right)^{nt} = \left(1 + \frac{0.056}{12}\right)^{(12 \cdot 1)} \approx 1.057459928$$

Compare the two compound amounts.
$1.059273739 > 1.057459928$
5.8% compounded quarterly has a higher yield than 5.6% compounded monthly.

EXCURSION EXERCISES, SECTION 11.3

1. a. Net capitalized cost = negotiated price – down payment – trade-in value
        = 28,990 – 2400 – 3850
        = $22,740

    b. Average monthly finance charge = (net capitalized cost + residual value) × money factor
        = (22,740 + 15,000) × 0.0027 = $101.90

    c. Average monthly depreciation
$$= \frac{\text{net capitalized cost} - \text{residual value}}{\text{term of the lease in months}}$$
$$= \frac{22,740 - 15,000}{4(12)} = \frac{7740}{48}$$
$$= \$161.25$$

    d. Monthly lease payment = average monthly finance charge + average monthly depreciation
        = 101.90 + 161.25 = $263.15

3. a. Net capitalized cost = negotiated price – down payment – trade-in value
        = 22,100 – 1000 – 0 = $21,100

    b. Money factor $= \dfrac{\text{Annual interest rate}}{24}$
$$= \frac{0.081}{24} = 0.003375$$

    c. Average monthly finance charge = (net capitalized cost + residual value) × money factor
        = (21,200 + 15,000) × 0.003375
        = $121.84

    d. Average monthly depreciation
$$= \frac{\text{net capitalized cost} - \text{residual value}}{\text{term of the lease in months}}$$
$$= \frac{21,100 - 15,000}{3(12)} = \frac{6100}{36}$$
$$= \$169.44$$

    e. Monthly lease payment = average monthly finance charge + average monthly depreciation
        = 121.84 + 169.4 = $291.28

5. Net capitalized cost = 31,900 – 0 – 0 = $31,900
    Residual value = 0.40(33,395) = $13,358
    Money factor $= \dfrac{0.08}{24} = 0.0\overline{3}$
    Average monthly finance charge
$$= (31,900 + 13,358) \times 0.0\overline{3} = \$150.86$$
    Average monthly depreciation
$$= \frac{31,900 - 13,358}{5(12)} = \frac{18,358}{60} = \$309.03$$
    Monthly lease payment $= 150.86 + 309.03$
$$= \$459.89$$

EXERCISE SET 11.3

1. $I = Prt = 118.72(0.0125)(1) \approx 1.48$
    The finance charge is $1.48.

3. $I = Prt = 10,1543.87(0.015)(1) \approx 152.32$
    The finance charge is $152.32.

5. Recording the balances:

Date	Payment or purchase	Balance	Days	Balance times Days
Mar. 5-11		$244	7	$1708
Mar. 12-27	$152	$396	16	$6336
Mar. 28-Apr. 4	–$100	$296	8	$2368
Total				$10,412

Average daily balance $= \dfrac{10,412}{31} \approx \$335.87$

7. Recording the balances:

Date	Payment or purchase	Balance	Days	Balance times Days
May 5-16		$944	12	$11,328
May 17-19	$255	$1199	3	$3597
May 20-Jun. 4	–$150	$1049	16	$16,784
Total				$31,709

Average daily balance $= \dfrac{31,709}{31} \approx \$1022.87$

$I = Prt = (1022.87)(0.015)(1) \approx 15.34$. The finance charge is $15.34.

9. Recording the balances:

Date	Payment or purchase	Balance	Days	Balance times Days
Aug. 10-14		$345	5	$1725
Aug. 15	$56	$401	0	$0
Aug. 15-26	– $75	$326	12	$3912
Aug. 27-Sep. 9	$157	$483	14	$6762
Total				$12,399

Average daily balance $= \dfrac{12,399}{31} \approx \$399.97$

$I = Prt = (399.97)(0.0125)(1) \approx 5.00$. The finance charge is $5.00.

11. Recording the balances:

Date	Payment or purchase	Balance	Days	Balance times Days
Aug. 15		$1236.43	1	$1236.43
Aug. 16	$125	$1361.43	1	$1361.43
Aug. 17	$23.56	$1384.99	1	$1384.99
Aug. 18-19	$53.45	$1438.44	2	$2876.88
Aug. 20-21	$41.36	$1479.80	2	$2959.60
Aug. 22-24	$223.65	$1703.45	3	$5110.35
Aug. 25-29	$310	$2013.45	5	$10,067.25
Aug. 30-31	$23.36	$2036.81	2	$4073.62
Sep. 1-11	$36.45	$2073.26	11	$22,804.76
Sep. 12	$41.25	$2114.41	1	$2114.41
Sep. 13-14	–$1345	$769341	2	$1538.82
Total				$55,528.34

Average daily balance $= \dfrac{55,528.34}{31} \approx \$1791.24$

$I = Prt = (1791.24)(0.015)(1) \approx 26.87$. The finance charge is $26.87.

13. Using the formula:

$\text{APR} \approx \dfrac{2nr}{n+1}$

$\approx \dfrac{2(3)(0.09)}{3+1} = \dfrac{0.54}{4} = 0.135$

The annual percentage rate on the loan is 13.5%.

15. Using the formula:

$\text{APR} \approx \dfrac{2nr}{n+1}$

$\approx \dfrac{2(24)(0.10)}{24+1} = \dfrac{4.8}{25} = 0.192$

The annual percentage rate on the loan is 19.2%.

17. Using:

$\dfrac{r}{n} = \dfrac{0.069}{12} = 0.00575$

Calculate the monthly payment. For a 1-year loan, $nt = 12(1) = 12$.

$PMT = A\left(\dfrac{\dfrac{r}{n}}{1-\left(1+\dfrac{r}{n}\right)^{-nt}}\right)$

$= 400\left(\dfrac{0.00575}{1-(1+0.00575)^{-12}}\right)$

≈ 34.59

The monthly payment is $34.59.

19. a.   Sales tax = 0.0725(649) = 47.0525
         The sales tax is $47.05.
         Total cost = 649 + 47.05 = 696.05
         The total cost is $696.05.

    b.   Down payment = 0.25(696.05) = 174.0125
         The down payment is $174.01.

    c.   Loan amount = total cost – down payment
         Loan amount = 696.05 – 174.01 = 522.04

         $$\frac{r}{n} = \frac{0.057}{12} = 0.00475$$

         Calculate the monthly payment. For a 6-month loan, $nt = 6$.

         $$PMT = A\left(\frac{\frac{r}{n}}{1-\left(1+\frac{r}{n}\right)^{-nt}}\right)$$

         $$= 522.04\left(\frac{0.00475}{1-(1+0.00475)^{-6}}\right)$$

         $$\approx 88.46$$

         The monthly payment is $88.46.

21. a.   Sales tax = 0.055(64,995) = 3574.725
         The sales tax is $3574.73.
         Total cost = 64,995 + 3574.73 = 68,569.73
         The total cost is $68,569.73.

    b.   Down payment = 0.20(68,569.73)
                      = 13,713.946
         The down payment is $13,713.95.

    c.   Loan amount = total cost – down payment
         Loan amount = 68,569.73 – 13,713.95
                     = 54,855.78

         $$\frac{r}{n} = \frac{0.0715}{12}$$

         Calculate the monthly payment. For a 10-year loan, $nt = 10(12) = 120$.

         $$PMT = A\left(\frac{\frac{r}{n}}{1-\left(1+\frac{r}{n}\right)^{-nt}}\right)$$

         $$= 54,855.78\left(\frac{0.0715/12}{1-(1+0.0715/12)^{-120}}\right)$$

         $$\approx 641.17$$

         The monthly payment is $641.17.

23. Sales tax = 0.06(42,600) = 2556
    License fee = 0.01(42,600) = 426
    Loan amount = 42,600 + 2556 + 426 = 45,582

    $$\frac{r}{n} = \frac{0.057}{12} = 0.00475$$

    Calculate the monthly payment. For a 5-year loan, $nt = 5(12) = 60$.

    $$PMT = A\left(\frac{\frac{r}{n}}{1-\left(1+\frac{r}{n}\right)^{-nt}}\right)$$

    $$= 45,582\left(\frac{0.00475}{1-(1+0.00475)^{-60}}\right)$$

    $$\approx 874.88$$

    The monthly payment is $874.88.

25. Sales tax = 0.055(24,500) = 1347.50
    Loan amount = purchase price – down payment
                + sales tax + license fee
    Loan amount = 24,500 – 3000
                + 1347.50 + 331 = 23,178.50

    $$\frac{r}{n} = \frac{0.085}{12}$$

    Calculate the monthly payment. For a 4-year loan, $nt = 4(12) = 48$.

    $$PMT = A\left(\frac{\frac{r}{n}}{1-\left(1+\frac{r}{n}\right)^{-nt}}\right)$$

    $$= 23,178.50\left(\frac{0.085/12}{1-(1+0.085/12)^{-48}}\right)$$

    $$\approx 571.31$$

    The monthly payment is $571.31.

27. a.   Using:

         $$\frac{r}{n} = \frac{0.08}{12}$$

         Calculate the monthly payment. For a 4-year loan, $nt = 4(12) = 48$.

         $$PMT = A\left(\frac{\frac{r}{n}}{1-\left(1+\frac{r}{n}\right)^{-nt}}\right)$$

         $$= 25,445\left(\frac{0.08/12}{1-(1+0.08/12)^{-48}}\right)$$

         $$\approx 621.19$$

         The monthly payment is $621.19.

    b.   There are 48 monthly payments of
         $621.19, so the total amount paid is
         48(621.19) = $29,817.12.
         Amount of interest paid = total amount
         paid – amount financed
             = 29,817.12 – 25,445 = 4372.12
         The total amount of interest paid is
         $4372.12.

29. She has owned the Jeep for 3 years, so she has 2 years, or 24 months, of payments remaining. Thus, $U = 24$.

$$\frac{r}{n} = \frac{0.063}{12} = 0.00525$$

$$A = PMT\left(\frac{1 - \left(1 + \frac{r}{n}\right)^{-U}}{\frac{r}{n}}\right)$$

$$= 603.50\left(\frac{1 - (1 + 0.000525)^{-24}}{0.00525}\right)$$

$$\approx 13{,}575.25$$

The loan payoff is $13,575.25.

31. You have owned the car for 3 years, so you have 1 year, or 12 months, of payments remaining. Thus, $U = 12$.

$$\frac{r}{n} = \frac{0.089}{12}$$

$$A = PMT\left(\frac{1 - \left(1 + \frac{r}{n}\right)^{-U}}{\frac{r}{n}}\right)$$

$$= 303.52\left(\frac{1 - (1 + 0.089/12)^{-12}}{0.089/12}\right)$$

$$\approx 3472.57$$

The loan payoff is $3472.57.

33. The monthly payment for the loan is the PMT. The interest rate per period, i, is the annual interest rate divided by 12. The term of the loan, n, is the number of years of the loan times 12. Substitute these values into the Payment Formula for an APR loan and solve for A, the selling price of the car. The selling price of the car is $9775.72.

35. The loan is non-subsidized, so simple interest will accrue for 4 years before graduation.

$$I = Prt = 17{,}000(0.062)(4) = 4216$$

Total loan amount after graduation:
$17{,}000 + 4216 = 21{,}216$
$A = 21{,}216, r = 0.062, n = 12, t = 5$

$$PMT = A\left(\frac{\frac{r}{n}}{1 - \left(1 + \frac{r}{n}\right)^{-nt}}\right)$$

$$= 21{,}216\left(\frac{0.062/12}{1 - (1 + 0.062/12)^{-(12\cdot5)}}\right)$$

$$\approx 412.14$$

The monthly payment is $412.14.

37. a. You have owned the car for 3 years, so you have 1 year, or 12 months, of payments remaining. Thus, $U = 12$.

$$\frac{r}{n} = \frac{0.084}{12} = 0.007$$

$$A = PMT\left(\frac{1 - \left(1 + \frac{r}{n}\right)^{-U}}{\frac{r}{n}}\right)$$

$$= 235.73\left(\frac{1 - (1 + 0.007)^{-12}}{0.007}\right)$$

$$\approx 2704.15$$

The loan payoff is $2704.15.

b. The loan payoff for the old car is added to the price of the new car.
$18{,}234 + 2704.15 = 20{,}938.15$
The actual amount you owe is $20,938.15.

c. Using:

$$\frac{r}{n} = \frac{0.084}{12} = 0.007$$

Calculate the monthly payment. For a 4-year loan, $nt = 4(12) = 48$.

$$PMT = A\left(\frac{\frac{r}{n}}{1 - \left(1 + \frac{r}{n}\right)^{-nt}}\right)$$

$$= 20{,}938.15\left(\frac{0.007}{1 - (1 + 0.007)^{-48}}\right)$$

$$\approx 515.10$$

The monthly payment is $515.10.

39. The loan is non-subsidized, so simple interest will accrue for 4 years before graduation. Each year 12,000 loan is taken out.

Year 1: (4 years until graduation)
$I = Prt = 12{,}000(0.05)(4) = 2400$

Loan amount after graduation:
$12{,}000 + 2400 = \$14{,}400$

Year 2: (3 years until graduation)
$I = Prt = 12{,}000(0.05)(3) = 1800$

Loan amount after graduation:
$12{,}000 + 1800 = \$13{,}800$

Year 3: (2 years until graduation)
$I = Prt = 12{,}000(0.05)(2) = 1200$

Loan amount after graduation:
$12{,}000 + 1200 = \$13{,}200$

Year 4: (1 year until graduation)
$I = Prt = 12{,}000(0.05)(1) = 600$

Loan amount after graduation:
$12{,}000 + 600 = \$12{,}600$

Total loan amount after graduation:
$14{,}400 + 13{,}800 + 13{,}200 + 12{,}600 = \$54{,}000$
$A = 54{,}000, r = 0.05, n = 12, t = 10$

$$PMT = A\left(\frac{r/n}{1 - (1 + r/n)^{-nt}}\right)$$

$$= 54{,}000\left(\frac{0.05/12}{1 - (1 + 0.05/12)^{-(12\cdot10)}}\right)$$

$$\approx 572.75$$

The monthly payment is $572.75.

EXCURSION EXERCISES, SECTION 11.4

1. First, compute the interest:

    $I = Prt = 30,000(0.0232)\left(\dfrac{182}{360}\right) \approx 351.87$

    Cost = (face value – interest) + service fee
    = 30,000 – 351.87 + 15 = $29,663.13

3. First, compute the interest:

    $I = Prt = 60,000(0.0228)\left(\dfrac{29}{360}\right) \approx 110.20$

    Cost = (face value – interest) + service fee
    = 60,000 – 110.20 + 35 = $59,924.80

EXERCISE SET 11.4

1. (1.02)(375) = $382.50

3. (0.63) (850) = $535.50

5. Solving:
    $I = Prt$
    $1.24 = 49.375r(1)$
    $0.0251 \approx r$
    The yield is 2.51%.

7. Solving:
    $I = Prt$
    $0.58 = 31.75r(1)$
    $0.0183 \approx r$
    The yield is 1.83%.

9. a. 158.83 – 115.02 = $43.81

    b. 4.36(750) = $3270.00

    c. 6,976,300 shares

    d. increase

    e. $141.13

11. Dividing:
    $\dfrac{5000}{98.802} \approx 50.6$
    You can buy 50 shares.

13. a. You paid:
        48.96(1000) = $48,960
        You sold the shares for:
        74.20(1000) = $74,200
        The profit is 74,200 – 48,960 = $25,240.

    b. 0.019(74,200) = $1409.80

15. a. You paid:
        92(800) = $73,600
        You sold the shares for:
        134.54(800) = $107,632
        The profit is 107,632 – 73,600 = $34,032.

    b. 0.023(107,632) = $2475.54

17. $I = Prt = 6000(0.042)(1) = $252

19. $I = Prt = 8000(0.035)(3) = $840

21. Using the formula:
    $NAV = \dfrac{50,000,000 - 5,000,000}{2,000,000} = 22.50

23. First find NAV:
    $NAV = \dfrac{15,000,000 - 1,000,000}{2,000,000} = 7
    Dividing: $\dfrac{5000}{7} \approx 714.3$. You can buy 714 shares.

25. First find total assets: 500,000,000 + 500,000
    + 1,000,000 = 501,500,000
    Find NAV:
    $NAV = \dfrac{501,500,000 - 2,000,000}{10,000,000} = 49.95
    Dividing: $\dfrac{12,000}{49.95} \approx 240.2$. You can buy 240 shares.

27. Load fund:
    First deduct the load:
    2500 – (0.04)(2500) = $2400
    At the end of the first year:
    2400 + (0.08)(2400) = $2592.00
    The management fee must be deducted:
    2592 – (0.00015)(2592) = $2591.61
    At the end of the second year:
    2591.61 + (0.08)(2591.61) = $2798.94
    The management fee must be deducted:
    2798.94 – (0.00015) (2798.94) = $2598.52
    No load fund:
    At the end of the first year:
    2500 + (0.06) (2500) = $2650.00
    The management fee must be deducted:
    2650–(0.0015)(2650)≈$2650 – $3.98 =
    $2646.02
    At the end of the second year:
    2646.02 + (0.06)(2646.02) = $2804.78
    The management fee must be deducted:
    2804.78–(0.0015)(2804.78) = $2800.57
    The no-load fund's value ($2800.57) is $2.05
    greater than the load fund's value ($2798.52).

29. Answers will vary.

EXCURSION EXERCISES, SECTION 11.5

1. First calculate $\frac{r}{n}$ and store the results.

$$\frac{r}{n} = \frac{0.065}{12}$$

Calculate the monthly payment. For a 30-year loan, $nt = 30(12) = 360$.

$$PMT = A \left(\frac{\dfrac{r}{n}}{1 - \left(1 + \dfrac{r}{n}\right)^{-nt}} \right)$$

$$= 285,000 \left(\frac{0.065 / 12}{1 - (1 + 0.065 / 12)^{-360}} \right)$$

$$\approx 1801.39$$

The monthly mortgage payment is $1801.39.

3. $285,000 + 4275 = 289,275$
The modified mortgage is $289,275.

5. For both options, the term of the loan is 30 years, so $nt = 30(12) = 360$.
Option 1

Calculate $\frac{r}{n}$ and store the results.

$$\frac{r}{n} = \frac{0.0725}{12}$$

Calculate the monthly payment.

$$PMT = A \left(\frac{\dfrac{r}{n}}{1 - \left(1 + \dfrac{r}{n}\right)^{-nt}} \right)$$

$$= 100,000 \left(\frac{0.0725 / 12}{1 - (1 + 0.0725 / 12)^{-360}} \right)$$

$$\approx 682.18$$

The monthly mortgage payment for Option 1 is $682.18.

Option 2

Calculate $\frac{r}{n}$ and store the results.

$$\frac{r}{n} = \frac{0.07}{12}$$

Calculate the monthly payment.

$$PMT = A \left(\frac{\dfrac{r}{n}}{1 - \left(1 + \dfrac{r}{n}\right)^{-nt}} \right)$$

$$= 100,000 \left(\frac{0.07 / 12}{1 - (1 + 0.07 / 12)^{-360}} \right)$$

$$\approx 665.30$$

The monthly mortgage payment for Option 2 is $665.30.

7. **Option 1**
In 2 years, there are 24 payments of $682.18.
$24(682.18) = 16,372.32$

Total paid = mortgage payments + points
$= 16,372.32 + 1500 = 17,872.32$
The total paid for Option 1, after 2 years, is $17,872.32.

Option 2
In 2 years, there are 24 payments of $665.30.
$24(665.30) = 15,967.20$
Total paid = mortgage payments + points
$= 15,967.20 + 2000 = 17,967.20$
The total paid for Option 2, after 2 years, is $17,967.20.

9. Option 1 is more cost effective if you stay in the home for 2 years or less. Option 2 is more cost effective if you stay in the home for 3 years or more.

EXERCISE SET 11.5

1. Down payment = 25% of $258,000
$= 0.25(258,000)$
$= 64,500$
The down payment is $64,500.
Mortgage = selling price – down payment
$= 258,000 - 64,500 = 193,500$
The mortgage is $193,500.

3. Points = 2.25% of $250,000
$= 0.0225(250,000) = 5625$
The charge for points is $5625.

5. Down payment = 30% of $309,000
$= 0.30(309,000)$
$= 92,700$
Mortgage = selling price – down payment
$= 309,000 - 92,700 = 216,300$
Points = 3% of $216,300
$= 0.03(216,300) = 6489$
Sum of down payment and closing costs
$= 92,700 + 350 + 6489 = 99539$
The total of the down payment and closing costs is $99,539.

7. Down payment = 25% of $121,500
$= 0.25(121,500)$
$= 30,375$
Mortgage = selling price – down payment
$= 121,500 - 30,375$
$= 91,125$

Points = $3\frac{1}{2}$% of $91,125

$= 0.035(91,125) = 3189.38$
Sum of down payment and closing costs
$= 30,375 + 725 + 3189.38 = 34,289.38$
The total of the down payment and closing costs is $34,289.38.

9. First calculate $\frac{r}{n}$ and store the results.

$$\frac{r}{n} = \frac{0.0775}{12}$$

Calculate the monthly payment. For a 25-year loan, $nt = 25(12) = 300$.

$$PMT = A\left(\frac{\frac{r}{n}}{1-\left(1+\frac{r}{n}\right)^{-nt}}\right)$$

$$= 129{,}000\left(\frac{0.0775/12}{1-(1+0.0775/12)^{-300}}\right)$$

$$\approx 974.37$$

The monthly mortgage payment is $974.37.

11. First calculate $\frac{r}{n}$ and store the results.

$$\frac{r}{n} = \frac{0.0815}{12}$$

Calculate the monthly payment. For a 15-year loan, $nt = 15(12) = 180$.

$$PMT = A\left(\frac{\frac{r}{n}}{1-\left(1+\frac{r}{n}\right)^{-nt}}\right)$$

$$= 223{,}500\left(\frac{0.0815/12}{1-(1+0.0815/12)^{-180}}\right)$$

$$\approx 2155.28$$

The monthly mortgage payment is $2155.28.

13. a. First calculate $\frac{r}{n}$ and store the results.

$$\frac{r}{n} = \frac{0.0775}{12}$$

Calculate the monthly payment. For a 30-year loan, $nt = 30(12) = 360$.

$$PMT = A\left(\frac{\frac{r}{n}}{1-\left(1+\frac{r}{n}\right)^{-nt}}\right)$$

$$= 152{,}000\left(\frac{0.0775/12}{1-(1+0.0775/12)^{-360}}\right)$$

$$\approx 1088.95$$

The monthly mortgage payment is $1088.95.

 b. Multiply the number of payments by the monthly payment.
 1088.95(360) = 392,022
 The total of payments over the life of the loan is $392,022.

 c. Subtract the mortgage from the total of the payments.
 392,022 − 152,000 = 240,022
 The amount of interest paid over the life of the loan is $240,022.

15. First calculate $\frac{r}{n}$ and store the results.

$$\frac{r}{n} = \frac{0.087}{12}$$

Calculate the monthly payment. For a 15-year loan, $nt = 15(12) = 180$.

$$PMT = A\left(\frac{\frac{r}{n}}{1-\left(1+\frac{r}{n}\right)^{-nt}}\right)$$

$$= 219{,}990\left(\frac{0.087/12}{1-(1+0.087/12)^{-180}}\right)$$

$$\approx 2192.20$$

The monthly mortgage payment is $2192.20.
Next, find the total paid over the life of the loan.
 2192.20(180) = 394,596
The total of payments over the life of the loan is $394,596.
Last, find the total amount of interest paid.
 394,596 − 219,990 = 174,606
The amount of interest paid over the life of the loan is $174,606.

17. Find the down payment.
 0.10(208,500) = 20,850
The down payment is $20,850.
Find the mortgage.
 208,500 − 20,850 = 187,650
The mortgage is $187,650.
Calculate the mortgage payment. First calculate $\frac{r}{n}$ and store the results.

$$\frac{r}{n} = \frac{0.09}{12}$$

Calculate the monthly payment. For a 15-year loan, $nt = 15(12) = 180$.

$$PMT = A\left(\frac{\frac{r}{n}}{1-\left(1+\frac{r}{n}\right)^{-nt}}\right)$$

$$= 187{,}650\left(\frac{0.09/12}{1-(1+0.09/12)^{-180}}\right)$$

$$\approx 1903.27$$

The monthly mortgage payment is $1903.27.
Find the amount of interest paid on the first mortgage payment.

$$I = Prt = 187{,}650(0.09)\left(\frac{1}{12}\right) \approx 1407.38$$

The interest paid is $1407.38.
Find the principal paid on the first mortgage payment.
 1903.27 − 1407.38 = 495.89
The principal paid is $495.89.

19. Find the down payment.
 0.30(185,000) = 55,500
The down payment is $55,500.

Find the mortgage.
$185{,}000 - 55{,}500 = 129{,}500$
The mortgage is $129,500.
Calculate the mortgage payment. First calculate
$\dfrac{r}{n}$ and store the results.

$\dfrac{r}{n} = \dfrac{0.125}{12}$

Calculate the monthly payment. For a 20-year
loan, $nt = 20(12) = 240$.

$$PMT = A\left(\dfrac{\dfrac{r}{n}}{1-\left(1+\dfrac{r}{n}\right)^{-nt}}\right)$$

$$= 129{,}500\left(\dfrac{0.125/12}{1-(1+0.125/12)^{-240}}\right)$$

$$\approx 1471.30$$

The monthly mortgage payment is $1471.30.
Find the amount of interest paid on the payment
for the first month.

$$I = Prt = 129{,}500(0.125)\left(\dfrac{1}{12}\right) \approx 1348.96$$

The interest paid on the first payment is
$1348.96.
Find the principal paid.
$1471.30 - 1348.96 = 122.34$
The principal paid on the first payment is
$122.34.
Find the balance on the loan after the first
mortgage payment.
$129{,}500 - 122.34 = 129{,}377.66$
The balance after the first payment is
$129,377.66.
Find the amount of interest paid on the
mortgage payment for the second month.

$$I = Prt = 129{,}377.6(0.125)\left(\dfrac{1}{12}\right)$$

$$\approx 1347.68$$

The interest paid on the second payment is
$1347.68.
Find the principal paid.
$1471.30 - 1347.68 = 123.62$
The principal paid on the second payment is
$123.62.

21. Use the APR Loan Payoff Formula. You have
been making payments for 6 years, or 72
months. There are 360 months in a 30-year
loan, so $U = 360 - 72 = 288$. $\dfrac{r}{n} = \dfrac{0.085}{12}$

$$A = PMT\left(\dfrac{1-\left(1+\dfrac{r}{n}\right)^{-U}}{\dfrac{r}{n}}\right)$$

$$= 913.10\left(\dfrac{1-(1+0.085/12)^{-288}}{0.085/12}\right)$$

$$\approx 112{,}025.49$$

The loan payoff is $112,025.49.

23. Use the APR Loan Payoff Formula. You have
been making payments for 4 years, or 48
months. There are 180 months in a 15-year
loan, so $U = 180 - 48 = 132$.

$$\dfrac{r}{n} = \dfrac{0.0725}{12}$$

$$A = PMT\left(\dfrac{1-\left(1+\dfrac{r}{n}\right)^{-U}}{\dfrac{r}{n}}\right)$$

$$= 672.39\left(\dfrac{1-(1+0.0725/12)^{-132}}{0.0725/12}\right)$$

$$\approx 61{,}039.75$$

The loan payoff is $61,039.75.

25. Find the monthly property tax bill.
$594 \div 12 = 49.50$
The monthly property tax bill is $49.50.
Find the monthly fire insurance bill.
$300 \div 12 = 25$
The monthly fire insurance bill is $25.
Find the total monthly payment.
$996.60 + 49.50 + 25 = 1071.10$
The total monthly payment for the mortgage,
property tax, and fire insurance is $1071.10.

27. Find the monthly mortgage payment. First
calculate $\dfrac{r}{n}$ and store the results.

$$\dfrac{r}{n} = \dfrac{0.0715}{12}$$

For a 25-year loan, $nt = 25(12) = 300$.

$$PMT = A\left(\dfrac{\dfrac{r}{n}}{1-\left(1+\dfrac{r}{n}\right)^{-nt}}\right)$$

$$= 259{,}500\left(\dfrac{0.0715/12}{1-(1+0.0715/12)^{-300}}\right)$$

$$\approx 1859.00$$

The monthly mortgage payment is $1859.00.
Find the monthly property tax bill.
$1320 \div 12 = 110$
The monthly property tax bill is $110.
Find the monthly fire insurance bill.
$642 \div 12 = 53.50$
The monthly fire insurance bill is $53.50.
Find the total monthly payment.
$1859.00 + 110 + 53.50 = 2022.50$
The total monthly payment for the mortgage,
property tax, and fire insurance is $2022.50.

29. a. Find the monthly mortgage payment for a
15-year loan. First calculate $\dfrac{r}{n}$ and store
the results.

$$\dfrac{r}{n} = \dfrac{0.08125}{12}$$

For a 15-year loan, $nt = 15(12) = 180$.

$$PMT = A \left(\frac{\frac{r}{n}}{1 - \left(1 + \frac{r}{n}\right)^{-nt}} \right)$$

$$= 150,000 \left(\frac{0.08125/12}{1 - (1 + 0.08125/12)^{-180}} \right)$$

$$\approx 1444.32$$

The monthly mortgage payment on a 15-year loan is $1444.32.

Find the monthly mortgage payment for a 30-year loan. The value of i is the same as for the 15-year loan.

For a 30-year loan, $n = 30(12) = 360$.

$$PMT = A \left(\frac{i}{1 - (1 + i)^{-nt}} \right)$$

$$= 150,000 \left(\frac{0.08125/12}{1 - (1 + 0.08125/12)^{-360}} \right)$$

$$\approx 1113.75$$

The monthly mortgage payment on a 30-year loan is $1113.75.

$$1444.32 - 1113.75 = 330.57$$

The payment on the 15-year loan is $330.57 greater than the payment on the 30-year loan.

b. Find the amount paid over the life of each loan.

For the 15-year loan, there are 180 payments of $1444.32.

$$180(1444.32) = 259,977.60$$

For the 30-year loan, there are 360 payments of $1113.75.

$$360(1113.75) = 400,950$$

$$400,950 - 259,977.60 = 140,972.40$$

The 15-year loan costs $140,972.40 less over the life of the loan than the 30-year loan.

31. a. Find the monthly mortgage payment for a 20-year loan. First calculate $\frac{r}{n}$ and store the results.

$$\frac{r}{n} = \frac{0.0675}{12}$$

For a 20-year loan, $nt = 20(12) = 240$.

$$PMT = A \left(\frac{\frac{r}{n}}{1 - \left(1 + \frac{r}{n}\right)^{-nt}} \right)$$

$$= 349,500 \left(\frac{0.0675/12}{1 - (1 + 0.0675/12)^{-240}} \right)$$

$$\approx 2657.47$$

The monthly mortgage payment on a 20-year loan is $2657.47.

Find the monthly mortgage payment for a 30-year loan. The value of $\frac{r}{n}$ is the same as for the 20-year loan.

For a 30-year loan, $nt = 30(12) = 360$.

$$PMT = A \left(\frac{\frac{r}{n}}{1 - \left(1 + \frac{r}{n}\right)^{-nt}} \right)$$

$$= 349,500 \left(\frac{0.0675/12}{1 - (1 + 0.0675/12)^{-360}} \right)$$

$$\approx 2266.85$$

The monthly mortgage payment on a 30-year loan is $2266.85.

$$2657.47 - 2266.85 = 390.62$$

The payment on the 20-year loan is $390.62 greater than the payment on the 30-year loan.

b. Find the amount paid over the life of each loan.

For the 20-year loan, there are 240 payments of $2657.47.

$$240(2657.47) = 637,792.80$$

For the 30-year loan, there are 360 payments of $2266.85.

$$360(2266.85) = 816,066$$

$$816,066 - 637,792.80 = 178,273.20$$

The 20-year loan costs $178,273.20 less over the life of the loan than the 30-year loan.

33. Let x = the price of the house.

20% of the price of the house = $25,000.

$$0.20x = 25,000$$

$$x = \frac{25,000}{0.20} = 125,000$$

The maximum price they can offer is $125,000.

35. Let x = the price of the house.

Then, down payment = $0.15x$, mortgage = $x - 0.15x$, 4 points = 4% of the mortgage = $0.04(x - 0.15x)$.

Down payment + closing costs + points = total.

$$0.15x + 380 + 0.04(x - 0.15x) = 39,400$$

$$0.15x + 380 + 0.04(0.85x) = 39,400$$

$$0.15x + 380 + 0.034x = 39,400$$

$$0.184x + 380 = 39,400$$

$$0.184x = 39,020$$

$$x = \frac{39,020}{0.184}$$

$$x = 212,065$$

The maximum price they can offer is $212,065.

37. No. The 260th payment remains the first payment in which the principal paid exceeds the amount of interest paid, regardless of the amount of the loan.

39. a.   Find the down payment.
$0.30(208,750) = 62,625$
The down payment is $62,625.
Find the mortgage.
$208,750 - 62,625 = 146,125$
The mortgage is $146,125.
Find the points.
Points = 1.5% of $146,125
$= 0.015(146,125) = 2191.88$
The charge for points is $2191.88.
Amount due at closing
= down payment + points + lender fees
$= 62,625 + 2191.88 + 825 = 65,641.88$
The amount due at closing is $65,641.88.

b.   First, find the monthly mortgage payment.

Calculate $\dfrac{r}{n}$ and store the results.

$\dfrac{r}{n} = \dfrac{0.0775}{12}$

For a 30-year loan, $nt = 30(12) = 360$.

$$PMT = A\left(\dfrac{\dfrac{r}{n}}{1-\left(1+\dfrac{r}{n}\right)^{-nt}}\right)$$

$$= 148,125\left(\dfrac{0.0775/12}{1-(1+0.0775/12)^{-360}}\right)$$

$$\approx 1046.86$$

The monthly mortgage payment is $1046.86.
Use the APR Loan Payoff Formula. You have been making payments for 5 years, or 60 months. There are 360 months in a 30-year loan, so $U = 360 - 60 = 300$.

$$A = PMT\left(\dfrac{1-(1+r/n)^{-U}}{r/n}\right)$$

$$= 1046.86\left(\dfrac{1-(1+0.0775/12)^{-300}}{0.0775/12}\right)$$

$$\approx 138,596.60$$

The loan payoff is $138,596.60.

c.   Selling fees = 6% of selling price
$= 0.06(248,000) = 14,880$
The selling fees are $14,880.
Proceeds of sale
= selling price – selling fee – loan payoff
– down payment – points paid
– lender fees
$= 248,000 - 14,880 - 138,596.60 - 62,625$
$- 2191.88 - 825$
$= 28,881.52$
The proceeds of the sale after deducting selling fees are $28,881.52.

d. $\dfrac{28,881.52}{65,640} \times 100 \approx 44\%$

The percent return on your investment is 44%.

CHAPTER 11 REVIEW EXERCISES

1.   Because the interest rate is an annual rate, the time must be measured in years:

$$t = \dfrac{4 \text{ months}}{1 \text{ year}} = \dfrac{4 \text{ months}}{12 \text{ months}} = \dfrac{4}{12}$$

$$I = Prt = 2750(0.0675)\left(\dfrac{4}{12}\right) \approx 61.88$$

The interest earned is $61.88.

2.   The interest rate is per month and the time period is in months.
$I = Prt = 8500(0.0115)(8) = 782$
The interest due is $782.

3.   The value for the time is

$$t = \dfrac{\text{number of days}}{360} = \dfrac{120}{360}$$

$$I = Prt = 4000(0.0675)\left(\dfrac{120}{360}\right) = 90$$

The interest earned is $90.

4.   Since the interest rate is per year, the time period of the loan is $t = \dfrac{108}{360}$ year.

$$A = P(1+rt) = 7000\left[1+0.104\left(\dfrac{108}{360}\right)\right]$$

$$= 7000(1.0312) = 7218.40$$

The maturity value of the loan is $7218.40.

5.   Solve the simple interest formula for r.
$I = Prt$

$$127.50 = 6800r\left(\dfrac{3}{12}\right)$$

$$127.50 = 1700r$$

$$\dfrac{127.50}{1700} = r$$

$$r = 7.5\%$$

The simple interest rate is 7.5%.

6. $P = 3000, r = 6.6\% = 0.066, n = 12, t = 3$

$$A = P\left(1+\dfrac{r}{n}\right)^{nt}$$

$$= 3000\left(1+\dfrac{0.066}{12}\right)^{(12 \cdot 3)}$$

$$= 3000(1.055)^{36} \approx 3654.90$$

The compound amount after 3 years is approximately $3654.90.

7. $P = 6400, r = 6\% = 0.06, n = 4, t = 10$

$$A = P\left(1+\dfrac{r}{n}\right)^{nt}$$

$$= 6400\left(1+\dfrac{0.06}{4}\right)^{(4 \cdot 10)}$$

$$= 6400(1.015)^{40} \approx 11,609.72$$

The compound amount after 10 years is approximately $11,609.72.

8. $P = 6000, r = 9\% = 0.09, n = 360, t = 3$

$$A = P\left(1+\frac{r}{n}\right)^{nt}$$

$$= 6000\left(1+\frac{0.09}{360}\right)^{(360\cdot3)}$$

$$= 6000(1.00025)^{1080} \approx 7859.52$$

The future value is approximately \$7859.52.

9. $P = 600, r = 7.2\% = 0.072, n = 360, t = 4$

$$A = P\left(1+\frac{r}{n}\right)^{nt}$$

$$= 600\left(1+\frac{0.072}{360}\right)^{(360\cdot4)}$$

$$= 600(1.0002)^{1440} \approx 800.23$$

Calculate the interest earned.
$I = A - P = 800.23 - 600 = 200.23$
The amount of interest earned is \$200.23.

10. $A = 18,500, r = 8\% = 0.08, n = 2, t = 7$

$$P = \frac{A}{\left(1+\frac{r}{n}\right)^{nt}}$$

$$= \frac{18,500}{\left(1+\frac{0.08}{2}\right)^{(2\cdot7)}}$$

$$= \frac{18,500}{(1.04)^{14}} \approx 10,683.29$$

\$10,683.29 should be invested in the account in order to have \$18,500 in 7 years.

11. a. $P = 8000, r = 7\% = 0.07, n = 4, t = 5$

$$A = P\left(1+\frac{r}{n}\right)^{nt}$$

$$= 8000\left(1+\frac{0.07}{4}\right)^{(4\cdot5)}$$

$$= 8000(1.0175)^{20} \approx 11,318.23$$

The maturity value of the loan is \$11,318.23.

b. $11,318.23 - 8000 = 3318.23$
The interest you are paying is \$3318.23.

12. $A = 80,000, r = 8\% = 0.08, n = 4, t = 18$

$$P = \frac{A}{\left(1+\frac{r}{n}\right)^{nt}}$$

$$= \frac{80,00}{\left(1+\frac{0.08}{4}\right)^{(4\cdot18)}}$$

$$= \frac{80,000}{(1.02)^{72}} \approx 19,225.50$$

\$19,225.50 should be invested in the account in order to have \$80,000 in 18 years.

13. Using the formula:
$I = Prt$
$0.66 = 60t(1)$
$0.66 = 60r$
$0.011 = r$
$r = 1.1\%$
The dividend yield is 1.1 %.

14. Find the annual interest payment.
$I = Prt = 20,000(0.045)(1) = 900$
Multiply the annual interest payment by the term of the bond.
$900(10) = 9000$
The total interest payments paid to the bondholder is \$9000.

15. The inflation rate is an annual rate, so $n = 1$.
$P = 0.89, r = 6\% = 0.06$

$$A = P\left(1+\frac{r}{n}\right)^{nt}$$

$$= 0.89\left(1+\frac{0.06}{1}\right)^{(1\cdot10)}$$

$$= 0.89(1.06)^{10} \approx 1.59$$

One pound of baking potatoes will cost approximately \$ 1.59 in 2021.

16. $A = 75,000, r = 7\% = 0.07$ and $t = 8$. The inflation rate is an annual rate, so $n = 1$.

$$P = \frac{A}{\left(1+\frac{r}{n}\right)^{nt}}$$

$$= \frac{75,000}{\left(1+\frac{0.07}{1}\right)^{(1\cdot8)}}$$

$$= \frac{75,000}{(1.07)^{8}} \approx 43,650.68$$

The purchasing power of the \$75,000 in will be approximately \$43,650.68 in 8 years.

17. Find the future value of \$100 after 1 year.
$r = 5.9\% = 0.059, n = 12$

$$A = P\left(1+\frac{r}{n}\right)^{nt}$$

$$= 100\left(1+\frac{0.059}{12}\right)^{(12\cdot1)}$$

$$\approx 106.06$$

Find the interest earned on the \$100.
$I = A - P = 106.06 - 100 = 6.06$. The effective interest rate is 6.06%.

18. Calculate $\left(1+\frac{r}{n}\right)^{nt}$ for each investment. 5.2% compounded quarterly: $n = 4, r = 5.2\% = 0.052, t = 1$

$$\left(1+\frac{r}{n}\right)^{nt} = \left(1+\frac{0.052}{4}\right)^{(4\cdot1)} \approx 1.053022817$$

5.4% compounded semiannually: $n = 2$,
$r = 5.4\% = 0.054$, $t = 1$

$$\left(1+\frac{r}{n}\right)^{nt} = \left(1+\frac{0.054}{2}\right)^{(2\cdot 1)} \approx 1.054729$$

Compare the two compound amounts.
  $1.054729 > 1.053022817$
So 5.4% compounded semiannually has a
higher yield than 5.2% compounded quarterly.

19. Recording the balances:

Date	Payment or purchase	Balance	Days	Balance times Days
Mar 11-17		$423.35	7	$2963.45
Mar 18-28	$145.50	$568.85	11	$6257.35
Mar 29-Apr10	−$250	$318.85	13	$4145.05
Total				$13,365.85

Average daily balance $= \dfrac{13,365.85}{31} \approx \$431.16$

20. Recording the balances:

Date	Payment or purchase	Balance	Days	Balance times Days
Sep 10-19		$450	10	$4500
Sep 20-24	$47	$497	5	$2485
Sep 25-27	$157	$654	3	$1962
Sep 28-Oct 9	−$175	$479	12	$5748
Total				$14,695

Average daily balance $= \dfrac{14,695}{30} \approx \$489.83$

$I = Prt = (489.83)(0.0125)(1) \approx 6.12$. The
finance charge is $6.12.

21. a.   Find the interest due.

$$I = Prt = 1500(0.075)\left(\frac{6}{12}\right) = 56.25$$

The total amount to be repaid to the bank is
$A = P + I = 1500 + 56.25 = 1556.25$

Monthly payment $= \dfrac{1556.25}{6} = 259.38$

The monthly payment is $259.38.

  b.   Using the formula:

$$\text{APR} \approx \frac{2nr}{n+1}$$

$$\approx \frac{2(6)(0.075)}{6+1} = \frac{0.9}{7} \approx 0.129$$

The annual percentage rate on the loan is
12.9%.

22. a.   Find the down payment.
Down payment $= 0.10(449) = 44.9$
Find the amount financed.
  $449 - 44.9 = 404.10$
Find the simple interest due.
  $I = Prt = 404.10(0.07)(1) \approx 28.29$
Find the total amount to be repaid to the
bank.
  $A = P + I = 404.10 + 28.29 = 432.39$
Find the monthly payment.

Monthly payment $= \dfrac{432.39}{12} = 36.03$

The monthly payment is $36.03.

  b.   Using the formula:

$$\text{APR} \approx \frac{2nr}{n+1}$$

$$\approx \frac{2(12)(0.07)}{12+1} = \frac{1.68}{13} = 0.129$$

The annual percentage rate on the loan is
12.9%.

23. Using:

$$\frac{r}{n} = \frac{0.085}{12}$$

Calculate the monthly payment. For a 2-year
loan, $nt = 2(12) = 24$.

$$PMT = A\left(\frac{\frac{r}{n}}{1-\left(1+\frac{r}{n}\right)^{-nt}}\right)$$

$$= 999\left(\frac{0.085/12}{1-(1+0.085/12)^{-24}}\right)$$

$$\approx 45.41$$

The monthly payment is $45.41.

24. a.   Sales tax $= 0.0625(9499) = 593.69$
The sales tax is $593.69.
Total cost $= 9499 + 593.69 = 10{,}092.69$
The total cost is $10,092.69.

  b.   Down payment $= 0.20(10{,}092.69) = 2018.54$
The down payment is $2018.54.

  c.   Loan amount $=$ total cost − down payment
Loan amount $= 10{,}092.69 - 2018.54$
$= 8074.15$

$$\frac{r}{n} = \frac{0.08}{12}$$

Calculate the monthly payment. For a 3-
year loan, $nt = 3(12) = 36$.

$$PMT = A\left(\frac{\frac{r}{n}}{1-\left(1+\frac{r}{n}\right)^{-nt}}\right)$$

$$= 8074.15\left(\frac{0.08/12}{1-(1+0.08/12)^{-36}}\right)$$

$$\approx 253.01$$

The monthly payment is $253.01.

25. Down payment = $(0.20)(28,450) = 5690$
Loan amount = purchase price – down payment
Loan amount = $28,450 - 5690 = 22,760$

$$\frac{r}{n} = \frac{0.0325}{12}$$

Calculate the monthly payment. For a 3-year loan, $nt = 3(12) = 36$.

$$PMT = A\left(\frac{\dfrac{r}{n}}{1-\left(1+\dfrac{r}{n}\right)^{-nt}}\right)$$

$$= 22,760\left(\frac{0.0325/12}{1-(1+0.0325/12)^{-36}}\right)$$

$$\approx 664.40$$

The monthly payment is $664.40.

26. a. Using the formula:

$$\frac{r}{n} = \frac{0.059}{12}$$

Calculate the monthly payment. For a 5-year loan, $nt = 5(12) = 60$.

$$PMT = A\left(\frac{\dfrac{r}{n}}{1-\left(1+\dfrac{r}{n}\right)^{-nt}}\right)$$

$$= 28,000\left(\frac{0.059/12}{1-(1+0.059/12)^{-60}}\right)$$

$$\approx 540.02$$

The monthly payment is $540.02.

b. He has owned the car for 3 years, so he has 2 years, or 24 months, of payments remaining. Thus, $U = 24$.

$$\frac{r}{n} = \frac{0.059}{12}$$

$$A = PMT\left(\frac{1-\left(1+\dfrac{r}{n}\right)^{-U}}{\dfrac{r}{n}}\right)$$

$$= 540.02\left(\frac{1-(1+0.059/12)^{-24}}{0.059/12}\right)$$

$$= 12,196.80$$

The loan payoff is $12,196.80.

27. a. You paid: $(500)(28.75) = 14,375$. You sold the stock for $(500)(39.40) = 19,700$.
The profit is $19,700 - 14,375 = \$5,325$.

b. $(0.013)(19,700) = \$256.10$

28. First find the NAV:

$$NAV = \frac{34,000,000 - 4,000,000}{2,000,000} = 15$$

You will buy: $\dfrac{3000}{15} = 200$ shares.

29. Down payment = 20% of $459,000
$$= 0.20(459,000)$$
$$= 91,800$$
Mortgage = selling price – down payment
$$= 459,000 - 91,800 = 367,200$$
Points = 1.75% of $367,200
$$= 0.0175(367,200) = 6426$$
Sum of down payment and closing costs
$$= 91,800 + 815 + 6426 = 99,041$$
The total of the down payment and closing costs is $99,041.

30. a. First calculate $\dfrac{r}{n}$ and store the results.

$$\frac{r}{n} = \frac{0.0675}{12}$$

Calculate the monthly payment. For a 30-year loan, $nt = 30(12) = 360$.

$$PMT = A\left(\frac{\dfrac{r}{n}}{1-\left(1+\dfrac{r}{n}\right)^{-nt}}\right)$$

$$= 255,800\left(\frac{0.0675/12}{1-(1+0.0675/12)^{-360}}\right)$$

$$\approx 1659.11$$

The monthly mortgage payment is $1659.11.

b. Multiply the number of payments by the monthly payment.
$1659.11(360) = 597,279.60$
The total of payments over the life of the loan is $597,279.60.

c. Subtract the mortgage from the total of the payments.
$597,279.60 - 255,800 = 341,479.60$
The amount of interest paid over the life of the loan is $341,479.60.

31. a. First calculate $\dfrac{r}{n}$ and store the results.

$$\frac{r}{n} = \frac{0.075}{12}$$

Calculate the monthly payment. For a 25-year loan, $nt = 25(12) = 300$.

$$PMT = A\left(\frac{\dfrac{r}{n}}{1-\left(1+\dfrac{r}{n}\right)^{-nt}}\right)$$

$$= 189,000\left(\frac{0.075/12}{1-(1+0.075/12)^{-300}}\right)$$

$$\approx 1396.69$$

The monthly mortgage payment is $1396.69.

b. Use the APR Loan Payoff Formula. He has been making payments for 10 years, or 120 months. There are 300 months in a 25-year loan, so $U = 300 - 120 = 180$.

$$\frac{r}{n} = \frac{0.075}{12}$$

$$A = PMT\left(\frac{1-\left(1+\frac{r}{n}\right)^{-U}}{\frac{r}{n}}\right)$$

$$= 1396.69\left(\frac{1-(1+0.075/12)^{-180}}{0.075/12}\right)$$

$$\approx 150,665.74$$

The loan payoff is \$150,665.74.

32. Find the monthly mortgage payment. First calculate $\frac{r}{n}$ and store the results.

$$\frac{r}{n} = \frac{0.07}{12}$$

Calculate the monthly payment. For a 15-year loan, $nt = 15(12) = 180$.

$$PMT = A\left(\frac{\frac{r}{n}}{1-\left(1+\frac{r}{n}\right)^{-nt}}\right)$$

$$= 278,950\left(\frac{0.07/12}{1-(1+0.07/12)^{-180}}\right)$$

$$\approx 2507.28$$

The monthly mortgage payment is \$2507.28.
Find the monthly property tax bill.
$1134 \div 12 = 94.50$
The monthly property tax bill is \$94.50.
Find the monthly fire insurance bill.
$681 \div 12 = 56.75$
The monthly fire insurance bill is \$56.75.
Find the total monthly payment.
$2507.28 + 94.50 + 56.75 = 2658.53$
The total monthly payment for the mortgage, property tax, and fire insurance is \$2658.53.

33. The loan is non-subsidized, so simple interest will accrue for 2 years before graduation.
$I = Prt = 17,000(0.04)(2) = 1394$

Total loan amount after graduation:
$17,000 + 1394 = 18,394$
$A = 18,394, r = 0.04, n = 12, t = 6$

$$PMT = A\left(\frac{\frac{r}{n}}{1-\left(1+\frac{r}{n}\right)^{-nt}}\right)$$

$$= 18,394\left(\frac{0.041/12}{1-(1+0.041/12)^{-(12\cdot6)}}\right)$$

$$\approx 288.62$$

The monthly payment is \$288.62.

CHAPTER 11 TEST

1. Because the interest rate is an annual rate, the time must be measured in years:

$$t = \frac{3\text{ months}}{1\text{ year}} = \frac{3\text{ months}}{12\text{ months}} = \frac{3}{12}$$

$$I = Prt = 5250(0.0825)\left(\frac{3}{12}\right) \approx 108.28$$

The interest earned is \$108.28.

2. The value for the time is

$$t = \frac{\text{number of days}}{360} = \frac{180}{360}$$

$$I = Prt = 6000(0.0675)\left(\frac{180}{360}\right) = 202.50$$

The interest earned is \$202.50.

3. Since the interest rate is per year, the time period of the loan is $t = \frac{200}{360}$ year.

$$A = P(1+rt)$$

$$= 8000\left[1+0.092\left(\frac{200}{360}\right)\right]$$

$$\approx 8000(1.051111) \approx 8408.89$$

The maturity value of the loan is \$8408.89.

4. Solve the simple interest formula for r.

$$I = Prt$$

$$114 = 7600r\left(\frac{2}{12}\right)$$

$$114 = \frac{7600}{6}r$$

$$\frac{6}{7600}(114) = r$$

$$r = 0.09 = 9\%$$

The simple interest rate is 9%.

5. $P = 44200, r = 7\% = 0.07, n = 12, t = 8$

$$A = P\left(1+\frac{r}{n}\right)^{nt}$$

$$= 4200\left(1+\frac{0.07}{12}\right)^{(12\cdot8)}$$

$$\approx 7340.87$$

The compound amount after 8 years is \$7340.87.

6. $P = 1500, r = 6.3\% = 0.063, n = 360, t = 3$

$$A = P\left(1+\frac{r}{n}\right)^{nt}$$

$$= 1500\left(1+\frac{0.063}{360}\right)^{(360\cdot3)}$$

$$= 1500(1.000175)^{1080} \approx 7340.87$$

Calculate the interest earned.
$I = A - P = 1812.03 - 1500 = 312.03$
The amount of interest earned is \$312.03.

7. a. $P = 10{,}500$, $r = 9.5\% = 0.095$, $n = 12$, $t = 4$

$$A = P\left(1 + \frac{r}{n}\right)^{nt}$$

$$= 10{,}500\left(1 + \frac{0.095}{12}\right)^{(12 \cdot 4)}$$

$$\approx 15{,}331.03$$

The maturity value of the loan is
$15,331.03.

b. $15{,}331.03 - 10{,}500 = 4831.03$
The interest you are paying is $4831.03.

8. $A = 30{,}000$, $r = 6.25\% = 0.0625$, $n = 360$, $t = 5$

$$P = A\left(1 + \frac{r}{n}\right)^{-nt}$$

$$= 30{,}000\left(1 + \frac{0.0625}{360}\right)^{-(360 \cdot 5)}$$

$$\approx 21{,}949.06$$

$21,949.06 should be invested in the account in
order to have $30,000 in 5 years.

9. Using the formula:
$$I = Prt$$
$$0.48 = 40r(1)$$
$$0.48 = 40r$$
$$r = 0.012 = 1.2\%$$
The dividend yield is 1.2%.

10. Find the annual interest payment.
$I = Prt = 5000(0.038)(1) = 190$
Multiply the annual interest payment by the
term of the bond.
$190(10) = 1900$
The total interest payments paid to the
bondholder is $1900.

11. The inflation rate is an annual rate, so $n = 1$.
$P = 224{,}000$, $r = 4.3\% = 0.043$,
$t = 2029 - 2016 = 13$

$$A = P\left(1 + \frac{r}{n}\right)^{nt}$$

$$= 224{,}000\left(1 + \frac{0.043}{1}\right)^{(1 \cdot 13)}$$

$$= 224{,}000(1.043)^{13} \approx 387{,}207.74$$

The median value of a single family house will
be approximately $387,207.74 in 2029.

12. Find the future value of $100 after 1 year.
$r = 6.25\%$, $n = 4$

$$A = P\left(1 + \frac{r}{n}\right)^{nt}$$

$$= 100\left(1 + \frac{0.0625}{4}\right)^{(4 \cdot 1)}$$

$$= 100(1.015625)^4 \approx 106.40$$

Find the interest earned on the $100.
$I = A - P = 106.40 - 100 = 6.40$
The effective interest rate is 6.40%.

13. Calculate $\left(1 + \frac{r}{n}\right)^{nt}$ for each investment. 0.4%
compounded monthly: $n = 12$,
$r = 0.4\% = 0.004$, $t = 1$

$$\left(1 + \frac{r}{n}\right)^{nt} = \left(1 + \frac{0.044}{12}\right)^{(12 \cdot 1)} \approx 1.044898269$$

4.6% compounded semiannually: $n = 2$,
$r = 4.6\% = 0.046$, $t = 1$

$$\left(1 + \frac{r}{n}\right)^{nt} = \left(1 + \frac{0.046}{2}\right)^{(2 \cdot 1)} \approx 1.046529$$

Compare the two compound amounts.
1.046529 > 1.044898269
4.6% compounded semiannually has a higher
yield than 4.4% compounded monthly.

14. Recording the balances:

Date	Payment or purchase	Balance	Days	Balance times Days
Oct 15-19		$515	5	$2575
Oct 20-27	$75	$590	8	$4720
Oct 28- Nov 14	-$250	$340	18	$6120
Total				$13,415

Average daily balance $= \dfrac{13{,}415}{31} \approx 432.74$

$I = Prt = (432.74)(0.018)(1) \approx 7.79$
The finance charge is $7.79.

15. a. Find the down payment.
Down payment $= 0.15(629) = 94.35$
Find the amount financed.
$629 - 94.35 = 534.65$
Find the simple interest due.
$I = Prt = 534.65(0.09)(1) \approx 48.12$
Find the total amount to be repaid to the
bank.
$A = P + I = 534.65 + 48.12 = 582.77$
Find the monthly payment.

Monthly payment $= \dfrac{582.77}{12} \approx 48.56$

The monthly payment is $48.56.

b. Using the formula:

$$\text{APR} \approx \frac{2nr}{n+1}$$

$$\approx \frac{2(12)(0.09)}{12+1} = \frac{2.16}{13} = 0.166$$

The annual percentage rate on the loan is
16.6%.

16. Using:
$$\frac{r}{n} = \frac{0.045}{12} = 0.00375$$

Calculate the monthly payment. For a 3-year
loan, $nt = 3(12) = 36$.

$$PMT = A\left(\frac{r/n}{1-(1+r/n)^{-nt}}\right)$$

$$= 1899\left(\frac{0.00375}{1-(1+0.00375)^{-36}}\right)$$

$$\approx 56.49$$

The monthly payment is $56.49.

17. a. You paid: $(800)(31.82) = 25{,}456$.
You sold the stock for
$(800)(25.70) = 20{,}560$.
The loss is $25{,}456 - 20{,}560 = \$4{,}896$.

b. $(0.011)(20560) = \$226.16$

18. First find the NAV:
$$NAV = \frac{42{,}000{,}000 - 6{,}000{,}000}{3{,}000{,}000} = 12$$

Dividing: $\frac{2500}{12} \approx 208.3$ You will buy 208

shares.

19. a. Sales tax $= 0.0625(6575) = 410.94$
The sales tax is $410.94.
Total cost $= 6575 + 410.94 = 6985.94$
The total cost is $6985.94.

b. Down payment $= 0.20(6985.94) = 1397.19$
The down payment is $1397.19.

c. Loan amount $=$ total cost $-$ down payment
Loan amount $= 6985.94 - 1397.19$
$$= 5588.75$$

$\dfrac{r}{n} = \dfrac{0.078}{12}$

Calculate the monthly payment. For a 3-year loan, $nt = 3(12) = 36$.

$$PMT = A\left(\frac{r/n}{1-(1+r/n)^{-nt}}\right)$$

$$= 5588.75\left(\frac{0.078/12}{1-(1+0.078/12)^{-36}}\right)$$

$$\approx 174.62$$

The monthly payment is $174.62

20. Down payment $= 20\%$ of $262{,}250$
$$= 0.20(262{,}250)$$
$$= 52{,}450$$

Mortgage $=$ selling price $-$ down payment
$$= 262{,}250 - 52{,}450$$
$$= 209{,}800$$

Points $= 3.25\%$ of $209{,}800$
$$= 0.0325(209{,}800) = 6818.50$$

Sum of down payment and closing costs
$= 52{,}450 + 815 + 6818.50 = 60{,}083.50$

The total of the down payment and closing costs is $60,083.50.

21. a. First calculate $\dfrac{r}{n}$ and store the results.

$$\frac{r}{n} = \frac{0.0675}{12}$$

Calculate the monthly payment. For a 30-year loan, $nt = 30(12) = 360$.

$$PMT = A\left(\frac{r/n}{1-(1+r/n)^{-nt}}\right)$$

$$= 236{,}000\left(\frac{0.0675/12}{1-(1+0.0675/12)^{-360}}\right)$$

$$\approx 1530.69$$

The monthly mortgage payment is $1530.69.

b. Use the APR Loan Payoff Formula. He has been making payments for 5 years, or 60 months. There are 360 months in a 30-year loan, so $U = 360 - 60 = 300$.

$$\frac{r}{n} = \frac{0.0675}{12}$$

$$A = PMT\left(\frac{1-(1+r/n)^{-U}}{r/n}\right)$$

$$= 1530.69\left(\frac{1-(1+0.0675/12)^{-300}}{0.0675/12}\right)$$

$$\approx 221{,}546.46$$

The loan payoff is $221,546.46.

22. Find the monthly mortgage payment. First calculate $\dfrac{r}{n}$ and store the results.

$$\frac{r}{n} = \frac{0.0725}{12}$$

Calculate the monthly payment. For a 20-year loan, $nt = 20(12) = 240$.

$$PMT = A\left(\frac{r/n}{1-(1+r/n)^{-nt}}\right)$$

$$= 312{,}000\left(\frac{0.0725/12}{1-(1+0.0725/12)^{-240}}\right)$$

$$\approx 2465.97$$

The monthly mortgage payment is $2465.97.
Find the monthly property tax bill.
$1044 \div 12 = 87$
The monthly property tax bill is $87.
Find the monthly fire insurance bill.
$516 \div 12 = 43$
The monthly fire insurance bill is $43.
Find the total monthly payment.
$2465.97 + 87 + 43 = 2595.97$
The total monthly payment for the mortgage, property tax, and fire insurance is $2595.97.

Chapter 12: Combinatorics and Probability

EXCURSION EXERCISES, SECTION 12.1

1.

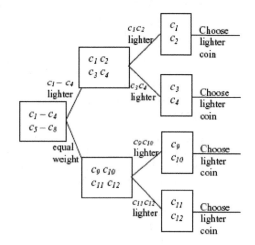

EXERCISE SET 12.1

1. $\{0, 2, 4, 6, 8\}$

3. {Monday, Tuesday, Wednesday, Thursday, Friday, Saturday, Sunday}

5.

	H	T
H	HH	HT
T	TH	TT

{HH, TT, HT, TH}

7.

	H	T
1	1H	1T
2	2H	2T
3	3H	3T
4	4H	4T
5	5H	5T
6	6H	6T

{1H, 2H, 3H, 4H, 5H, 6H,1T, 2T, 3T, 4T, 5T, 6T}

9. $\{S_1E_1D_1, S_1E_1D_2, S_1E_2D_1, S_1E_2D_2, S_1E_3D_1, S_1E_3D_2, S_2E_1D_1, S_2E_1D_2, S_2E_2D_1, S_2E_2D_2, S_2E_3D_1, S_2E_3D_2\}$

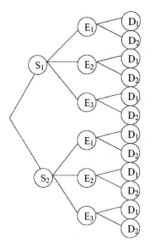

11. {ABCD, ABDC, ACBD, ACDB, ADBC, ADCB}

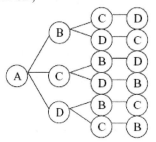

13. Any of the 4 numbers can be used in the first digit, so $n_1 = 4$. Because each number can be used only once, $n_2 = 3$. $4 \cdot 3 = 12$ two digit numbers.

15. There are four choices on each question, so there are
$4 \cdot 4 \cdot 4 \cdot 4 \cdot 4 \cdot 4 \cdot 4 \cdot 4 \cdot 4 \cdot 4 \cdot 4 \cdot 4 \cdot 4 \cdot 4 \cdot 4 \cdot 4 \cdot 4 \cdot 4 \cdot 4 \cdot 4 = 4^{20}$
possible ways to complete the test.

17. There are 7 numbers that can be used in the first digit, so $n_1 = 7$. There are 10 numbers that can be used in the last three digits, so $n_2 = 10$, $n_3 = 10$ and $n_4 = 10$. By the counting principle, there are $7 \cdot 10 \cdot 10 \cdot 10 \cdot 7000$ possible four-digit numbers.

19. Because a number starting with 0 is not two-digits, there are 9 numbers that can be used in the first digit, so $n_1 = 9$. There are 10 numbers that can be used in the last digit, so $n_2 = 10$. By the counting principle, there are $9 \cdot 10 = 90$ possible two-digit numbers.

21. Because a number starting with 0 is not two- digits, there are 9 numbers that can be used in the first digit, so $n_1 = 9$. Because a number divisible by 5 must end in 0 or 5, there are 2 numbers that can be used in the last digit, so $n_2 = 2$. By the counting principle, there are $9 \cdot 2 = 18$ possible two-digit numbers divisible by 5.

23. Because a number with 4 or greater in the first place is greater than 37, there are 6 numbers that can be used in the first digit, so $n_1 = 6$. For numbers starting with 4 or greater, any number may be in the second digit, so $n_2 = 10$. Additionally, some numbers may start with 3, so for these cases $n_1 = 1$. When paired with 3, 8 and 9 are the only numbers that make a number greater than 37, so $n_2 = 2$. There are $6 \cdot 10 = 60$ numbers and $1 \cdot 2 = 2$ numbers or $60 + 2 = 62$ possible two-digit numbers greater than 37.

25. Any one of the cards can be placed in box A, so $n_1 = 4$. Because a card cannot be placed in more than one box, there are 3 cards that can be placed in box B, so $n_2 = 3$. Similarly, there are 2 cards that can be placed in box C and 1 card in box D. By the counting principle, there are $4 \cdot 3 \cdot 2 \cdot 1 = 24$ possible placements.

27. Using the sample space found in Exercise 26, there are 15 elements where at least one card is in the box with the corresponding letter.

29. There are thirteen choices in each position, so there are $13 \cdot 13 \cdot 13 \cdot 13 = 13^4$ possible ways to choose the cards.

31. There is 1 choice for each position, so n_1, n_2, n_3, and $n_4 = 1$. By the counting principle, there are $1 \cdot 1 \cdot 1 \cdot 1 = 1$ possible ways to choose the cards.

33. Answers may vary.

35. There are 10 numbers that I can equal, so $n_1 = 10$. There are 15 numbers that J can equal, so $n_2 = 15$. By the counting principle, the instruction will have been executed $10 \cdot 15 = 150$ times. The last number for I is 10 and the last number for J is 15. Therefore, the final value will be $10 + 15 = 25$.

EXCURSION EXERCISES, SECTION 12.2

1. Using the formula:
$$C(60, 4) = \frac{60!}{4!(60-4)!} = \frac{60!}{4!56!}$$
$$= \frac{60 \cdot 59 \cdot 58 \cdot 57 \cdot 56!}{4!56!}$$
$$= \frac{60 \cdot 59 \cdot 58 \cdot 57}{4 \cdot 3 \cdot 2 \cdot 1} = 487,635$$

3. Using the formula:
$$C(20, 2) \cdot C(60, 2) = \frac{20!}{2!(20-2)!} \cdot \frac{60!}{2!(60-2)!}$$
$$= \frac{20!}{2!18!} \cdot \frac{60!}{2!58!}$$
$$= \frac{20 \cdot 19 \cdot 18!}{2!18!} \cdot \frac{60 \cdot 59 \cdot 58!}{2!58!}$$
$$= \frac{20 \cdot 19}{2 \cdot 1} \cdot \frac{60 \cdot 59}{2 \cdot 1}$$
$$= 190 \cdot 1770 = 336,300$$

5. Using the formula:
$$C(20, 4) = \frac{20!}{4!(20-4)!} = \frac{20!}{4!16!}$$
$$= \frac{20 \cdot 19 \cdot 18 \cdot 17 \cdot 16!}{4!16!}$$
$$= \frac{20 \cdot 19 \cdot 18 \cdot 17}{4 \cdot 3 \cdot 2 \cdot 1} = 4845$$

EXERCISE SET 12.2

1. $8! = 8 \cdot 7 \cdot 6 \cdot 5 \cdot 4 \cdot 3 \cdot 2 \cdot 1 = 40,320$

3. Evaluating:
$9! - 5! = (9 \cdot 8 \cdot 7 \cdot 6 \cdot 5 \cdot 4 \cdot 3 \cdot 2 \cdot 1) - (5 \cdot 4 \cdot 3 \cdot 2 \cdot 1)$
$= 362,880 - 120 = 362,760$

5. $(8-3)! = 5! = 5 \cdot 4 \cdot 3 \cdot 2 \cdot 1 = 120$

7. Evaluating:
$$P(8, 5) = \frac{8!}{(8-5)!} = \frac{8!}{3!} = \frac{8 \cdot 7 \cdot 6 \cdot 5 \cdot 4 \cdot 3!}{3!}$$
$$= 8 \cdot 7 \cdot 6 \cdot 5 \cdot 4 = 6720$$

9. Evaluating:
$$P(9, 7) = \frac{9!}{(9-7)!} = \frac{9!}{2!} = \frac{9 \cdot 8 \cdot 7 \cdot 6 \cdot 5 \cdot 4 \cdot 3 \cdot 2!}{2!}$$
$$= 9 \cdot 8 \cdot 7 \cdot 6 \cdot 5 \cdot 4 \cdot 3 = 181,440$$

11. $P(8, 0) = \dfrac{8!}{(8-0)!} = \dfrac{8!}{8!} = 1$

13. $P(8, 8) = \dfrac{8!}{(8-8)!} = \dfrac{8!}{0!} = \dfrac{8 \cdot 7 \cdot 6 \cdot 5 \cdot 4 \cdot 3 \cdot 2 \cdot 1}{1}$
$= 8 \cdot 7 \cdot 6 \cdot 5 \cdot 4 \cdot 3 \cdot 2 \cdot 1 = 40,320$

15. Evaluating:
$$P(8, 2) \cdot P(5, 3) = \frac{8!}{(8-2)!} \cdot \frac{5!}{(5-3)!}$$
$$= \frac{8 \cdot 7 \cdot 6!}{6!} \cdot \frac{5 \cdot 4 \cdot 3 \cdot 2!}{2!}$$
$$= 56 \cdot 60 = 3360$$

17. Evaluating:
$$\frac{P(6, 0)}{P(6, 6)} = \frac{\frac{6!}{(6-0)!}}{\frac{6!}{(6-6)!}} = \frac{\frac{6!}{6!}}{\frac{6!}{0!}} = \frac{1}{\frac{6 \cdot 5 \cdot 4 \cdot 3 \cdot 2 \cdot 1}{1}}$$
$$= \frac{1}{720}$$

19. Evaluating:
$$C(9, 2) = \frac{9!}{2!(9-2)!} = \frac{9!}{2!7!}$$
$$= \frac{9 \cdot 8 \cdot 7!}{2!7!} = \frac{9 \cdot 8}{2 \cdot 1} = 36$$

21. Evaluating:
$$C(12,\ 0) = \frac{12!}{0!(12-0)!} = \frac{12!}{0!12!} = \frac{1}{1} = 1$$

23. Evaluating:
$$C(6,\ 2) \cdot C(7,\ 3) = \frac{6!}{2!(6-2)!} \cdot \frac{7!}{3!(7-3)!}$$
$$= \frac{6!}{2!4!} \cdot \frac{7!}{3!4!}$$
$$= \frac{6 \cdot 5 \cdot 4!}{2!4!} \cdot \frac{7 \cdot 6 \cdot 5 \cdot 4!}{3!4!}$$
$$= \frac{6 \cdot 5}{2 \cdot 1} \cdot \frac{7 \cdot 6 \cdot 5}{3 \cdot 2 \cdot 1}$$
$$= 15 \cdot 35 = 525$$

25. Evaluating:
$$\frac{C(10,\ 4) \cdot C(5,\ 2)}{C(15,\ 6)} = \frac{\frac{10!}{4!(10-4)!} \cdot \frac{5!}{2!(5-2)!}}{\frac{15!}{6!(15-6)!}}$$
$$= \frac{\frac{10!}{4!6!} \cdot \frac{5!}{2!3!}}{\frac{15!}{6!9!}}$$
$$= \frac{\frac{10 \cdot 9 \cdot 8 \cdot 7 \cdot 6!}{2!18!} \cdot \frac{5 \cdot 4 \cdot 3!}{2!3!}}{\frac{15 \cdot 14 \cdot 13 \cdot 12 \cdot 11 \cdot 10 \cdot 9!}{6!9!}}$$
$$= \frac{\frac{10 \cdot 9 \cdot 8 \cdot 7}{4 \cdot 3 \cdot 2 \cdot 1} \cdot \frac{5 \cdot 4}{2 \cdot 1}}{\frac{15 \cdot 14 \cdot 13 \cdot 12 \cdot 11 \cdot 10}{6 \cdot 5 \cdot 4 \cdot 3 \cdot 2 \cdot 1}}$$
$$= \frac{210 \cdot 10}{5005} = \frac{60}{143}$$

27. Evaluating:
$$\frac{C(9,\ 7) \cdot C(5,\ 3)}{C(14,\ 10)} = \frac{\frac{9!}{7!(9-7)!} \cdot \frac{5!}{3!(5-3)!}}{\frac{14!}{10!(14-10)!}}$$
$$= \frac{\frac{9!}{7!2!} \cdot \frac{5!}{3!2!}}{\frac{14!}{10!4!}}$$
$$= \frac{\frac{9 \cdot 8 \cdot 7!}{7!2!} \cdot \frac{5 \cdot 4 \cdot 3!}{3!2!}}{\frac{14 \cdot 13 \cdot 12 \cdot 11 \cdot 10!}{10!4!}}$$
$$= \frac{\frac{9 \cdot 8}{2 \cdot 1} \cdot \frac{5 \cdot 4}{2 \cdot 1}}{\frac{14 \cdot 13 \cdot 12 \cdot 11}{4 \cdot 3 \cdot 2 \cdot 1}}$$
$$= \frac{36 \cdot 10}{1001} = \frac{360}{1001}$$

29. Evaluating:
$$4! \cdot C(10,\ 3) = 4! \cdot \frac{10!}{3!(10-3)!} = 4! \cdot \frac{10!}{3!7!}$$
$$= 4! \cdot \frac{10 \cdot 9 \cdot 8 \cdot 7!}{3!7!}$$
$$= 4 \cdot 10 \cdot 9 \cdot 8 = 2880$$

31. Using the formula:
$$C(7,\ 5) = \frac{7!}{5!(7-5)!} = \frac{7!}{5!2!}$$
$$= \frac{7 \cdot 6 \cdot 5!}{5!2!} = \frac{7 \cdot 6}{2 \cdot 1} = 21$$

33. Using the formula:
$$C(12,\ 7) = \frac{12!}{7!(12-7)!} = \frac{12!}{7!5!}$$
$$= \frac{12 \cdot 11 \cdot 10 \cdot 9 \cdot 8 \cdot 7!}{7!5!}$$
$$= \frac{12 \cdot 11 \cdot 10 \cdot 9 \cdot 8}{5 \cdot 4 \cdot 3 \cdot 2 \cdot 1} = 792$$

35. No. If there are only 7 items, there is no way to choose 9 of them.

37. The order in which the songs are played is important, so the number of ways to order the songs is
$$P(5,\ 5) = \frac{5!}{(5-5)!} = \frac{5!}{0!} = \frac{5 \cdot 4 \cdot 3 \cdot 2 \cdot 1}{1}$$
$$= 5 \cdot 4 \cdot 3 \cdot 2 \cdot 1 = 120$$

39. Because each officer position is different, the order of the selection is important, so the number of ways to make the selection is
$$P(16,\ 4) = \frac{16!}{(16-4)!} = \frac{16!}{12!}$$
$$= \frac{16 \cdot 15 \cdot 14 \cdot 13 \cdot 12!}{12!}$$
$$= 16 \cdot 15 \cdot 14 \cdot 13 = 43,680$$

41. Because band arrangements have a definite order, they are permutations.
   a. If there are no restrictions on the band order, then the number of different band orders is:
$$P(9,\ 9) = \frac{9!}{(9-9)!} = \frac{9!}{0!} = 9! = 362,880$$
   b. There are 6! ways to arrange the country bands and 3! ways to arrange the rock bands. We must also consider that either the country bands or the rock bands could be arranged at the beginning of the show. There are two ways to do this. By the counting principle, there are 2·6!·3! = 8640 arrangements for the bands to play.

43. We are looking for the number of permutations of the characters SOCCER77.
   With $n = 8$ (number of characters),
   $k_1 = 1$ (number of S's), $k_2 = 1$ (number of O's),
   $k_3 = 2$ (number of C's), $k_4 = 1$ (number of E's),
   $k_5 = 1$ (number of R's), and
   $k_6 = 2$ (number of 7's), we have
$$\frac{8!}{1!1!2!1!1!2!} = 10,080$$
   There are 10,080 possible passwords.

45. The order in which the firefighters are selected is not important, so the number of different teams is

$$C(24, 8) = \frac{24!}{8!(24-8)!} = \frac{24!}{8!16!}$$
$$= \frac{24 \cdot 23 \cdot 22 \cdot 21 \cdot 20 \cdot 19 \cdot 18 \cdot 17 \cdot 16!}{8!16!}$$
$$= \frac{24 \cdot 23 \cdot 22 \cdot 21 \cdot 20 \cdot 19 \cdot 18 \cdot 17}{8 \cdot 7 \cdot 6 \cdot 5 \cdot 4 \cdot 3 \cdot 2 \cdot 1}$$
$$= 735,471$$

47. The order of the questions on the test is not important, so the number of different exams is

$$C(7, 3) = \frac{7!}{3!(7-3)!} = \frac{7!}{3!4!}$$
$$= \frac{7 \cdot 6 \cdot 5 \cdot 4!}{3!4!} = \frac{7 \cdot 6 \cdot 5}{3 \cdot 2 \cdot 1} = 35$$

49. The order of the choices is not important, so the number of possible ways to choose the women, is $C(8, 3)$ and the number of ways to choose the men is $C(8, 3)$. Therefore, by the counting principle, there are $C(8,3) \cdot C(8,3)$ different committees.

$$C(8, 3) \cdot C(8, 3) = \frac{8!}{3!(8-3)!} \cdot \frac{8!}{3!(8-3)!}$$
$$= \frac{8!}{3!5!} \cdot \frac{8!}{3!5!}$$
$$= \frac{8 \cdot 7 \cdot 6 \cdot 5!}{5!3!} \cdot \frac{8 \cdot 7 \cdot 6 \cdot 5!}{3!5!}$$
$$= \frac{8 \cdot 7 \cdot 6}{3 \cdot 2 \cdot 1} \cdot \frac{8 \cdot 7 \cdot 6}{3 \cdot 2 \cdot 1}$$
$$= 56 \cdot 56 = 3136$$

51. Because the choice of team A playing team B is the same as the choice of team B playing team A, the order of selection is not important, so the number of different ways to match the teams is

$$C(4, 2) = \frac{4!}{2!(4-2)!} = \frac{4!}{2!2!}$$
$$= \frac{4 \cdot 3 \cdot 2!}{2!2!} = \frac{4 \cdot 3}{2 \cdot 1} = 6$$

Since the teams play each other twice, the number of games is $6 \cdot 2 = 12$.

53. There are 6 points and we must choose 2 to form a diagonal. Because order is not important, we find $C(6, 2)$.

$$C(6, 2) = \frac{6!}{2!(6-2)!} = \frac{6!}{2!4!}$$
$$= \frac{6 \cdot 5 \cdot 4!}{2!4!} = \frac{6 \cdot 5}{2 \cdot 1} = 15$$

Because 6 of the lines made are the sides of the hexagon, the number of diagonals is $15 - 6 = 9$.

55. The order in which the players are selected is not important, so the number of different teams

is

$$C(18, 9) = \frac{18!}{9!(18-9)!} = \frac{18!}{9!9!}$$
$$= \frac{18 \cdot 17 \cdot 16 \cdot 15 \cdot 14 \cdot 13 \cdot 12 \cdot 11 \cdot 10!}{9!9!}$$
$$= \frac{18 \cdot 17 \cdot 16 \cdot 15 \cdot 14 \cdot 13 \cdot 12 \cdot 11 \cdot 10}{9 \cdot 8 \cdot 7 \cdot 6 \cdot 5 \cdot 4 \cdot 3 \cdot 2 \cdot 1}$$
$$= 48,620$$

57. The order in which the flags are raised is important, so the number of different signals possible is

$$P(4, 4) = \frac{4!}{(4-4)!} = \frac{4!}{0!} = \frac{4 \cdot 3 \cdot 2 \cdot 1}{1}$$
$$= 4 \cdot 3 \cdot 2 \cdot 1 = 24$$

59. We are looking for the number of permutations of the letters *committee*. With
$n = 9$ (number of letters),
$k_1 = 1$ (number of *c*'s), $k_2 = 1$ (number of *o*'s),
$k_3 = 2$ (number of *m*'s), $k_4 = 1$ (number of *i*'s),
$k_5 = 2$ (number of *t*'s), $k_6 = 2$ (number of *e*'s),
we have

$$\frac{9!}{1!1!2!1!2!2!} = 45,360$$

There are 45,360 different arrangements possible.

61. The order in which the heads and tails are chosen is not important, so the number of distinct arrangements is

$$C(10, 5) = \frac{10!}{5!(10-5)!} = \frac{10!}{5!5!}$$
$$= \frac{10 \cdot 9 \cdot 8 \cdot 7 \cdot 6 \cdot 5!}{5!5!}$$
$$= \frac{10 \cdot 9 \cdot 8 \cdot 7 \cdot 6}{5 \cdot 4 \cdot 3 \cdot 2 \cdot 1} = 252$$

63. Because the group arrangements have a definite order, they are permutations.
There are 3! ways to arrange the first group,
4! ways to arrange the second group, and
2! ways to arrange the third group. We must also consider that either the first, second, or third group could be arranged at the beginning of the show. There are 3! ways to do this.
By the counting principle, there are
$3! \cdot 3! \cdot 4! \cdot 2! = 1728$ possible orders of the groups.

65. There are $C(4, 4)$ ways of choosing four aces from four aces and 48 choices for the fifth card in the hand. By the counting principle, the number of hands containing 4 aces is

$$C(4, 4) \cdot 48 = \frac{4!}{4!(4-4)!} \cdot 48 = \frac{4!}{4!0!} \cdot 48$$
$$= \frac{1}{1} \cdot 48 = 48$$

67. There are $C(4, 3)$ ways of choosing three jacks from four jacks and $C(48, 2)$ ways of choosing the other 2 cards from the remaining 48 cards. By the counting principle, the number of hands containing exactly three jacks is

$$C(4, 3) \cdot C(48, 2) = \frac{4!}{3!(4-3)!} \cdot \frac{48!}{2!(48-2)!}$$

$$= \frac{4!}{3!1!} \cdot \frac{48!}{2!46!}$$

$$= \frac{4 \cdot 3!}{3!1!} \cdot \frac{48 \cdot 47 \cdot 46!}{2!46!}$$

$$= \frac{4}{1} \cdot \frac{48 \cdot 47}{2 \cdot 1}$$

$$= 4 \cdot 1128 = 4512$$

69. There are $C(4, 2)$ ways of choosing two sevens from four sevens and $C(48, 3)$ ways of choosing the other 3 cards from the remaining 48 cards. By the counting principle, the number of hands containing exactly two sevens is

$$C(4, 2) \cdot C(48, 3) = \frac{4!}{2!(4-2)!} \cdot \frac{48!}{3!(48-3)!}$$

$$= \frac{4!}{2!2!} \cdot \frac{48!}{3!45!}$$

$$= \frac{4 \cdot 3 \cdot 2!}{2!2!} \cdot \frac{48 \cdot 47 \cdot 46 \cdot 45!}{2!45!}$$

$$= \frac{4 \cdot 3}{2 \cdot 1} \cdot \frac{48 \cdot 47 \cdot 46}{3 \cdot 2 \cdot 1}$$

$$= 6 \cdot 17,296 = 103,776$$

71. Answers may vary.

73. a. The number of possible Hamiltonian circuits is
$$\frac{(8-1)!}{2} = \frac{7!}{2} = \frac{7 \cdot 6 \cdot 5 \cdot 4 \cdot 3 \cdot 2 \cdot 1}{2} = 2520$$

   b. There are 6 vertices so the number of possible Hamiltonian circuits is
$$\frac{(6-1)!}{2} = \frac{5!}{2} = \frac{5 \cdot 4 \cdot 3 \cdot 2 \cdot 1}{2} = 60$$

   c. The number of possible Hamiltonian circuits is
$$\frac{(20-1)!}{2} = \frac{19!}{2} \approx \frac{1.21645 \times 10^{17}}{2}$$
$$\approx 6.0823 \times 10^{16}$$
The number of seconds needed to analyze them is
$6.0823 \times 10^{16} \div 1,000,000 = 6.0823 \times 10^{10}$
The years needed to analyze them is
$6.0823 \times 10^{10} \div 60 \div 60 \div 24 \div 365 \approx 1929$
Approximately 1929 years.

   d. The number of possible Hamiltonian circuits is
$$\frac{(40-1)!}{2} = \frac{39!}{2} \approx \frac{2.0398 \times 10^{46}}{2}$$
$$\approx 1.0199 \times 10^{46}$$

The number of seconds needed to analyze them is
$1.0199 \times 10^{46} \div 1 \times 10^{12} = 1.0199 \times 10^{34}$
The number of years needed to analyze them is
$1.0199 \times 10^{34} \div 60 \div 60 \div 24 \div 365$
$\approx 3.2341 \times 1026$
Approximately 3.23×10^{26} years.

EXCURSION EXERCISES, SECTION 12.3

1. Answers will vary.

3. Answers will vary.

5. Answers will vary.

EXERCISE SET 12.3

1. $S = \{$HHH, HHT, HTH, HTT, THH, THT, TTH, TTT$\}$

3. $S = \{$Nov. 1, Nov. 2, Nov. 3, Nov. 4, Nov. 5, Nov. 6, Nov. 7, Nov. 8, Nov. 9, Nov. 10, Nov. 11, Nov. 12, Nov. 13, Nov. 14$\}$

5. $S = \{$Alaska, Alabama, Arizona, Arkansas$\}$

7. $S = \{$BBB, BBG, BGB, BGG, GBB, GBG, GGB, GGG$\}$

9. $S = \{$BGG, GBG, GGB, GGG$\}$

11. $S = \{$BBG, BGB, BGG, GBB, GBG, GGB, GGG$\}$

13. The sample space was found in Exercise 7. The elements in the event that the Lins will have at least two girls was found in Exercise 9. Then the probability is $P(E) = \frac{n(E)}{n(S)} = \frac{4}{8} = \frac{1}{2}$.

15. The sample space was found in Exercise 7. The elements in the event that the Lins will have at least one girl are $\{$BBG, BGB, GBB, BGG, GBG, GGB, GGG$\}$. Then the probability is
$P(E) = \frac{n(E)}{n(S)} = \frac{7}{8}$.

17. The sample space for tossing a coin four times is $\{$HHHH, HHHT, HHTH, HHTT, HTHH, HTHT, HTTH, HTTT, THHH, THHT, THTH, THTT, TTHH, TTHT, TTTH, TTTT$\}$. The elements in the event that one head and three tails are tossed are $\{$HTTT, THTT, TTHT, TTTH$\}$. Then the probability is
$P(E) = \frac{n(E)}{n(S)} = \frac{4}{16} = \frac{1}{4}$.

19. The sample space for tossing a dodecahedral die is $\{1, 2, 3, 4, 5, 6, 7, 8, 9, 10, 11, 12\}$. The element in the event that the number on the upward face is 12 is $\{12\}$. Then the probability is

    $$P(E) = \frac{n(E)}{n(S)} = \frac{1}{12}.$$

21. The sample space for tossing a dodecahedral die was found in Exercise 19. The elements in the event that the number on the upward face is divisible by 4 are $\{4, 8, 12\}$. Then the probability is

    $$P(E) = \frac{n(E)}{n(S)} = \frac{3}{12} = \frac{1}{4}.$$

23. The sample space for tossing two six-sided dice was found in the text. The dice must be considered as distinct, so there are 36 possible outcomes. The elements in the event that the sum of the pips on the upward faces is 6 are

    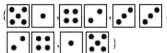

    Then the probability is

    $$P(E) = \frac{n(E)}{n(S)} = \frac{5}{36}.$$

25. The sample space for tossing two six-sided dice was found in the text. The dice must be considered as distinct, so there are 36 possible outcomes. The element in the event that the sum of the pips on the upward faces is 2 is

    $\{\boxed{\bullet \mid \bullet}\}$.

    Then the probability is $P(E) = \frac{n(E)}{n(S)} = \frac{1}{36}.$

27. The sample space for tossing two six-sided dice was found in the text. The dice must be considered as distinct, so there are 36 possible outcomes. There are no elements in the event that the sum of the pips on the upward faces is 1.

    Then the probability is $P(E) = \frac{n(E)}{n(S)} = \frac{0}{36} = 0.$

29. The sample space for tossing two six-sided dice was found in the text. The dice must be considered as distinct, so there are 36 possible outcomes. The elements in the event that the sum of the pips on the upward faces is at least 10 are (6 and 6, 6 and 5, 6 and 4, 5 and 6, 5 and 5, 4 and 6). Then the probability is

    $$P(E) = \frac{n(E)}{n(S)} = \frac{6}{36} = \frac{1}{6}.$$

31. The sample space for tossing two six-sided dice was found in the text. The dice must be considered as distinct, so there are 36 possible outcomes. The elements in the event that the sum of the pips on the upward faces is an even number are $\{6$ and 6, 6 and 4, 6 and 2, 5 and 5, 5 and 3, 5 and 1, 4 and 6, 4 and 4, 4 and 2, 3 and 5, 3 and 3, 3 and 1, 2 and 6, 2 and 4, 2 and 2, 1 and 5, 1 and 3, 1 and $1\}$. Then the probability is

    $$P(E) = \frac{n(E)}{n(S)} = \frac{18}{36} = \frac{1}{2}.$$

33. The sample space for tossing two six-sided dice was found in the text. The dice must be considered as distinct, so there are 36 possible outcomes. The elements in the event that both dice show the same number of pips are $\{6$ and 6, 5 and 5, 4 and 4, 3 and 3, 2 and 2, 1 and $1\}$. Then the probability is

    $$P(E) = \frac{n(E)}{n(S)} = \frac{6}{36} = \frac{1}{6}.$$

35. Because there are 52 cards in a standard deck of playing cards, there are 52 items in the sample space. The elements in the event that a nine is drawn are $\{9$ hearts, 9 diamonds, 9 spades, 9 clubs$\}$. Then the probability is

    $$P(E) = \frac{n(E)}{n(S)} = \frac{4}{52} = \frac{1}{13}.$$

37. Because there are 52 cards in a standard deck of playing cards, there are 52 items in the sample space. The elements in the event that a card between 5 and 9 inclusive is drawn, which includes the 5, 6, 7, 8, and 9 in each of four suits. Then the probability is

    $$P(E) = \frac{n(E)}{n(S)} = \frac{15}{52} = \frac{5}{13}.$$

39. Let E be the event that the voter is a Democrat. Then $P(E) = \frac{1267}{3228} \approx 0.39.$

41. Let E be the event that the voter is 50 years old or older. Then

    $$P(E) = \frac{804}{3228} = \frac{67}{269} \approx 0.25.$$

43. Let E be the event that the voter is an Independent. Then

    $$P(E) = \frac{150}{3228} = \frac{25}{538} \approx 0.05.$$

45. Let E be the event that the respondent did not complete high school. Then

    $$P(E) = \frac{52}{850} = \frac{26}{425}.$$

47. Let E be the event that the respondent has a Ph.D. or professional degree. Then
$$P(E) = \frac{36}{850} = \frac{18}{425}.$$

49. Let E be the event that the respondent earns from \$36,000 to \$45,999 annually. Then
$$P(E) = \frac{58}{293}.$$

51. Let E be the event that the respondent earns at least \$80,000 annually. There are $22 + 14 = 36$ people in the survey that earn at least \$80,000 annually. Then
$$P(E) = \frac{36}{293}.$$

53. To have white flowers, the plant must be *rr*. From the table, only one of the four possible genotypes is *rr*, so the probability that an offspring will have white flowers is $\frac{1}{4}$.

55. Make a Punnett square.

Parents	E	E
e	Ee	Ee
e	Ee	Ee

To have red eyes, the mouse must be *ee*. From the table, none of the four possible genotypes is *ee*, so the probability that an offspring will have red eyes is 0.

57. Answers may vary.

59. Use the formula for calculating the probability of an event when the odds in favor are known.
$$P(E) = \frac{a}{a+b} = \frac{1}{1+2} = \frac{1}{3}$$

61. Use the formula for calculating the probability of an event when the odds in favor are known.
$$P(E) = \frac{a}{a+b} = \frac{3}{3+7} = \frac{3}{10}$$

63. Use the formula for calculating the probability of an event when the odds in favor are known.
$$P(E) = \frac{a}{a+b} = \frac{8}{8+5} = \frac{8}{13}$$

65. The odds in favor are
$$\frac{P(E)}{1-P(E)} = \frac{0.2}{1-0.2} = \frac{0.2}{0.8} = \frac{1}{4}$$
The odds in favor are 1 to 4.

67. The odds in favor are
$$\frac{P(E)}{1-P(E)} = \frac{0.375}{1-0.375} = \frac{0.375}{0.625} = \frac{3}{5}$$
The odds in favor are 3 to 5.

69. The odds in favor are
$$\frac{P(E)}{1-P(E)} = \frac{0.55}{1-0.55} = \frac{0.55}{0.45} = \frac{11}{9}$$
The odds in favor are 11 to 9.

71. There are 30 answers total. Two are desirable and the other 28 are undesirable. The odds of choosing a Daily Double are 2:28 or 1:14, or 1 to 14.

73. Let E be the event of rolling an even number. Because there are three ways to roll an even number, there are three favorable outcomes, leaving three unfavorable outcomes. Odds in favor of $E = \frac{3}{3} = \frac{1}{1}$.
The odds in favor are 1 to 1.

75. Let E be the event of tossing heads four times. Refer to Exercise 16. Because there is one way to toss head four times, there is one favorable outcome, leaving 15 unfavorable outcomes.
Odds against $E = \frac{15}{1}$
The odds against are 15 to 1.

77. Let E be the event of an earthquake of magnitude 7.5 or greater occurring in California within any one year
Because the odds against are 48 to 49, the odds in favor are 49 to 48.
Use the formula for calculating the probability of an event when the odds in favor are known.
$$P(E) = \frac{a}{a+b} = \frac{49}{49+48} = \frac{49}{97}$$

79. Because the odds against are 8 to 3, the odds in favor are 3 to 8. Use the formula for calculating the probability of an event when the odds in favor are known.
$$P(E) = \frac{a}{a+b} = \frac{3}{3+8} = \frac{3}{11}$$

81. By making a tree diagram, we can find the sample space for the possible arrangements of the cards to be {ABCD, ABDC, ACBD, ACDB, ADBC, ADCB, BACD, BADC, BCAD, BCDA, BDAC, BDCA, CABD, CADB, CBAD, CBDA, CDAB, CDBA, DABC, DACB, DBAC, DBCA, DCAB, DCBA}. Assuming the boxes are in the order ABCD, the elements in the event that no card will be in a box with the same letter are {BADC, BCDA, BDAC, CADB, CDAB, CDBA, DABC, DCAB, DCBA}. Then the probability is
$$P(E) = \frac{n(S)}{n(S)} = \frac{9}{24} = \frac{3}{8}.$$

83. Answers may vary.

85. There are $C(52, 5)$ different hands of five cards possible.

$$C(52,\ 5) = \frac{52!}{5!(52-5)!} = \frac{52!}{5!47!}$$
$$= \frac{52\cdot51\cdot50\cdot49\cdot48\cdot47!}{5!47!}$$
$$= \frac{52\cdot51\cdot50\cdot49\cdot48}{5\cdot4\cdot3\cdot2\cdot1} = 2,598,960$$

There are $C(4, 3)$ ways of choosing three jacks from four jacks and $C(4, 2)$ ways of choosing two queens from four queens. By the counting principle, the number of hands containing three jacks and two queens is

$$C(4,\ 3)\cdot C(4,\ 2) = \frac{4!}{3!(4-3)!}\cdot\frac{4!}{2!(4-2)!}$$
$$= \frac{4!}{3!1!}\cdot\frac{4!}{2!2!}$$
$$= \frac{4\cdot3!}{3!1!}\cdot\frac{4\cdot3\cdot2!}{2!2!}$$
$$= \frac{4}{1}\cdot\frac{4\cdot3}{2\cdot1}$$
$$= 4\cdot6 = 24$$

Then the probability is

$$P(E) = \frac{n(S)}{n(S)} = \frac{24}{2,598,960} = \frac{1}{108,290}.$$

87. Because there are 38 slots, there are 38 items in the sample space. The elements in the event that the ball will stop on an odd number slot include 18 slots. Then the probability is

$$P(E) = \frac{n(S)}{n(S)} = \frac{18}{38} = \frac{9}{19}.$$

89. Because there are 38 slots, there are 38 items in the sample space. The element in the event that the ball will stop in a particular number includes only one slot. Then the probability is

$$P(E) = \frac{n(S)}{n(S)} = \frac{1}{38}.$$

91. Because there are 38 slots, there are 38 items in the sample space. The elements in the event that the ball will stop in one of 6 numbers include 6 slots. Then the probability is

$$P(E) = \frac{n(S)}{n(S)} = \frac{6}{38} = \frac{3}{19}.$$

EXCURSION EXERCISES, SECTION 12.4

1. Let A be the event of matching at least one lucky number. Then E^C is the event of matching none.

$$P(E) = 1 - P(E^C)$$
$$P(E) = 1 - \frac{C(60,\ 5)}{C(80,\ 5)} = 1 - \frac{5,461,512}{24,040,016}$$
$$\approx 0.773$$

The probability is 77.3%.

3. Thirteen out of the twenty numbers chosen are correct and two out of the sixty numbers not chosen are incorrect. Therefore, there are
$$C(20,\ 13)\cdot C(60,\ 2) = 77,520\cdot1770$$
$$= 137,210,400$$
ways to choose 13 correct numbers. There are $C(80, 15) = 6.636 \times 10^{15}$ ways to choose 15 numbers. The probability is
$$\frac{n(E)}{n(S)} = \frac{137,210,400}{6.636\times10^{15}} \approx 0.0000000207 \text{ or}$$
0.00000207%.

5. In the first event, none of the twenty numbers chosen are correct and twenty out of the sixty numbers not chosen are incorrect. Therefore, there are $C(60, 20) \approx 4.192 \times 10^{15}$ ways to choose no correct numbers. In the other event, five out of the twenty numbers chosen are correct and fifteen out of the sixty numbers not chosen are incorrect. Therefore, there are
$$C(20,\ 5)\cdot C(60,\ 15) = 15,504\cdot5.319\times10^{13}$$
$$\approx 8.247\times10^{17}$$
ways to choose 5 correct numbers. There are $C(80, 20) = 3.535 \times 10^{18}$ ways to choose 20 numbers. The probability of matching none of the numbers is
$$\frac{n(E)}{n(S)} = \frac{4.192\times10^{15}}{3.535\times10^{18}} \approx 0.0012 \text{ or } 0.12\%.$$
The probability of matching 5 of the lucky numbers is
$$\frac{n(E)}{n(S)} = \frac{8.247\times10^{17}}{3.535\times10^{18}} \approx 0.233 \text{ or } 23.3\%.$$

EXERCISE SET 12.4

1. Answers may vary.

3. Let $A = \{\heartsuit4,\ \heartsuit4,\ \diamondsuit4,\ \clubsuit4\}$ and $B = \{\spadesuit A,\ \heartsuit A,\ \diamondsuit A,\ \clubsuit A\}$. The events in A and B cannot occur at the same time, so they are mutually exclusive. There are 52 cards in a standard deck, thus $n(S) = 52$. The probability is
$$P(A \text{ or } B) = P(A) + P(B) = \frac{1}{13} + \frac{1}{13} = \frac{2}{13}.$$

5. Let A be the event of rolling a 2, and let B be the event of rolling a 10. The events in A and B cannot occur at the same time, so they are mutually exclusive. From the sample space given in the text,
$$P(A) = \frac{1}{36} \text{ and } P(B) = \frac{3}{36} = \frac{1}{12}.$$
Because A and B are mutually exclusive events,
$$P(A \text{ or } B) = P(A) + P(B) = \frac{1}{36} + \frac{1}{12} = \frac{4}{36} = \frac{1}{9}.$$

7. $P(A \text{ or } B) = P(A) + P(B) - P(A \text{ and } B)$
$$= 0.2 + 0.5 - 0.1$$
$$= 0.6$$

9. $P(A \cap B) = P(A) + P(B) - P(A \cup B)$
$$= 0.3 + 0.8 - 0.9$$
$$= 0.2$$

11. Let A be the event of the number being more than 6, and let B be the event of the number being odd. The sample space consists of 10 numbers. $n(A) = 4$, $n(B) = 5$ and $n(A) = 4$, $n(B) = 5$ and $n(A \text{ and } B) = 2$.
$P(A \text{ or } B) = P(A) + P(B) - P(A \text{ and } B)$
$$= \frac{4}{10} + \frac{5}{10} - \frac{2}{10} = \frac{7}{10}.$$

13. Let A be the event of the number being even, and let B be the event of the number being prime. The sample space consists of 10 numbers. $n(A) = 5$, $n(B) = 4$ and $n(A \text{ and } B) = 1$.
$P(A \text{ or } B) = P(A) + P(B) - P(A \text{ and } B)$
$$= \frac{4}{10} + \frac{5}{10} - \frac{1}{10} = \frac{4}{5}.$$

15. Let A be the event of rolling a 6, and let B be the event of rolling doubles. From the figure in in the text, there are 36 elements in the sample space. $n(A) = 5$, $n(B) = 6$ and $n(A \text{ and } B) = 1$.
$P(A \text{ or } B) = P(A) + P(B) - P(A \text{ and } B)$
$$= \frac{5}{36} + \frac{6}{36} - \frac{1}{36} = \frac{10}{36} = \frac{5}{18}.$$

17. Let A be the event of rolling an even number, and let B be the event of rolling doubles. From the figure in the text, there are 36 elements in the sample space. $n(A) = 18$, $n(B) = 6$ and $n(A \text{ and } B) = 6$.
$P(A \text{ or } B) = P(A) + P(B) - P(A \text{ and } B)$
$$= \frac{18}{36} + \frac{6}{36} - \frac{6}{36} = \frac{18}{36} = \frac{1}{2}.$$

19. Let A be the event of rolling an odd number, and let B be the event of rolling a number less than 6. From the figure in the text, there are 36 elements in the sample space. $n(A) = 18$, $n(B) = 10$ and $n(A \text{ and } B) = 6$.
$P(A \text{ or } B) = P(A) + P(B) - P(A \text{ and } B)$
$$= \frac{18}{36} + \frac{10}{36} - \frac{6}{36} = \frac{22}{36} = \frac{11}{18}.$$

21. Let A be the event of drawing an 8, and let B be the event of drawing a spade. The sample space consists of 52 cards. $n(A) = 4$, $n(B) = 13$ and $n(A \text{ and } B) = 1$.
$P(A \text{ or } B) = P(A) + P(B) - P(A \text{ and } B)$
$$= \frac{4}{52} + \frac{13}{52} - \frac{1}{52} = \frac{16}{52} = \frac{4}{13}.$$

23. Let A be the event of drawing a jack, and let B be the event of drawing a face card. The sample space consists of 52 cards. $n(A) = 4$, $n(B) = 12$ and $n(A \text{ and } B) = 4$.
$P(A \text{ or } B) = P(A) + P(B) - P(A \text{ and } B)$
$$= \frac{4}{52} + \frac{12}{52} - \frac{4}{52} = \frac{12}{52} = \frac{3}{13}$$

25. Let A be the event of drawing a diamond, and let B be the event of drawing a black card. The sample space consists of 52 cards. $n(A) = 13$, $n(B) = 26$ and $n(A \text{ and } B) = 0$.
$P(A \text{ or } B) = P(A) + P(B)$
$$= \frac{13}{52} + \frac{26}{52} = \frac{39}{52} = \frac{3}{4}$$

27. Let $A = \{\text{people aged 26-34}\}$, and let $B = \{\text{people employed part-time}\}$. Then, from the table, $n(A) = 348 + 67 + 27 = 442$, $n(B) = 164 + 203 + 67 + 179 + 162 = 775$ and $n(A \text{ and } B) = 67$. The total number of people represented in the table is 3179.
$P(A \text{ or } B) = P(A) + P(B) - P(A \text{ and } B)$
$$= \frac{442}{3179} + \frac{775}{3179} - \frac{67}{3179} = \frac{1150}{3179}$$

29. Let $A = \{\text{people under 18}\}$, and let $B = \{\text{people employed part-time}\}$. Then, from the table, $n(A) = 24 + 164 + 371 = 559$, $n(B) = 164 + 203 + 67 + 179 + 162 = 775$ and $n(A \text{ and } B) = 164$. The total number of people represented in the table is 3179.
$P(A \text{ or } B) = P(A) + P(B) - P(A \text{ and } B)$
$$= \frac{559}{3179} + \frac{775}{3179} - \frac{164}{3179} = \frac{1170}{3179}$$

31. If E is the event of winning the contest, then E^C is the event of not winning the contest, and $P(E^C) = 1 - P(E) = 1 - 0.04 = 0.96$

33. If E is the event of a senior being drafted by a NBA team, then E^C is the event of not being drafted on a NBA team, and
$$P(E^C) = 1 - P(E) = 1 - \frac{1}{75} = \frac{74}{75} \approx 98.7\%$$

35. If E is the event of tossing a 7, then E^C is the event of not tossing a 7, and
$$P(E^C) = 1 - P(E) = 1 - \frac{6}{36} = \frac{30}{36} = \frac{5}{6}$$

37. If E is the event of tossing a sum less than 4, then E^C is the event of tossing a sum of at least four, and
$$P(E^C) = 1 - P(E) = 1 - \frac{3}{36} = \frac{33}{36} = \frac{11}{12}$$

39. If E is the event of drawing an ace, then E^C is the event of not drawing an ace, and
$$P(E^C) = 1 - P(E) = 1 - \frac{4}{52} = \frac{48}{52} = \frac{12}{13}$$

41. If E is the event of getting no tails, then E^C is the event of getting at least one tail, and
$$P(E^C) = 1 - P(E) = 1 - \frac{1}{16} = \frac{15}{16}$$

43. Let $E = \{$at least one 1$\}$; then $E^C = \{$no 1's$\}$. Because on each toss of the die there are six possible outcomes, $n(S) = 6 \cdot 6 \cdot 6 = 216$. On each toss of the die there are five numbers that are not 1's. Therefore, $n(E^C) = 5 \cdot 5 \cdot 5 = 125$.
$$P(E^C) = 1 - P(E) = 1 - \frac{125}{216} = \frac{91}{216} \approx 0.421$$
The probability is 42.1%.

45. Let $E = \{$at least one roll of sum of 8$\}$; then $E^C = \{$no sum of 8 is rolled$\}$. Using the table in the text, there are 36 possibilities for each toss of the dice. Thus, $n(S) = 36 \cdot 36 \cdot 36 = 46,656$. On each toss of the dice there are 31 numbers that do not total 8, so
$$n(E^C) = 31 \cdot 31 \cdot 31 = 29,791$$
$$P(E) = 1 - P(E^C) = 1 - \frac{29,791}{46,656} = \frac{16,865}{46,656}$$
$$\approx 0.361$$
The probability is 36.1%.

47. Let $E = \{$at least one face card is drawn$\}$; then $E^C = \{$no face card is drawn$\}$. There are 52 possibilities for each card drawn. Thus, $n(S) = 52 \cdot 52 \cdot 52 = 140,608$. Each time a card is drawn, there are 40 cards that are not face cards, so $n(E^C) = 40 \cdot 40 \cdot 40 = 64,000$.
$$P(E) = 1 - P(E^C) = 1 - \frac{64,000}{140,608}$$
$$= \frac{76,608}{140,608} \approx 0.545$$
The probability is 54.5%.

49. Let $E = \{$at least one e-reader is defective$\}$; then $E^C = \{$no e-readers are defective$\}$. The number of elements in the sample space is
$$n(S) = C(30, 12) = \frac{30!}{12!(30-12)!}$$
$$= \frac{30!}{12!18!} = 86,493,225.$$
To find the number of outcomes that contain no defective e-readers , all of them must come from the 26 nondefective e-readers . The number of elements is
$$n(E^C) = C(26, 12) = \frac{26!}{12!(26-12)!}$$
$$= \frac{26!}{12!14!} = 9,657,700$$

$$P(E) = 1 - P(E^C) = 1 - \frac{9,657,700}{86,493,225}$$
$$= \frac{76,835,525}{86,493,22} \approx 0.888$$
The probability is 88.8%.

51. Let $E = \{$at least one employee wins a prize$\}$; then $E^C = \{$no employee wins a prize$\}$. The number of elements in the sample space is
$$n(S) = C(42, 3) = \frac{42!}{3!(42-3)!}$$
$$= \frac{42!}{3!39!} = 11,480.$$
To find the number of outcomes where no employee wins, all 3 must come from the 39 non-employee cards. The number of elements is
$$n(E^C) = C(39, 3) = \frac{39!}{3!(39-3)!}$$
$$= \frac{39!}{3!36!} = 9139.$$
$$P(E) = 1 - P(E^C) = 1 - \frac{9139}{11,480}$$
$$= \frac{2341}{11,480} \approx 0.204$$
The probability is 20.4%.

53. No, the ratio of the cards is still the same.

55. There are 2 possible settings for each of the 8 digits, so by the counting principle, there are $2 \cdot 2 \cdot 2 \cdot 2 \cdot 2 \cdot 2 \cdot 2 \cdot 2 = 256$ possible settings. Because there are less than 300 settings possible, the probability that at least two homeowners will set their door openers to the same code is 100%.

57. Answers will vary.

EXCURSION EXERCISES, SECTION 12.5

1. Using the Product Rule:
$$P(E^C) = \frac{364}{365} \cdot \frac{363}{365} \cdot \frac{362}{365} \cdot \frac{361}{365} \cdot \frac{360}{365} \cdot \frac{357}{365} \cdot \frac{358}{365}$$
$$\approx 0.9257$$
$$P(E) = 1 - P(E^C) \approx 0.0743$$
There is a 7.43% chance.

3. The pattern above continues for 22 times.
$$P(E^C) = \frac{364}{365} \cdot \frac{363}{365} \cdots \frac{344}{365} \cdot \frac{343}{365}$$
$$\approx 0.493$$
$$P(E) = 1 - P(E^C) \approx 0.507$$
There is a 50.7% chance.

EXERCISE SET 12.5

1. Answers will vary.

3. $P(A \mid B) = \dfrac{P(A \cap B)}{P(B)} = \dfrac{0.25}{0.4} = 0.625$

    $P(B \mid A) = \dfrac{P(A \cap B)}{P(A)} = \dfrac{0.25}{0.7} \approx 0.357$

5. $P(A \mid B) = \dfrac{P(A \cap B)}{P(B)} = \dfrac{0.07}{0.18} \approx 0.389$

    $P(B \mid A) = \dfrac{P(A \cap B)}{P(A)} = \dfrac{0.07}{0.61} \approx 0.115$

7. Let B = {people employed part-time} and
    A = {people between 35 and 49}.
    From the table, $n(A \cap B) = 179$,
    $n(A) = 581 + 179 + 104 = 864$, and $n(S) = 3179$.
    We have

    $P(B \mid A) = \dfrac{P(A \cap B)}{P(A)} = \dfrac{\frac{179}{3179}}{\frac{864}{3179}} = \dfrac{179}{864}$.

9. Let B = {people employed full-time} and
    A = {people between 18 and 49}. From the
    table, $n(A \cap B) = 185 + 348 + 581 = 1114$,
    $n(A) = 185 + 203 + 148 + 348 + 67 + 27 + 581$
    $+ 179 + 104 = 1842$, and $n(S) = 3179$. We have

    $P(B \mid A) = \dfrac{P(A \cap B)}{P(A)} = \dfrac{\frac{1114}{3179}}{\frac{1842}{3179}} = \dfrac{1114}{1842} = \dfrac{557}{921}$.

11. There are 2000 people in the sample space. The
    number of elements in the event that a
    participant prefers the Nintendo Wii system is
    607. Then the probability is

    $P(E) = \dfrac{n(E)}{n(S)} = \dfrac{607}{2000} \approx 0.30$.

13. Let B = {participants who prefer Sega
    Dreamcast} and A = {participants between 13
    and 24}. From the table,
    $n(A \cap B) = 92 + 83 = 175$,
    $n(A) = 105 + 139 + 92 + 113 + 248 + 217 + 83$
    $+ 169 = 1166$, and $n(S) = 2000$. We have

    $P(B \mid A) = \dfrac{P(A \cap B)}{P(A)} = \dfrac{\frac{175}{2000}}{\frac{1166}{2000}} \approx 0.15$.

15. Let B = {the sum is 8} and A = {the sum is
    even}. From the table in the text, there are five
    possible rolls of the dice for which the sum is 8
    and the sum is even. So $P(A \cap B) = \dfrac{5}{36}$. There
    are 18 possibilities for which the sum is even,

so $P(A) = \dfrac{18}{36} = \dfrac{1}{2}$. We have

$P(B \mid A) = \dfrac{P(A \cap B)}{P(A)} = \dfrac{\frac{5}{36}}{\frac{1}{2}} = \dfrac{5}{18}$.

17. Let B = {doubles} and A = {sum less than 7}.
    From the table in the text, there are three
    possible rolls of the dice for which the roll is
    doubles and the sum is less than 7. So
    $P(A \cap B) = \dfrac{3}{36} = \dfrac{1}{12}$. There are 15 possibilities
    for which the sum is less than 7, so
    $P(A) = \dfrac{15}{36} = \dfrac{5}{12}$. We have

    $P(B \mid A) = \dfrac{P(A \cap B)}{P(A)} = \dfrac{\frac{1}{12}}{\frac{5}{12}} = \dfrac{1}{5}$.

19. Let A = {a face card is drawn first} and B = {a
    face card is drawn second}. On the first draw,
    there are 12 face cards in the deck of 52 cards.
    Therefore, $P(A) = \dfrac{12}{52} = \dfrac{3}{13}$. On the second
    draw, there are only 51 cards remaining and
    only 11 face cards. Therefore, $P(B \mid A) = \dfrac{11}{51}$.
    We have
    $P(A \cap B) = P(A) \cdot P(B \mid A)$
    $\qquad = \dfrac{3}{13} \cdot \dfrac{11}{51} = \dfrac{33}{663} \approx 0.050$.

21. Let A = {a spade is drawn first} and B = {a red
    card is drawn second}. On the first draw, there
    are 13 spades in the deck of 52 cards.
    Therefore, $P(A) = \dfrac{13}{52} = \dfrac{1}{4}$. On the second
    draw, there are only 51 cards remaining and 26
    red cards. Therefore, $P(B \mid A) = \dfrac{26}{51}$. We have
    $P(A \cap B) = P(A) \cdot P(B \mid A)$
    $\qquad = \dfrac{1}{4} \cdot \dfrac{26}{51} = \dfrac{26}{204} \approx 0.127$.

23. Let A = {a blue is selected first},
    B = {a red is selected second} and
    C = {a green is selected third}. Then
    $P(A$ followed by B followed by $C)$
    $= P(A) \cdot P(B \mid A)P(C \mid A$ and $B)$
    $= \dfrac{12}{57} \cdot \dfrac{12}{56} \cdot \dfrac{7}{55} = \dfrac{6}{1045}$.

25. Let A = {a green is selected first}, B = {a green is selected second} and C = {a red is selected third}. Then

    $P(A$ followed by B followed by $C)$
    $= P(A) \cdot P(B \mid A) P(C \mid A$ and $B)$
    $= \dfrac{7}{57} \cdot \dfrac{6}{56} \cdot \dfrac{12}{55} = \dfrac{3}{1045}$.

27. Let A = {a red card is dealt first}, B = {a black card is dealt second} and C = {a red card is dealt third}. Then

    $P(A$ followed by B followed by $C)$
    $= P(A) \cdot P(B \mid A) P(C \mid A$ and $B)$
    $= \dfrac{26}{52} \cdot \dfrac{26}{51} \cdot \dfrac{25}{50} = \dfrac{13}{102}$.

29. Let A = {an ace is dealt first}, B = {a face card is dealt second} and C = {an 8 is dealt third}. Then

    $P(A$ followed by B followed by $C)$
    $= P(A) \cdot P(B \mid A) P(C \mid A$ and $B)$
    $= \dfrac{4}{52} \cdot \dfrac{12}{51} \cdot \dfrac{4}{50} = \dfrac{8}{5525}$.

31. Let A = {absent the first day}, B = {absent the second day} and C = {absent the third day}. Then

    $P(A$ followed by B followed by $C)$
    $= P(A) \cdot P(B \mid A) \cdot P(C \mid A$ and $B)$
    $= 0.04 \cdot 0.11 \cdot 0.11 = 0.000484.$

33. Let A = {absent the first day}, B = {not absent the second day} and C = {absent the third day}. Then

    $P(A$ followed by B followed by $C)$
    $= P(A) \cdot P(B \mid A) \cdot P(C \mid A$ and $B)$
    $= 0.04 \cdot 0.89 \cdot 0.04 = 0.001424.$

35. The outcome of the first roll has no effect on the outcome of the second roll so the events are independent.

37. The probability of numbers on the second slip of paper depends on the result of the first slip of paper so the events are not independent.

39. The rolls of a pair of dice are independent. Let A = {sum of 8 on the first roll} and B = {sum of 8 on the second roll}.

    $P(A \cap B) = P(A) \cdot P(B)$
    $= \dfrac{5}{36} \cdot \dfrac{5}{36} = \dfrac{25}{1296}$

41. The rolls of a pair of dice are independent. Let A = {sum of at least 10 on the first roll} and B = {sum of at least 11 on the second roll}.

    $P(A \cap B) = P(A) \cdot P(B)$
    $= \dfrac{6}{36} \cdot \dfrac{3}{36} = \dfrac{18}{1296} = \dfrac{1}{72}$

43. The rolls of a pair of dice are independent. Let A = {sum of even number on the first roll} and B = {sum of even number on the second roll}.

    $P(A \cap B) = P(A) \cdot P(B)$
    $= \dfrac{18}{36} \cdot \dfrac{18}{36} = \dfrac{324}{1296} = \dfrac{1}{4}$

45. Each flip of a coin is independent. Let A = {heads}, B = {heads}, C = {tails} and D = {tails}.

    $P(A \cap B \cap C \cap D) = P(A) \cdot P(B) \cdot P(C) \cdot P(D)$
    $= \dfrac{1}{2} \cdot \dfrac{1}{2} \cdot \dfrac{1}{2} \cdot \dfrac{1}{2} = \dfrac{1}{16}$

47. The rolls of a pair of dice are independent. Let A = {sum of at least 10 on the first roll}, B = {sum of at least 10 on the second roll} and C = {sum of at least 10 on the third roll}.

    $P(A \cap B \cap C) = P(A) \cdot P(B) \cdot P(C)$
    $= \dfrac{6}{36} \cdot \dfrac{6}{36} \cdot \dfrac{6}{36} = \dfrac{216}{46{,}656} = \dfrac{1}{216}$

49. The choices of a card are independent. Let A = {ace on the first draw} and B = {ace on the second draw}.

    $P(A \cap B) = P(A) \cdot P(B)$
    $= \dfrac{4}{52} \cdot \dfrac{4}{52} = \dfrac{16}{2704} = \dfrac{1}{169}$

51. The choices of a card are independent. Let A = {spade on the first draw} and B = {diamond on the second draw}.

    $P(A \cap B) = P(A) \cdot P(B)$
    $= \dfrac{13}{52} \cdot \dfrac{13}{52} = \dfrac{169}{2704} = \dfrac{1}{16}$

53. The choices of a card are independent. Let A = {heart on the first draw} and B = {spade on the second draw}.

    $P(A \cap B) = P(A) \cdot P(B)$
    $= \dfrac{13}{52} \cdot \dfrac{13}{52} = \dfrac{169}{2704} = \dfrac{1}{16}$

55. a. The choices of a card with replacement are independent Let
    A = {diamond on the first draw}, B = {diamond on the second draw} and C = {black card on the third draw}.

    $P(A \cap B \cap C) = P(A) \cdot P(B) \cdot P(C)$
    $= \dfrac{13}{52} \cdot \dfrac{13}{52} \cdot \dfrac{26}{52}$
    $= \dfrac{4394}{140{,}608} = \dfrac{1}{32}$

    b. The choices of a card without replacement are not independent. Let
    A = {diamond card on the first draw}, B = {diamond card on the second draw} and C = {black card on the third draw}.

$P(A \text{ followed by } B \text{ followed by } C)$
$= P(A) \cdot P(B \mid A) \cdot P(C \mid A \text{ and } B)$
$= \dfrac{13}{52} \cdot \dfrac{12}{51} \cdot \dfrac{26}{50} = \dfrac{4056}{132,600} = \dfrac{13}{425}$

57. a. The choices of a marble with replacement are independent. Let
$A = \{\text{red marble chosen}\}$,
$B = \{\text{red marble chosen}\}$, and
$C = \{\text{green marble chosen}\}$.
$P(A \cap B \cap C) = P(A) \cdot P(B) \cdot P(C)$
$\qquad = \dfrac{5}{17} \cdot \dfrac{5}{17} \cdot \dfrac{4}{17} = \dfrac{100}{4913}$

b. The choices of a marble without replacement are not independent. Let
$A = \{\text{red marble chosen}\}$,
$B = \{\text{red marble chosen}\}$, and
$C = \{\text{green marble chosen}\}$.
$P(A \text{ followed by } B \text{ followed by } C)$
$= P(A) \cdot P(B \mid A) \cdot P(C \mid A \text{ and } B)$
$= \dfrac{5}{17} \cdot \dfrac{4}{16} \cdot \dfrac{4}{15} = \dfrac{80}{4080} = \dfrac{1}{51}$

59. Let D be the event that a person has been using the drug and let T be the event that the test is positive. The probability we wish to determine is $P(D \mid T)$.

$$P(D \mid T) = \frac{P(D \text{ and } T)}{P(T)}$$

A tree diagram will help us compute the needed probabilities.

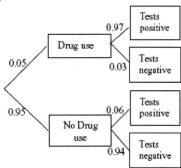

From the diagram,
$P(D \text{ and } T) = P(D) \cdot P(T \mid D) = (0.5)(0.97)$.

To compute $P(T)$ we need to combine two branches from the diagram, one corresponding to a correct positive test result when the person has drug use, and one corresponding to a false positive result when the person has no drug use:
$P(T) = (0.05)(0.97) + (0.95)(0.06)$.

$$P(D \mid T) = \frac{P(D \text{ and } T)}{P(T)}$$
$$= \frac{(0.05)(0.97)}{(0.05)(0.97) + (0.95)(0.06)} \approx 0.46$$

61. Let D be the event that a person has the disease and let T be the event that the test is positive. The probability we wish to determine is $P(D \mid T)$.

$$P(D \mid T) = \frac{P(D \text{ and } T)}{P(T)}$$

A tree diagram will help us compute the needed probabilities.

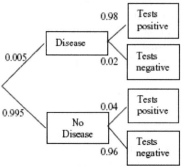

From the diagram,
$P(D \text{ and } T) = P(D) \cdot P(T \mid D) = (0.005)(0.98)$.

To compute $P(T)$ we need to combine two branches from the diagram, one corresponding to a correct positive test result when the person has the disease, and one corresponding to a false positive result when the person does not have the disease:
$P(T) = (0.005)(0.98) + (0.995)(0.04)$.

$$P(D \mid T) = \frac{P(D \text{ and } T)}{P(T)}$$
$$= \frac{(0.005)(0.98)}{(0.005)(0.98) + (0.995)(0.04)} \approx 0.11$$

63. a. If the grand prize is behind door 1, and the contestant chooses door 1, Monte Hall can open either door 2 or 3. The probability that he will open door 2 is $\dfrac{1}{2}$.

b. If the grand prize is behind door 2, and the contestant chooses door 1, Monte Hall will not open door 2. The probability that he will open door 2 is 0.

c. If the grand prize is behind door 3, and the contestant chooses door 1, Monte Hall will open door 2. The probability that he will open door 2 is 1.

d. $P(A \mid \overline{B})$

$$= \frac{P(\overline{B} \mid A) P(A)}{P(\overline{B} \mid A) P(A) + P(\overline{B} \mid B) P(B) + P(\overline{B} \mid C) P(C)}$$

$$= \frac{\dfrac{1}{2}\left(\dfrac{1}{3}\right)}{\dfrac{1}{2}\left(\dfrac{1}{3}\right) + 0\left(\dfrac{1}{3}\right) + 1\left(\dfrac{1}{3}\right)}$$

$$= \frac{\dfrac{1}{6}}{\dfrac{1}{2}} = \frac{1}{3}$$

e. $P(C|\overline{B})$

$$= \frac{P(\overline{B}|C)P(C)}{P(\overline{B}|A)P(A)+P(\overline{B}|B)P(B)+P(\overline{B}|C)P(C)}$$

$$= \frac{1\left(\dfrac{1}{3}\right)}{\dfrac{1}{2}\left(\dfrac{1}{3}\right)+0\left(\dfrac{1}{3}\right)+1\left(\dfrac{1}{3}\right)}$$

$$= \frac{\dfrac{1}{3}}{\dfrac{1}{2}} = \frac{2}{3}$$

f.   Yes.

EXCURSION EXERCISES, SECTION 12.6

1. The sample space for tossing three dice has $6 \cdot 6 \cdot 6 = 216$ items in it. There is one element in the event that three fives are tossed. Then the probability is

$$P(E) = \frac{n(E)}{n(S)} = \frac{1}{216}.$$

3. The sample space for tossing three dice has $6 \cdot 6 \cdot 6 = 216$ items in it. The number of elements in the event that only 2 die will show a five is $3(1 \cdot 1 \cdot 5) = 15$. Then the probability is

$$P(E) = \frac{n(E)}{n(S)} = \frac{15}{216} = \frac{5}{72}.$$

5. Let S_1 be the event that one die matches, in which case the player wins $1. $P(S_1) = \dfrac{25}{72}$.

Let S_2 be the event that two dice match, in which case the player wins $2. $P(S_2) = \dfrac{5}{72}$.

Let S_3 be the event that three dice match, in which case the player wins $10. $P(S_3) = \dfrac{1}{216}$.

Let S_4 be the event that no prize is won in which case the player loses $1. $P(S_4) = \dfrac{125}{216}$.

Expectation =
$$= P(S_1) \cdot S_1 + P(S_2) \cdot S_2 + P(S_3) \cdot S_3 + P(S_4) \cdot S_4$$
$$= \frac{25}{72}(1) + \frac{5}{72}(2) + \frac{1}{216}(10) + \frac{125}{216}(-1)$$
$$\approx -0.05$$
Expectation is −5 cents.

7. A sum less than 11 is the complement of a sum more than 10, therefore,

$$P(E) = 1 - P(E^C) = 1 - \frac{108}{216} = \frac{108}{216}$$

The expectation is
$$\frac{108}{216}(1) + \frac{108}{216}(-1) = 0.$$

EXERCISE SET 12.6

1. Expectation
$$= P(S_1) \cdot S_1 + P(S_2) \cdot S_2 + P(S_3) \cdot S_3$$
$$+ P(S_4) \cdot S_4 + P(S_5) \cdot S_5$$
$$= 30 \cdot 0.15 + 40 \cdot 0.2 + 50 \cdot 0.4 + 60 \cdot 0.05 + 70 \cdot 0.2$$
$$= 49.5$$

3. Let S_1 be the event that the roulette ball lands on a black number, in which case the player wins $1. There are 38 possible numbers, so $P(S_1) = \dfrac{18}{38}$. Let S_2 be the event that a black number does not come up, in which case the player loses $1. Then $P(S_2) = 1 - \dfrac{18}{38} = \dfrac{20}{38}$.

Expectation $= P(S_1) \cdot S_1 + P(S_2) \cdot S_2$
$$= \frac{18}{38}(1) + \frac{20}{38}(-1) \approx -0.05$$
Expectation is −5 cents.

5. Let S_1 be the event that the wheel stops on Joker ($40), in which case the player wins $40. There are 54 possible numbers, so $P(S_1) = \dfrac{1}{54}$. Let S_2 be the event that the wheel does not stop on Joker, in which case the player loses $1. Then

$$P(S_2) = 1 - \frac{1}{54} = \frac{53}{54}.$$

Expectation $= P(S_1) \cdot S_1 + P(S_2) \cdot S_2$
$$= \frac{1}{54}(40) + \frac{53}{54}(-1) \approx -0.24$$
Expectation is −24 cents.

7. Let S_1 be the event that the wheel stops on $5, in which case the player wins $5. There are 54 possible numbers, so $P(S_1) = \dfrac{7}{54}$. Let S_2 be the event that the wheel does not stop on $5, in which case the player loses $1. Then

$$P(S_2) = 1 - \frac{7}{54} = \frac{47}{54}.$$

Expectation $= P(S_1) \cdot S_1 + P(S_2) \cdot S_2$
$$= \frac{7}{54}(5) + \frac{47}{54}(-1) \approx -0.22$$
Expectation is −22 cents.

9. Let S_1 be the event that the person dies within one year. Then $P(S_1) = 0.000487$ and the company must pay out $25,000. Since the company charged $75 for the policy, the company's actual loss is $24,925. Let S_2 be the event that the policy holder does not die during the year of the policy. Then $P(S_2) = 0.999513$ and the company keeps the premium of $75. Expectation

$$= P(S_1) \cdot S_1 + P(S_2) \cdot S_2$$
$$= 0.000487(-24,925) + 0.999513(75) = 62.83$$
Expectation is $62.83.

11. Let S_1 be the event that the person dies within one year. Then $P(S_1) = 0.070471$ and the company must pay out \$10,000. Since the company charged \$495 for the policy, the company's actual loss is \$9505. Let $S2$ be the event that the policy holder does not die during the year of the policy. Then $P(S_2) = 0.929529$ and the company keeps the premium of \$495.

    Expectation
    $$= P(S_1) \cdot S_1 + P(S_2) + S_2$$
    $$= 0.070471(-9505) + 0.929529(495)$$
    $$= -209.71$$
    Expectation is $-\$209.71$.

13. Let S_1 be the event that the person dies within one year. Then $P(S_1) = 0.001406$ and the company must pay out \$30,000. If the company charges \$$x$ for the policy, the company's actual loss is \$$(30,000 - x)$. Let S_2 be the event that the policyholder does not die during the year of the policy. Then $P(S_2) = 0.998594$ and the company keeps the premium of \$$x$.

    Expectation $= P(S_1) \cdot S_1 + P(S_2) \cdot S_2$
    $$0 = 0.001406(-(30,000 - x)) + 0.998594x$$
    $$0 = -42.18 + 0.001406x + 0.998594x$$
    $$42.18 = x$$
    The company should charge more than \$42.18.

15. Expectation
    $$= 0.10(100,000) + 0.40(60,000) + 0.25(30,000)$$
    $$+ 0.15(0) + 0.08(-20,000) + 0.02(-40,000)$$
    $$= 10,000 + 24,000 + 7500 + 0 - 1600 - 800$$
    $$= 39,100$$
    The expected profit is \$39,100.

17. Expectation
    $$= 0.05(40,000) + 0.2(30,000) + 0.5(20,000)$$
    $$+ 0.2(10,000) + 0.05(5,000)$$
    $$= 2000 + 6000 + 10,000 + 2000 + 250$$
    $$= 20,250$$
    The expected profit is \$20,250.

19. Possible outcomes can be found on the table in the text. Let $S_1 = \{$ sum of $2\}$, $S_2 = \{$ sum of $3\}$, $S_3 = \{$ sum of $4\}$, $S_4 = \{$ sum of $5\}$, $S_5 = \{$ sum of $6\}$, $S_6 = \{$ sum of $7\}$, $S_7 = \{$ sum of $8\}$, $S_8 = \{$ sum of $9\}$, $S_9 = \{$ sum of $10\}$, $S_{10} = \{$ sum of $11\}$, and $S_{11} = \{$ sum of $12\}$. Then probabilities are:

    $$P(S_1) = \frac{1}{36}, \ P(S_2) = \frac{2}{36}, \ P(S_3) = \frac{3}{36},$$
    $$P(S_4) = \frac{4}{36}, \ P(S_5) = \frac{5}{36}, \ P(S_6) = \frac{6}{36},$$
    $$P(S_7) = \frac{5}{36}, \ P(S_8) = \frac{4}{36}, \ P(S_9) = \frac{3}{36},$$
    $$P(S_{10}) = \frac{2}{36}, \ P(S_{11}) = \frac{1}{36}$$

Expectation
$$= P(S_1) \cdot S_1 + P(S_2) \cdot S_2 + P(S_3) \cdot S_3$$
$$+ P(S_4) \cdot S_4 + P(S_5) \cdot S_5 + P(S_6) \cdot S_6$$
$$+ P(S_7) \cdot S_7 + P(S_8) \cdot S_8 + P(S_9) \cdot S_9$$
$$+ P(S_{10}) \cdot S_{10} + P(S_{11}) \cdot S_{11}$$
$$= \frac{1}{36}(2) + \frac{2}{36}(3) + \frac{3}{36}(4) + \frac{4}{36}(5)$$
$$+ \frac{5}{36}(6) + \frac{6}{36}(7) + \frac{5}{36}(8) + \frac{4}{36}(9)$$
$$+ \frac{3}{36}(10) + \frac{2}{36}(11) + \frac{1}{36}(12)$$
$$= 7$$
The expected sum is 7.

21. The sample space can be found using a table.

	1	2	2	3	3	4
1	11	12	12	13	13	14
3	31	32	32	33	33	34
4	41	42	42	43	43	44
5	51	52	52	53	53	54
6	61	62	62	63	63	64
8	81	82	82	83	83	84

Possible sums are:
$\{2, 3, 4, 5, 6, 7, 8, 9, 10, 11, 12\}$
Probabilities are:

$$P(2) = \frac{1}{36}, \ P(3) = \frac{2}{36}, \ P(4) = \frac{3}{36},$$
$$P(5) = \frac{4}{36}, \ P(6) = \frac{5}{36}, \ P(7) = \frac{6}{36},$$
$$P(8) = \frac{5}{36}, \ P(9) = \frac{4}{36}, \ P(10) = \frac{3}{36},$$
$$P(11) = \frac{2}{36}, \ P(12) = \frac{1}{36}$$

Expectation
$$= \frac{1}{36}(2) + \frac{2}{36}(3) + \frac{3}{36}(4) + \frac{4}{36}(5)$$
$$+ \frac{5}{36}(6) + \frac{6}{36}(7) + \frac{5}{36}(8) + \frac{4}{36}(9)$$
$$+ \frac{3}{36}(10) + \frac{2}{36}(11) + \frac{1}{36}(12)$$
$$= 7$$
The expected sum is 7.

23. The sample space of the red dice can be found using a table.

	0	0	4	4	4	4
2	20	20	24	24	24	24
3	30	30	34	34	34	34
3	30	30	34	34	34	34
9	90	90	94	94	94	94
10	100	100	104	104	104	104
11	110	110	114	114	114	114

Possible sums are:
$\{2, 3, 6, 7, 9, 10, 11, 13, 14, 15\}$

Probabilities are:

$$P(2) = \frac{2}{36}, \ P(3) = \frac{4}{36}, \ P(6) = \frac{4}{36},$$

$$P(7) = \frac{8}{36}, \ P(9) = \frac{2}{36}, \ P(10) = \frac{2}{36},$$

$$P(11) = \frac{2}{36}, \ P(13) = \frac{4}{36}, \ P(14) = \frac{4}{36},$$

$$P(15) = \frac{4}{36}$$

Expectation

$$= \frac{2}{36}(2) + \frac{4}{36}(3) + \frac{4}{36}(6) + \frac{8}{36}(7)$$

$$+ \frac{2}{36}(9) + \frac{2}{36}(10) + \frac{2}{36}(11) + \frac{4}{36}(13)$$

$$+ \frac{4}{36}(14) + \frac{4}{36}(15)$$

$$= 9$$

The expected sum for the red dice is 9.

The sample space of the green dice can be found using a table.

	3	3	3	3	3	3
0	03	03	03	03	03	03
1	13	13	13	13	13	13
7	73	73	73	73	73	73
8	83	83	83	83	83	83
8	83	83	83	83	83	83
8	83	83	83	83	83	83

Possible sums are: {3, 4, 10, 11}
Probabilities are:

$$P(3) = \frac{6}{36}, \ P(4) = \frac{6}{36}, \ P(10) = \frac{6}{36},$$

$$P(11) = \frac{18}{36}$$

Expectation

$$= \frac{6}{36}(3) + \frac{6}{36}(4) + \frac{6}{36}(10) + \frac{18}{36}(11)$$

$$\approx 8.3$$

The expected sum for the green dice is 8.3.

The red dice has a higher expected sum, so it would be the better choice.

CHAPTER 12 REVIEW EXERCISES

1. {11, 12, 13, 21, 22, 23, 31, 32, 33}

2. {26, 28, 62, 68, 82, 86}

3.

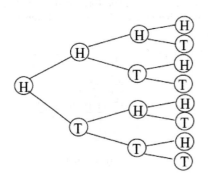

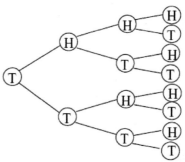

{HHHH, HHHT, HHTH, HHTT, HTHH, HTHT, HTTH, HTTT, THHH, THHT, THTH, THTT, TTHH, TTHT, TTTH, TTTT}

4.

	7	8	9
A	7A	8A	9A
B	7B	8B	9B

{7A, 8A, 9A, 7B, 8B, 9B}

5. There are 3 choices for style, so $n_1 = 3$. There are 4 choices for color, so $n_2 = 4$. There are 6 choices for sizes, so $n_3 = 6$. By the counting principle, there are $3 \cdot 4 \cdot 6 = 72$ shoes available.

6. There are 10 numbers that can be used in each position, so n_1, n_2, n_3 and $n_4 = 10$. By the counting principle, there are $10 \cdot 10 \cdot 10 \cdot 10 = 10,000$ different lock combinations possible.

7. There are 2 letters that can be used in the first position, so $n_1 = 2$. There are 12 letters that can be used in the second position, so $n_2 = 12$. There are 10 digits that can be used in the third and fourth positions, so $n_3 = 10$ and $n_4 = 10$. By the counting principle, there are $2 \cdot 12 \cdot 10 \cdot 10 = 2400$ different four-character sequences possible.

8. There are 7 binary digits, the first two of which are required to be different. There are two choices for the first digit, $n_1 = 2$, and only one choice for the second, $n_2 = 1$, since it must be different from the first. There are two choices for each of the five remaining digits, $n_3 = n_4 = n_5 = n_6 = n_7 = 2$. The counting principle then says there are $2 \cdot 1 \cdot 2 \cdot 2 \cdot 2 \cdot 2 \cdot 2 = 64$ biquinary numbers.

9. $7! = 7 \cdot 6 \cdot 5 \cdot 4 \cdot 3 \cdot 2 \cdot 1 = 5040$

10. Evaluating:
$$8! - 4! = (8 \cdot 7 \cdot 6 \cdot 5 \cdot 4 \cdot 3 \cdot 2 \cdot 1) - (4 \cdot 3 \cdot 2 \cdot 1)$$
$$= 40,320 - 24 = 40,296$$

11. $\dfrac{9!}{2!3!4!} = \dfrac{9 \cdot 8 \cdot 7 \cdot 6 \cdot 5 \cdot 4!}{2 \cdot 1 \cdot 3 \cdot 2 \cdot 1 \cdot 4!} = 1260$

12. Evaluating:

$$P(10,\ 6) = \dfrac{10!}{(10-6)!} = \dfrac{10!}{4!}$$
$$= \dfrac{10 \cdot 9 \cdot 8 \cdot 7 \cdot 6 \cdot 5 \cdot 4!}{4!}$$
$$= 10 \cdot 9 \cdot 8 \cdot 7 \cdot 6 \cdot 5 = 151,200$$

13. Evaluating:

$$P(8,\ 3) = \dfrac{8!}{(8-3)!} = \dfrac{8!}{5!} = \dfrac{8 \cdot 7 \cdot 6 \cdot 5!}{5!}$$
$$= 8 \cdot 7 \cdot 6 = 336$$

14. Evaluating:

$$\dfrac{C(6,\ 2) \cdot C(8,\ 3)}{C(14,\ 5)} = \dfrac{\dfrac{6!}{2!(6-2)!} \cdot \dfrac{8!}{3!(8-3)!}}{\dfrac{14!}{5!(14-5)!}}$$
$$= \dfrac{\dfrac{6!}{2!4!} \cdot \dfrac{8!}{3!5!}}{\dfrac{14!}{5!9!}}$$
$$= \dfrac{\dfrac{6 \cdot 5 \cdot 4!}{2!4!} \cdot \dfrac{8 \cdot 7 \cdot 6 \cdot 5!}{3!5!}}{\dfrac{14 \cdot 13 \cdot 12 \cdot 11 \cdot 10 \cdot 9!}{5!9!}}$$
$$= \dfrac{\dfrac{6 \cdot 5}{2 \cdot 1} \cdot \dfrac{8 \cdot 7 \cdot 6}{3 \cdot 2 \cdot 1}}{\dfrac{14 \cdot 13 \cdot 12 \cdot 11 \cdot 10}{5 \cdot 4 \cdot 3 \cdot 2 \cdot 1}}$$
$$= \dfrac{15 \cdot 56}{2002} = \dfrac{840}{2002} = \dfrac{60}{143}$$

15. The order in which the people arrange themselves is important, so the number of ways to receive service is

$$P(7,\ 7) = \dfrac{7!}{(7-7)!} = \dfrac{7!}{0!} = \dfrac{7 \cdot 6 \cdot 5 \cdot 4 \cdot 3 \cdot 2 \cdot 1}{1}$$
$$= 7 \cdot 6 \cdot 5 \cdot 4 \cdot 3 \cdot 2 \cdot 1 = 5040$$

16. The order in which the answers are chosen is important, so the number of matches possible is

$$P(7,\ 7) = \dfrac{7!}{(7-7)!} = \dfrac{7!}{0!} = \dfrac{7 \cdot 6 \cdot 5 \cdot 4 \cdot 3 \cdot 2 \cdot 1}{1}$$
$$= 7 \cdot 6 \cdot 5 \cdot 4 \cdot 3 \cdot 2 \cdot 1 = 5040$$

17. The order in which the answers are chosen is important, so the number of matches possible is

$$P(7,\ 5) = \dfrac{7!}{(7-5)!} = \dfrac{7!}{2!} = \dfrac{7 \cdot 6 \cdot 5 \cdot 4 \cdot 3 \cdot 2 \cdot 1}{2 \cdot 1}$$
$$= 2520$$

18. We are looking for the number of permutations of the letters *letter*. With $n = 6$ (number of letters), $k_1 = 1$ (number of l's), $k_2 = 2$ (number of e's), $k_3 = 2$ (number of t's)

and $k_4 = 1$ (number of r's), we have

$$\dfrac{6!}{1!2!2!1!} = 180$$

There are 180 different arrangements possible.

19. The order in which the heads and tails are chosen is not important, so the number of distinct arrangements is

$$C(12,\ 4) = \dfrac{12!}{4!(12-4)!} = \dfrac{12!}{4!8!}$$
$$= \dfrac{12 \cdot 11 \cdot 10 \cdot 9 \cdot 8!}{4!8!}$$
$$= \dfrac{12 \cdot 11 \cdot 10 \cdot 9}{4 \cdot 3 \cdot 2 \cdot 1} = 495$$

20. Because each shift is different, the order of the assignments to shifts is important, so the number of different assignments possible is

$$P(5,\ 3) = \dfrac{5!}{(5-3)!} = \dfrac{5!}{2!} = \dfrac{5 \cdot 4 \cdot 3 \cdot 2!}{2!}$$
$$= 5 \cdot 4 \cdot 3 = 60.$$

21. The order in which the selections are made is not important, so the number of sets that can be chosen is

$$C(25,\ 10) = \dfrac{25!}{10!(25-10)!} = \dfrac{25!}{10!15!}$$
$$= \dfrac{25 \cdot 24 \cdot 23 \cdot 22 \cdot 21 \cdot 20 \cdot 19 \cdot 18 \cdot 17 \cdot 16 \cdot 15!}{10!15!}$$
$$= \dfrac{25 \cdot 24 \cdot 23 \cdot 22 \cdot 21 \cdot 20 \cdot 19 \cdot 18 \cdot 17 \cdot 16}{10 \cdot 9 \cdot 8 \cdot 7 \cdot 6 \cdot 5 \cdot 4 \cdot 3 \cdot 2 \cdot 1}$$
$$= 3,268,760.$$

22. The order in which the stocks are selected is not important, so the number of different portfolios is

$$C(11,\ 3) = \dfrac{11!}{3!(11-3)!} = \dfrac{11!}{3!8!}$$
$$= \dfrac{11 \cdot 10 \cdot 9 \cdot 8!}{3!8!}$$
$$= \dfrac{11 \cdot 10 \cdot 9}{3 \cdot 2 \cdot 1} = 165.$$

23. The order of the choices is not important, so the number of possible ways to choose the nondefective monitors, is $C(12, 3)$ and the number of ways to choose the defective monitors is $C(3,2)$. Therefore, by the counting principle, there are $C(12,3) \cdot C(3,2)$ different sets.

$$C(12,\ 3) \cdot C(3,\ 2) = \dfrac{12!}{3!(12-3)!} \cdot \dfrac{3!}{2!(3-2)!}$$
$$= \dfrac{12!}{3!9!} \cdot \dfrac{3!}{2!1!}$$
$$= \dfrac{12 \cdot 11 \cdot 10 \cdot 9!}{3!9!} \cdot \dfrac{3 \cdot 2!}{2!1!}$$
$$= \dfrac{12 \cdot 11 \cdot 10}{3 \cdot 2 \cdot 1} \cdot \dfrac{3}{1}$$
$$= 220 \cdot 3 = 660$$

24. Call the two people that refuse to sit next to each other A and B. Consider two cases. In the first case, A sits at an end of the row. There are 2 ways A can do this. Once A is seated, there are 7 places where B could sit and not be next to A. Then there are 7! ways the remaining people can be seated, for a total of $2 \cdot 7 \cdot 7!$ ways to seat the group if A sits at an end of the row. In the second case, A does not sit at the end of the row. There are 7 ways A can do this. Once A is seated, there are 6 places where B could sit and not be next to A. Then there are 7! ways the remaining people can be seated, for a total of $7 \cdot 6 \cdot 7!$. Summing the totals from case A and case B gives 282,240 ways.

25. There are $C(4, 4)$ ways of choosing four aces from four aces, $C(4, 4)$ ways of choosing four kings from four kings, and $C(4, 4)$ ways of choosing each of the other possible cards of which there are 13 total. There are 48 choices for the fifth card in the hand. By the counting principle, the number of hands containing four of a kind is

$$13 \cdot C(4, \ 4) \cdot 48 = 13 \cdot \frac{4!}{4!(4-4)!} \cdot 48$$
$$= 13 \cdot \frac{4!}{4!0!} \cdot 48 = 13 \cdot \frac{1}{1} \cdot 48$$
$$= 624.$$

26. The sample space is {BBBB, BBBG, BBGB, BBGG, BGBB, BGBG, BGGB, BGGG, GBBB, GBBG, GBGB, GBGG, GGBB, GGBG, GGGB, GGGG}. The elements in the event that a family will have one boys and three girls is {BGGG, GBGG, GGBG, GGGB}. Then the probability is

$$P(E) = \frac{n(E)}{n(S)} = \frac{4}{16} = \frac{1}{4}.$$

27. The sample space is {HHH, HHT, HTH, HTT, THH, THT, TTH, TTT}. The elements in the event of getting one head and two tails is {HTT, THT, TTH}. Then the probability is

$$P(E) = \frac{n(E)}{n(S)} = \frac{3}{8}.$$

28. Let E be the event that the employee is a woman.

$$P(E) = \frac{7920}{5739 + 7290} = \frac{7290}{13,029} \approx 0.56$$

29. The total number of students is 2495. Let E be the event that the student is a junior or senior. There are $483 + 445 = 928$ students that are juniors or seniors.

$$P(E) = \frac{928}{2495} \approx 0.37$$

30. The total number of students in 2495. Let E be the event that the student is not a graduate student. There are $642 + 549 + 483 + 445 = 2119$ students that are not graduate students.

$$P(E) = \frac{2119}{2495} \approx 0.85$$

31. The sample space for tossing two six-sided dice was found in the text. The dice must be considered as distinct, so there are 36 possible outcomes. The elements in the event that the sum of the pips on the upward faces is 9 are {3 6, 4 5, 5 4, 6 3}. Then the probability is

$$P(E) = \frac{n(E)}{n(S)} = \frac{4}{36} = \frac{1}{9}.$$

32. If E is the event of tossing an 11, then E^C is the event of not tossing an 11, and

$$P(E^C) = 1 - P(E) = 1 - \frac{2}{36} = \frac{34}{36} = \frac{17}{18}.$$

33. If E is the event of tossing a sum less than 10, then E^C is the event of tossing a sum of at least ten, and

$$P(E^C) = 1 - P(E) = 1 - \frac{30}{36} = \frac{6}{36} = \frac{1}{6}.$$

34. Let A be the event of rolling an even number, and let B be the event of rolling a number less than 5. From the figure in the text, there are 36 elements in the sample space. $n(A) = 18$, $n(B) = 6$ and $n(A \text{ and } B) = 4$.

$$P(A \text{ or } B) = P(A) + P(B) - P(A \text{ and } B)$$
$$= \frac{18}{36} + \frac{6}{36} - \frac{4}{36} = \frac{20}{36} = \frac{5}{9}$$

35. Let $B = \{\text{the sum is 9}\}$ and $A = \{\text{the sum is odd}\}$. From the table in the text, there are four possible rolls of the dice for which the sum is 9 and the sum is odd. So $P(A \cap B) = \frac{4}{36} = \frac{1}{9}$. There are 18 possibilities for which the sum is odd, so $P(A) = \frac{18}{36} = \frac{1}{2}$. We have

$$P(B \mid A) = \frac{P(A \cap B)}{P(A)} = \frac{\frac{1}{9}}{\frac{1}{2}} = \frac{2}{9}.$$

36. Let $B = \{\text{the sum is 8}\}$ and $A = \{\text{doubles}\}$. From the table in the text, there is one possible roll of the dice for which the sum is 8 and doubles.

So $P(A \cap B) = \frac{1}{36}$. There are 6 possibilities for which the roll is doubles, so $P(A) = \frac{6}{36} = \frac{1}{6}.$

We have

$$P(B \mid A) = \frac{P(A \cap B)}{P(A)} = \frac{\frac{1}{36}}{\frac{1}{6}} = \frac{1}{6}.$$

37. Let A be the event of drawing a heart, and let B be the event of drawing a black card. The sample space consists of 52 cards. $n(A) = 13$, $n(B) = 26$ and $n(A$ and $B) = 0$.
$P(A$ or $B) = P(A) + P(B)$
$$= \frac{13}{52} + \frac{26}{52} = \frac{39}{52} = \frac{3}{4}$$

38. Let A be the event of drawing a heart, and let B be the event of drawing a jack. The sample space consists of 52 cards. $n(A) = 13$, $n(B) = 4$ and $n(A$ and $B) = 1$.
$P(A$ or $B) = P(A) + P(B) - P(A$ and $B)$
$$= \frac{13}{52} + \frac{4}{52} - \frac{1}{52} = \frac{16}{52} = \frac{4}{13}$$

39. If E is the event of drawing a three, then E^C is the event of not drawing a three, and
$$P(E^C) = 1 - P(E) = 1 - \frac{4}{52} = \frac{48}{52} = \frac{12}{13}.$$

40. Let $B = \{$the card is red$\}$ and $A = \{$the card is not a club$\}$. There are 26 cards that are red and not clubs. So $P(A \cap B) = \frac{26}{52} = \frac{1}{2}$. There are 39 possibilities for cards that are not clubs, so $P(A) = \frac{39}{52} = \frac{3}{4}$. We have
$$P(B \mid A) = \frac{P(A \cap B)}{P(A)} = \frac{\frac{1}{2}}{\frac{3}{4}} = \frac{2}{3}.$$

41. Let E be the event of rolling a sum of 6. Refer to the text. Because there are five ways to roll a sum of 6, there are five favorable outcomes, leaving 31 unfavorable outcomes.
Odds in favor of $E = \frac{5}{31}$
The odds in favor are 5 to 31.

42. Let E be the event of pulling a heart. Because there are thirteen ways to draw a heart, there are thirteen favorable outcomes, leaving 39 unfavorable outcomes.
Odds in favor of $E = \frac{13}{39} = \frac{1}{3}$
The odds in favor are 1 to 3.

43. Because the odds against are 4 to 5, the odds in favor are 5 to 4. Use the formula for calculating the probability of an event when the odds in favor are known.
$$P(E) = \frac{a}{a+b} = \frac{5}{5+4} = \frac{5}{9}$$

44. Make a Punnett square.

Parents	H	h
h	Hh	hh
h	Hh	hh

To have short hair, the rodent must have two h. From the table, two of the four possible genotypes have two h's, so the probability that an offspring will have short hair is $\frac{2}{4} = \frac{1}{2}$.

45. Let A be the event of exactly one card being an ace and let B be the event of exactly one card being a face card. Then,
$P(A$ and $B) = P(A) + P(B) - P(A$ or $B)$
$$= 0.145 + 0.362 - 0.471$$
$$= 0.036$$

46. Let $A = \{$people who like cheese flavored$\}$, and let $B = \{$people who like jalapeno-flavored$\}$. Then, $n(A) = 642$, $n(B) = 487$ and $n(A$ and $B) = 302$. The total number of people represented is 1000.
$P(A$ or $B) = P(A) + P(B) - P(A$ and $B)$
$$= \frac{642}{1000} + \frac{487}{1000} - \frac{302}{1000} = \frac{827}{1000}$$
If E is the event of a person liking one of the two flavors, then E^C is the event of not liking either of the two flavors, and
$$P(E^C) = 1 - P(E) = 1 - \frac{827}{1000} = \frac{173}{1000}.$$

47. Because there are 24 chips in the box, there are 24 items in the sample space. Let E be the event of choosing a yellow or white chip. There are 7 yellow chips and 8 white chips, so there are $7 + 8 = 15$ items in E. Then the probability is
$$P(E) = \frac{n(E)}{n(S)} = \frac{15}{24} = \frac{5}{8}.$$

48. Let E be the event of getting a red chip. Because there are 4 red chips, there are 4 favorable outcomes, leaving 20 unfavorable outcomes.
Odds in favor of $E = \frac{4}{20} = \frac{1}{5}$
The odds in favor are 1 to 5.

49. Let B = {the chip is yellow} and A = {the chip is not white}. There are seven chips that are yellow and not white. So $P(A \cap B) = \dfrac{7}{24}$.

There are 16 chips that are not white, so $P(A) = \dfrac{16}{24} = \dfrac{2}{3}$. We have

$$P(B \mid A) = \frac{P(A \cap B)}{P(A)} = \frac{\dfrac{7}{24}}{\dfrac{2}{3}} = \frac{7}{16}.$$

50. Let E = {no red chips are drawn}. There are 24 possibilities for the first chip drawn, 23 possibilities for the second chip drawn, 22 for the third, 21 for the fourth and 20 for the fifth. Thus, $n(S)$ = 24·23·22·21·20 = 5,100,480. The first time a chip is drawn, there are 20 chips that are not red, the second time there are 19 chips that are not red, 18 chips the third time 17 chips the fourth time, and 16 chips the fifth time, so $n(E)$ = 20·19·18·17·16 = 1,860,480.

$$P(E) = \frac{1,860,480}{5,100,480} = \frac{646}{1771}$$

51. Let A = {a yellow is selected first}, B = {a white is selected second} and C = {a yellow is selected third}. Then
$P(A$ followed by B followed by $C)$
$= P(A) \cdot P(B \mid A) \cdot P(C \mid A$ and $B)$
$= \dfrac{7}{24} \cdot \dfrac{8}{23} \cdot \dfrac{6}{22} = \dfrac{336}{12,144} = \dfrac{7}{253}$

52. $n(S)$ = 23,500. Let E be the event that the person voted against the proposition. Then
$$P(E) = \frac{2527 + 5370 + 712}{23,500} = \frac{8609}{23,500} \approx 0.37.$$

53. $n(S)$ = 23,500. Let A = {Democrats}, and let B = {Independents}. Then, from the table, $n(A)$ = 8452 + 2527 + 894 = 11,873, $n(B)$ = 1225 + 712 + 686 = 2623 and $n(A$ and $B)$ = 0.
$P(A$ or $B) = P(A) + P(B) - P(A$ and $B)$
$$= \frac{11,873}{23,500} + \frac{2623}{23,500} - 0$$
$$= \frac{14,496}{23,500} \approx 0.62$$

54. $n(S)$ = 23,500. Let E be the event that the person abstained from voting and is not a Republican. Then
$$P(E) = \frac{894 + 686}{23,500} = \frac{1580}{23,500} \approx 0.07.$$

55. $n(S)$ = 23,500. Let B = {voted for the proposition} and A = {registered Independent}. From the table, $n(A \cap B)$ = 1225, $n(A)$ = 1225 + 712 + 686 = 2623, and $n(S)$ =23,500. We have
$$P(B \mid A) = \frac{P(A \cap B)}{P(A)} = \frac{\dfrac{1225}{23,500}}{\dfrac{2623}{23,500}} \approx 0.47.$$

56. $n(S)$ = 23,500. Let B = {registered Democrat} and A = {voted against the proposition}. From the table, $n(A \cap B)$ = 2527 and $n(A)$ = 2527 + 5370 + 712 = 8609. We have
$$P(B \mid A) = \frac{P(A \cap B)}{P(A)} = \frac{\dfrac{2327}{23,500}}{\dfrac{8609}{23,500}} \approx 0.29.$$

57. The rolls of a pair of dice are independent. Let A ={6 on the first roll}, B = {6 on the second roll} and C = {6 on the third roll}. Then
$P(A \cap B \cap C) = P(A) \cdot P(B) \cdot P(C)$
$$= \frac{1}{6} \cdot \frac{1}{6} \cdot \frac{1}{6} = \frac{1}{216}$$

58. Let E = {at least one 6}; then E^C = {no 6's}. Because on each toss of the die there are six possible outcomes, $n(S)$ = 6·6·6·6·6 = 7776. On each toss of the die there are five numbers that are not 6's. Therefore, $n(E^C)$ = 5·5·5·5·5 =3125.
$$P(E) = 1 - P(E^C) = 1 - \frac{3125}{7776} = \frac{4651}{7776} \approx 0.60$$

59. Let E = {exactly two 6's}. Because on each toss of the die there are six possible outcomes, $n(S)$ = 6·6·6·6 = 7776. On each toss of the die there are five numbers that are not 6's and one number that is a 6. The number of possible arrangements of the two sixes is $C(5, 2)$ = 10. Therefore, $n(E)$ = 10·5·5·5·1·1= 1250.
$$P(E) = \frac{1250}{7776} \approx 0.16$$

60. Let E = {at least one spade is drawn}; then E^C = {no spade is drawn}. There are 52 possibilities for each card drawn. Thus, $n(S)$ = 52·52·52·52 = 7,311,616. Each time a card is drawn, there are 39 cards that are not spades, so $n(E^C)$ = 39·39·39·39 = 2,313,441.
$$P(E) = 1 - P(E^C) = 1 - \frac{2,313,441}{7,311,616}$$
$$= \frac{4,998,175}{7,311,616} = \frac{175}{256}$$

61. Let D be the event that a dog has the disease and let T be the event that the test is positive. The probability we wish to determine is $P(D/T)$. Then

    $$P(D \mid T) = \frac{P(D \text{ and } T)}{P(T)}.$$

    A tree diagram will help us compute the needed probabilities.

    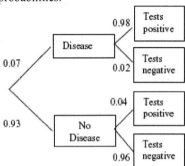

    From the diagram,
    $P(D \text{ and } T) = P(D) \cdot P(T \mid D) = (0.07)(0.98)$. To compute $P(T)$ we need to combine two branches from the diagram, one corresponding to a correct positive test result when the dog has the disease, and one corresponding to a false positive result when the dog does not have the disease:
    $P(T) = (0.07)(0.98) + (0.93)(0.04)$. Then

    $$P(D \mid T) = \frac{P(D \text{ and } T)}{P(T)}$$
    $$= \frac{(0.07)(0.98)}{(0.07)(0.98) + (0.93)(0.04)} \approx 0.648$$

62. Let $A = \{\text{rain the first day}\}$, $B = \{\text{rain the second day}\}$, $C = \{\text{no rain the third day}\}$ and $D = \{\text{no rain the fourth day}\}$. Then
    $P(A \text{ followed by } B \text{ followed by } C \text{ followed by } D)$
    $= P(A) \cdot P(B \mid A) \cdot P(C \mid A \text{ and } B)$
    $\qquad \cdot P(D \mid A, B \text{ and } C)$
    $= 0.15 \cdot 0.65 \cdot 0.35 \cdot 0.85$
    ≈ 0.029

63. $1 - (0.988)^{12} \approx 0.135$

64. Let S_1 be the event that a 1 is rolled, in which case the player loses $3. There is 1 possible number, so $P(S_1) = \frac{1}{6}$. Let S_2 be the event that a 2 is rolled, in which case the player loses $2. $P(S_2) = \frac{1}{6}$. Let S_3 be the event that a 3 is rolled, in which case the player loses $1. There is 1 possible number, so $P(S_3) = \frac{1}{6}$. Let S_4 be the event that a 4 is rolled, in which case the player wins $0. $P(S_4) = \frac{1}{6}$. Let S_5 be the event

that a 5 is rolled, in which case the player wins $1. There is 1 possible number, so $P(S_5) = \frac{1}{6}$. Let S_6 be the event that a 6 is rolled, in which case the player wins $2. $P(S_6) = \frac{1}{6}$.

Expectation
$= P(S_1) \cdot S_1 + P(S_2) \cdot S_2 + P(S_3) \cdot S_3$
$\quad + P(S_4) \cdot S_4 + P(S_5) \cdot S_5 + P(S_6) \cdot S_6$
$= \frac{1}{6}(-3) + \frac{1}{6}(-2) + \frac{1}{6}(-1)$
$\quad + \frac{1}{6}(0) + \frac{1}{6}(1) + \frac{1}{6}(2)$
$= -\frac{1}{2}$

65. Let S_1 be the event that both coins are tails, in which case the player wins $5. There is 1 outcome in the sample space of 4, so

    $$P(S_1) = \frac{1}{4}.$$

    Let S_2 be the event that one coin is heads and one is tails, in which case the player loses $2. There are two outcomes in the sample space of 4, so $P(S_2) = \frac{2}{4} = \frac{1}{2}$. Let S_3 be the event that both coins are heads, in which case the player neither wins nor loses money. There is 1 outcome in the sample space of 4, so

    $$P(S_3) = \frac{1}{4}.$$

    Expectation
    $= P(S_1) \cdot S_1 + P(S_2) \cdot S_2 + P(S_3) \cdot S_3$
    $= \frac{1}{4}(5) + \frac{1}{2}(-2) + \frac{1}{4}(0)$
    $= 0.25$
    Expectation is 25 cents.

66. Let S_1 be the event that the first prize is won, in which case the player wins $200. $P(S_1) = \frac{1}{800}$.
    Let S_2 be the event that the second prize is won, in which case the player wins $75.

    $$P(S_2) = \frac{4}{800}. \text{ It costs \$1 to buy a ticket.}$$

    Expectation
    $= P(S_1) \cdot S_1 + P(S_2) \cdot S_2 - 1$
    $= \frac{1}{800}(200) + \frac{4}{800}(75) - 1$
    ≈ -0.37
    Expectation is -37 cents.

67. The sample space for tossing two six-sided dice is found in the text. There are 36 possible outcomes. There are 3 elements in the event that the sum of the pips on the upward faces is

4. Then the probability is

$$P(E) = \frac{n(E)}{n(S)} = \frac{3}{36} = \frac{1}{12}.$$

Multiply the probability by 65 and find that we can expect a sum of 4 about 5.4 times.

68. Let S_1 be the event that the person dies within one year. Then $P(S_1) = 0.000595$ and the company must pay out \$40,000. Since the company charged \$320 for the policy, the company's actual loss is \$39,680. Let S_2 be the event that the policy holder does not die during the year of the policy. Then $P(S_2) = 0.999405$ and the company keeps the premium of \$320.
Expectation
$= P(S_1) \cdot S_1 + P(S_2) \cdot S_2$
$= 0.000595(-39,680) + 0.999405(320)$
$= 296.20$
Expectation is \$296.20 .

69. Let S_1 be the event that the person dies within one year. Then $P(S_1) = 0.001188$ and the company must pay out \$25,000. Since the company charged \$795 for the policy, the company's actual loss is \$24,205. Let S_2 be the event that the policy holder does not die during the year of the policy. Then $P(S_2) = 0.998812$ and the company keeps the premium of \$795.
Expectation
$= P(S_1) \cdot S_1 + P(S_2) \cdot S_2$
$= 0.001188(-24,205) + 0.998812(795)$
$= 765.30$
Expectation is \$765.30.

70. Expectation
$= 0.20(25,000) + 0.25(15,000) + 0.20(10,000)$
$\quad + 0.15(5,000) + 0.10(0) + 0.10(-5000)$
$= 5000 + 3750 + 2000 + 750 + 0 - 500$
$= 11,000$
The expected profit is \$11,000.

CHAPTER 12 TEST

1. {A2, D2, G2, K2, A3, D3, G3, K3, A4, D4, G4, K4}

2. There are 3 choices for processors, so $n_1 = 3$.
There are 4 choices for disk drives, so $n_2 = 4$.
There are 3 choices for monitors, so $n_3 = 3$.
There are 2 choices for graphics cards, so $n_4 = 2$.
By the counting principle, there are
$3 \cdot 4 \cdot 3 \cdot 2 = 72$ computer systems possible.

3. The order in which the instructions is sent is important, so the number of ways to send the instructions is

$$P(10,\ 4) = \frac{10!}{(10-4)!} = \frac{10!}{6!} = \frac{10 \cdot 9 \cdot 8 \cdot 7 \cdot 6!}{6!}$$
$$= 10 \cdot 9 \cdot 8 \cdot 7 = 5040.$$

4. The order in which the words are chosen is important, so the number of pairs possible is

$$P(15,\ 10) = \frac{15!}{(15-10)!} = \frac{15!}{5!}$$
$$= \frac{15 \cdot 14 \cdot 13 \cdot 12 \cdot 11 \cdot 10 \cdot 9 \cdot 8 \cdot 7 \cdot 6 \cdot 5!}{5!}$$
$$= 15 \cdot 14 \cdot 13 \cdot 12 \cdot 11 \cdot 10 \cdot 9 \cdot 8 \cdot 7 \cdot 6$$
$$\approx 1.09 \times 10^{10}.$$

5. Because the choice of team A playing team B is the same as the choice of team B playing team A, the order of the selection is not important, so the number of different games is

$$C(8,\ 2) = \frac{8!}{2!(8-2)!} = \frac{8!}{2!6!}$$
$$= \frac{8 \cdot 7 \cdot 6!}{2!6!} = \frac{8 \cdot 7}{2 \cdot 1} = 28.$$

6. Let A be the event of tossing a head, and let B be the event of rolling a 5. There are 2 outcomes possible for the coin and 6 outcomes possible for the die, so by the counting principle, there are $2 \cdot 6 = 12$ items in the sample space. $n(A) = 6$, $n(B) = 2$ and $n(A \text{ and } B) = 1$.
$$P(A \text{ or } B) = P(A) + P(B) - P(A \text{ and } B)$$
$$= \frac{6}{12} + \frac{2}{12} - \frac{1}{12} = \frac{7}{12}$$

7. Let $E = \{$at least one ace is drawn$\}$; then $E^C = \{$no ace is drawn$\}$. There are 52 possibilities for each card drawn. Thus, $n(S) = 52 \cdot 52 \cdot 52 \cdot 52 = 7,311,616$. Each time a card is drawn; there are 48 cards that are not aces, so $n(E^C) = 48 \cdot 48 \cdot 48 \cdot 48 = 5,308,416$.
$$P(E^C) = 1 - P(E)$$
$$= 1 - \frac{5,308,416}{7,311,616} = \frac{2,003,200}{7,311,616}$$
$$= \frac{7825}{28,561} \approx 27.4\%$$

8. Let $A = \{$a heart is drawn first$\}$ and $B = \{$a heart is drawn second$\}$. On the first draw, there are 13 hearts in the deck of 52 cards. Therefore,
$$P(A) = \frac{13}{52} = \frac{1}{4}.$$ On the second draw, there are only 51 cards remaining and only 12 hearts. Therefore, $P(B \mid A) = \frac{12}{51} = \frac{4}{17}$. We have
$$P(A \cap B) = P(A) \cdot P(B \mid A) = \frac{1}{4} \cdot \frac{4}{17} = \frac{1}{17}.$$

9. Let $E = \{$no 9 is drawn$\}$. There are 52 possibilities for the first card drawn, 51 possibilities for the second card drawn, 50 for the third and 49 for the fourth. Thus, $n(S) = 52 \cdot 51 \cdot 50 \cdot 49 = 6,497,400$. The first time a card is drawn, there are 48 cards that are not 9's, the second time there are 47 cards that are not 9's, 46 cards the third time and 45 cards the fourth time, so
    $n(E) = 48 \cdot 47 \cdot 46 \cdot 45 = 4,669,920$.
    $$P(E) = \frac{4,669,920}{6,497,400} \approx 0.72.$$

10. The odds in favor are
    $$\frac{P(E)}{1 - P(E)} = \frac{\frac{1}{8}}{1 - \frac{1}{8}} = \frac{\frac{1}{8}}{\frac{7}{8}} = \frac{1}{7}$$
    The odds in favor are 1 to 7.

11. Let D be the event that a person has the disease and let T be the event that the test is positive. The probability we wish to determine is $P(D|T)$. Then
    $$P(D \mid T) = \frac{P(D \text{ and } T)}{P(T)}.$$
    A tree diagram will help us compute the needed probabilities.

    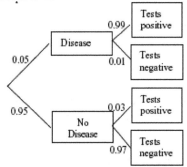

    From the diagram,
    $P(D \text{ and } T) = P(D) \cdot P(T \mid D) = (0.05)(0.99)$. To compute $P(T)$ we need to combine two branches from the diagram, one corresponding to a correct positive test result when the person has the disease, and one corresponding to a false positive result when the person does not have the disease:
    $P(T) = (0.05)(0.99) + (0.95)(0.03)$. Then
    $$P(D \mid T) = \frac{P(D \text{ and } T)}{P(T)}$$
    $$= \frac{(0.05)(0.99)}{(0.05)(0.99) + (0.95)(0.03)} \approx 0.635.$$

12. Let $B = \{$person is a woman$\}$ and $A = \{$responded negatively$\}$. From the table, $n(A \cap B) = 642$, $n(A) = 1378$, and $n(S) = 2815$. We have
    $$P(B \mid A) = \frac{P(A \cap B)}{P(A)} = \frac{\frac{642}{2815}}{\frac{1378}{2815}} \approx 0.466.$$

13. Make a Punnett square.

Parents	S	s
s	Ss	ss
s	Ss	ss

    To have curly hair, the hamster must have two s's. From the table, two of the four possible genotypes have two s's, so the probability that an offspring will have curly hair is $\frac{2}{4} = \frac{1}{2}$ or 50%.

14. The expected profit is
    Expectation
    $= 0.18(75,000) + 0.36(50,000)$
    $\quad + 0.31(25,000) + 0.08(0)$
    $\quad + 0.05(-10,000) + 0.02(-20,000)$
    $= 13,500 + 18,000 + 7750 + 0 - 500 - 400$
    $= 38,350$
    The expected profit is \$38,350.

Chapter 13: Statistics

EXCURSION EXERCISES, SECTION 13.1

1. $1.972 - 1.856 = 0.1160$
   $0.7 \cdot (0.1160) = 0.0812$
   $1.856 + 0.0812 = 1.9372$

3. $1.732 - 1.414 = 0.318$
   $0.4 \cdot (0.318) = 0.1272$
   $1.414 + 0.1272 = 1.5412 \approx 1.541$

5. The player needs to lose $325 - 290 = 35$
   pounds. He needs to lose $\dfrac{25}{90}$ of the weight in
   25 days, so he must lose $\dfrac{25}{90} \cdot (35) \approx 9.7$
   pounds. In 25 days, he should weigh
   approximately $325 - 9.7 = 315.3$ pounds.

EXERCISE SET 13.1

1. $\bar{x} = \dfrac{\sum x}{n} = \dfrac{2+7+5+7+14}{5} = 7$

   Ranking the numbers from smallest to largest
   gives 2, 5, 7, 7, 14. The middle number is
   7.Thus 7 is the median.
   The number 7 occurs more often than other
   numbers. Thus 7 is the mode.

3. $\bar{x} = \dfrac{11+8+2+5+17+39+52+42}{8} = 22$

   Ranking the numbers from smallest to largest
   gives 2, 5, 8, 11, 17, 39, 42, 52. The two middle
   numbers are 11 and 17. The mean of 11 and 17
   is 14. Thus 14 is the median. No number occurs
   more often than the others, so there is no mode.

5. $\bar{x} = \dfrac{\sum x}{n} = \dfrac{2.1+4.6+8.2+3.4+5.6+8.0+9.4+12.2+56.1+78.2}{10} \approx 18.8$

   Ranking the numbers from smallest to largest gives 2.1, 3.4, 4.6, 5.6, 8.0, 8.2, 9.4, 12.2, 56.1, 78.2. The two
   middle numbers are 8.0 and 8.2. The mean of 8.0 and 8.2 is 8.1. Thus 8.1 is the median. No number occurs
   more often than the others, so there is no mode.

7. $\bar{x} = \dfrac{\sum x}{n} = \dfrac{255+178+192+145+202+188+178+201}{8} \approx 192.4$

   Ranking the numbers from smallest to largest gives 145, 178, 178, 188, 192, 201, 202, 255. The two middle
   numbers are 188 and 192. The mean of 188 and 192 is 190. Thus 190 is the median. The number 178 occurs
   more often than the other numbers. Thus 178 is the mode.

9. $\bar{x} = \dfrac{\sum x}{n} = \dfrac{-12+(-8)+(-5)+(-5)+(-3)+0+4+9+21}{9} \approx 0.1$

   Ranking the numbers from smallest to largest gives –12, –8, –5, –5, –3, 0, 4, 9, 21. The middle number is –
   3.Thus –3 is the median. The number –5 occurs more often than the other numbers. Thus –5 is the mode.

11. a. Yes. The mean is computed by using the sum of all the data.

    b. No. The median is not affected unless the middle value, or one of the two middle values, in a data set is
       changed.

13. $\bar{x} = \dfrac{\sum x}{n} = \dfrac{1439}{36} \approx 40.0$

    Ranking the numbers from smallest to largest shows the two middle numbers to be 35 and 35. Thus 35 is
    the median.

15. a. Answers will vary.

    b. Answers will vary.

17. The A is worth 4 points with a weight of 3; the B is worth 3 points with a weight of 4; the C^+ is
    worth 2.33 with a weight of 3; and the B^- is worth 2.67 with a weight of 2. The sum of all the weights
    is $3 + 3 + 4 + 3 + 2$, or 15.

    Weighted mean $= \dfrac{(4 \times 3)+(4 \times 3)+(3 \times 4)+(2.33 \times 3)+(2.67 \times 2)}{15} = \dfrac{48.33}{15} \approx 3.22$

    Jerry's GPA for the fall semester is 3.22.

19. Find the weighted mean of both semesters. The sum of all the weights is 46 + 12, or 58.

$$\text{Weighted mean} = \frac{(3.24 \times 46) + (3.86 \times 12)}{58}$$

$$= \frac{195.36}{58} \approx 3.37$$

Tessa's cumulative GPA is 3.37.

21. $\text{Weighted Mean} = \dfrac{\sum (x \cdot w)}{\sum w}$

$$= \frac{(70 \times 10\%) + (65 \times 10\%) + (82 \times 10\%) + (94 \times 10\%) + (85 \times 10\%) + (92 \times 20\%) + (80 \times 30\%)}{15} = \frac{82}{1} = 82$$

23. $\text{Players slugging average} = \dfrac{s + 2d + 3t + 4h}{n}$

$$= \frac{73 + 2(36) + 3(9) + 4(54)}{458}$$

$$= \frac{388}{458} \approx 0.847$$

25. $\text{Players slugging average} = \dfrac{s + 2d + 3t + 4h}{n}$

$$= \frac{94 + 2(33) + 3(1) + 4(49)}{535}$$

$$= \frac{359}{535} \approx 0.671$$

27. $\text{Mean} = \dfrac{\sum (x \cdot f)}{\sum f} = \dfrac{(2 \cdot 6) + (4 \cdot 5) + (5 \cdot 6) + (9 \cdot 3) + (10 \cdot 1) + (14 \cdot 2) + (19 \cdot 1)}{24} = \dfrac{146}{24} \approx 6.1 \text{ points}$

If the data were written in a list (6 twos, 5 fours, 6 fives, 3 nines, 1 ten, 2 fourteens, 1 nineteen), the middle number would be 5. Thus 5 points is the median. 2 points and 5 points occur most. Thus 2 and 5 points are the mode.

29. $\text{Mean} = \dfrac{\sum (x \cdot f)}{\sum f} = \dfrac{(2 \cdot 1) + (4 \cdot 2) + (6 \cdot 7) + (7 \cdot 12) + (8 \cdot 10) + (9 \cdot 4) + (10 \cdot 3)}{39} = \dfrac{282}{39} \approx 7.2$

The middle score is 7. Thus 7 is the median. The score of 7 occurs more often than any other score. Thus 7 is the mode.

31. $\text{Midrange} = \dfrac{56° + 76°}{2} = 64°$

33. $\text{Midrange} = \dfrac{-56°F + 44°F}{2} = -6°F$

35. Let x be the score Ruben needs on the next test. The mean score has a weight of 6 and the new score has a weight of 1, so

$$80 = \frac{(78 \times 6) + (x \times 1)}{7}$$

$$560 = 468 + x$$

$$92 = x$$

37. a. $\bar{x} = \dfrac{0.336 + 0.213}{2} \approx 0.275$

b. Batting average

$$\frac{92 + 60}{274 + 282} \approx 0.273$$

c. No.

39. Yes. In the first month, Dawn's average is 0.4 and Joanne's average is 0.3973. In the second month, Dawn's average is 0.3878 and Joanne's average is 0.3875. For both months, Dawn has a total of 21 hits at 54 at-bats and an average of 0.3889, while Joanne has a total of 60 hits at 153 at-bats and an average of 0.3922. Joanne has a smaller average for the first month and the second month, but she has a larger average for both months combined.

EXCURSION EXERCISES, SECTION 13.2

1.   a., b.

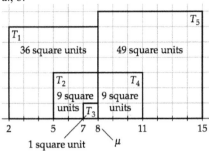

c.   104 square units

d.   Using the formula:
$$\frac{36+9+1+9+49}{5}=\frac{104}{5}$$
$$= 20.8 \text{ square units}$$

e.

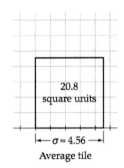

Average tile

f.   The variance is 20.8. It is the area of the average tile shown in part e.

g.   The standard deviation is $\sqrt{20.8} \approx 4.56$. It is the width of the average tile shown in part e.

EXERCISE SET 13.2

1.   The highest temperature is 63°F and the lowest temperature is –21°F. The range is 63°F – (–21°F)= 84°F.

3.   The greatest data value is 22 and the least data value is 1. The range is 22 – 1= 21.
$$\overline{x}=\frac{1+2+5+7+8+19+22}{7}=\frac{64}{7}\approx 9.1429$$

x	$x-\overline{x}$	$(x-\overline{x})^2$
1	$1-9.1429$	$(-8.1429)^2 = 66.3068$
2	$2-9.1429$	$(-7.1429)^2 = 51.0210$
5	$5-9.1429$	$(-4.1429)^2 = 17.1636$
7	$7-9.1429$	$(-2.1429)^2 = 4.5920$
8	$8-9.1429$	$(-1.1429)^2 = 1.3062$
19	$19-9.1429$	$(9.8571)^2 = 97.1624$
22	$22-9.1429$	$(12.8571)^2 = 165.3050$
		402.857

$$\frac{\text{sum of the (deviations)}^2}{n-1}=\frac{402.857}{7-1}=\frac{402.857}{6}$$
$$\approx 67.1428$$
The square root of the quotient is
$$s = \sqrt{67.1428} \approx 8.2$$
The variance is the square of the standard deviation.
$$s^2 = (\sqrt{67.1428})^2 \approx 67.1$$

5.   The greatest data value is 4.8 and the least data value is 1.5. The range is 4.8 – 1.5 = 3.3.
$$\overline{x}=\frac{2.1+3.0+1.9+1.5+4.8}{5}=\frac{13.3}{5}=2.66$$

x	$x-\overline{x}$	$(x-\overline{x})^2$
2.1	$2.1-2.66$	$(-0.56)^2 = 0.3136$
3.0	$3.0-2.66$	$(0.34)^2 = 0.1156$
1.9	$1.9-2.66$	$(-0.76)^2 = 0.5776$
1.5	$1.5-2.66$	$(-1.16)^2 = 1.3456$
4.8	$4.8-2.66$	$(2.14)^2 = 4.5796$
		6.932

$$\frac{\text{sum of the (deviations)}^2}{n-1}=\frac{6.932}{5-1}=\frac{6.932}{4}$$
$$= 1.733$$
The square root of the quotient is
$$s = \sqrt{1.733} \approx 1.3.$$
Variance is the square of the standard deviation.
Thus the variance is $s^2 = (\sqrt{1.733})^2 \approx 1.7$.

7.   The greatest data value is 100 and the least data value is 48. The range is 100 – 48 = 52.
$$\overline{x}=\frac{48+91+87+93+59+68+92+100+81}{9}$$
$$=\frac{719}{9}=79.889$$

x	$x-\overline{x}$	$(x-\overline{x})^2$
48	$48-79.889$	$(-31.889)^2 = 1016.90$
91	$91-79.889$	$(11.111)^2 = 123.454$
87	$87-79.889$	$(7.111)^2 = 50.566$
93	$93-79.889$	$(13.11)^2 = 171.898$
59	$59-79.889$	$(-20.889)^2 = 436.350$
68	$68-79.889$	$(-11.889)^2 = 141.348$
92	$92-79.889$	$(12.111)^2 = 146.676$
100	$100-79.889$	$(20.111)^2 = 404.452$
81	$81-79.889$	$(1.111)^2 = 1.234$
		2492.878

$$\frac{\text{sum of the (deviations)}^2}{n-1}=\frac{2492.878}{9-1}$$
$$=\frac{2492.878}{8}=311.610$$

The square root of the quotient is
$$s = \sqrt{311.610} \approx 17.7.$$
Variance is the square of the standard deviation.
Thus the variance is
$$s^2 = (\sqrt{311.610})^2 \approx 311.6.$$

9. All of the data values are 4. Thus, the range is $4 - 4 = 0$.
    $$\bar{x} = \frac{4+4+4+4+4+4+4+4+4+4+4+4+4+4+4+4+4}{17} = 4$$

x	$x - \bar{x}$	$(x - \bar{x})^2$
4	$4 - 4$	$(0)^2 = 0$
4	$4 - 4$	$(0)^2 = 0$
4	$4 - 4$	$(0)^2 = 0$
4	$4 - 4$	$(0)^2 = 0$
4	$4 - 4$	$(0)^2 = 0$
4	$4 - 4$	$(0)^2 = 0$
etc.	etc.	etc.
		0

$$\frac{\text{sum of the (deviations)}^2}{n-1} = \frac{0}{14-1} = \frac{0}{13} = 0.$$

The square root of the quotient is $s = \sqrt{0} = 0$.

Variance is the square of the standard deviation. Thus the variance is $s^2 = (\sqrt{0})^2 = 0$.

11. The greatest data value is 11 and the least data value is –12. The range is $11 - (-12) = 23$.
    $$\bar{x} = \frac{-8+(-5)+(-12)+(-1)+4+7+11}{7} = \frac{-4}{7} \approx -0.571$$

x	$x - \bar{x}$	$(x - \bar{x})^2$
–8	$-8 - (-0.571)$	$(-7.429)^2 = 55.190$
–5	$-5 - (-0.571)$	$(-4.429)^2 = 19.616$
–12	$-12 - (-0.571)$	$(-11.429)^2 = 130.622$
–1	$-1 - (-0.571)$	$(-0.429)^2 = 0.184$
4	$4 - (-0.571)$	$(4.571)^2 = 20.894$
7	$7 - (-0.571)$	$(7.571)^2 = 57.320$
11	$11 - (-0.571)$	$(11.571)^2 = 133.888$
		417.714

$$\frac{\text{sum of the (deviations)}^2}{n-1} = \frac{417.714}{7-1} = \frac{417.714}{6} = 69.619$$

The square root of the quotient is $s = \sqrt{69.619} \approx 8.3$.

Variance is the square of the standard deviation. Thus the variance is $s^2 = (\sqrt{69.619})^2 \approx 69.6$.

13. Opinions will vary. However, many climbers would consider rope B to be safer because of its smaller standard deviation in breaking strength.

15. The students in the college statistics course because the range of weights is greater.

17. Using a calculator, the mean, to the nearest hundredth, is $\bar{x} \approx 23.67$ mpg.

    The sample standard deviation (given as Sx on a TI-83/84 calculator), to the nearest hundredth, is $s \approx 3.92$ mpg.

19. Using a calculator, the mean, to the nearest hundredth, is $\bar{x} \approx 493.60$ cal.

    The sample standard deviation (given as Sx on a TI-83/84 calculator), to the nearest hundredth, is $s \approx 20.30$ cal.

21. Using a calculator, the mean, to the nearest hundredth, is $\bar{x} \approx 4.27$ h.

    The sample standard deviation (given as Sx on a TI-83/84 calculator), to the nearest hundredth, is $s \approx 0.69$ h.

23. a. Using a calculator, the mean $(\bar{x})$, to the nearest second, is 210 sec, or 3 min 30 sec, and the sample standard deviation (Sx), to the nearest second, is 26 sec.

    b. Yes.
        Add and Subtract from the mean:
        210 + 26 = 236 sec = 3:56 (min:sec)
        210 – 26 = 184 sec = 3:04 (min:sec)
        Therefore, the song length greater than 3:56 min:sec and lower than 3:04 min:sec are 4:02 min:sec, 3:01 min:sec, and 2:27 min:sec.

25. a. Answer will vary.

    b. The standard deviation of the new data is k times the standard deviation of the original data.

27. Because variance is the square of the standard deviation, if the standard deviation is 0, then $s^2 = 0^2 = 0$. If the standard deviation is 1, then $s^2 = 1^2 = 1$. The variance is 0 or 1.

EXCURSION EXERCISES, SECTION 13.3

1. Ages of Customers who Purchased a Cruise

Stems	Leaves
2	1 3 8
3	2 3 3 5 8
4	0 1 2 4 5 5 6
5	0 1 1 2 2 5
6	1 2 4 6 8
7	2 7

   Legend:
   5 | 2 represents 52 years old

3.

Before	Stem	After
	10	7
6	11	7 9
8 8 8	12	1 3 5 6 8 9
9 9 6 6 4 4 1	13	3 4 4 8 9 9
8 4 1	14	0 0 9
8 6 6 1	15	0
1	16	4
4	17	

   Legend: Legend:
   8 | 13 represents   13 | 3 represents
   128 beats per minute   133 beats per minute

EXERCISE SET 13.3

1. a. $z_{85} = \dfrac{85 - 75}{11.5} \approx 0.87$

   b. $z_{95} = \dfrac{95 - 75}{11.5} \approx 1.74$

   c. $z_{50} = \dfrac{50 - 75}{11.5} \approx -2.17$

   d. $z_{75} = \dfrac{75 - 75}{11.5} = 0$

3. a. $z_{6.2} = \dfrac{6.2 - 6.8}{1.9} \approx -0.32$

   b. $z_{7.2} = \dfrac{7.2 - 6.8}{1.9} \approx 0.21$

   c. $z_{9.0} = \dfrac{9.0 - 6.8}{1.9} \approx 1.16$

   d. $z_{5.0} = \dfrac{5.0 - 6.8}{1.9} \approx -0.95$

5. a. $z_{110.5} = \dfrac{110.5 - 119.4}{13.2} \approx -0.67$

   b. Solving for x:
   $$z_x = \frac{x - 119.4}{13.2} = 2.15$$
   $$x - 119.4 = 28.4$$
   $$x = 147.78$$
   Her systolic blood pressure reading was 147.78 mm Hg.

7. a. $z_{214} = \dfrac{214 - 182}{44.2} \approx 0.72$

   b. Solving for x:
   $$z_x = \frac{x - 182}{44.2} = -1.58$$
   $$x - 182 = -69.836$$
   $$x \approx 112.16$$
   His blood cholesterol level is approximately 112.16 mg/dl.

9. a. $z_{65} = \dfrac{65 - 72}{8.2} \approx -0.85$

   b. $z_{102} = \dfrac{102 - 130}{18.5} \approx -1.51$

   c. $z_{605} = \dfrac{605 - 720}{116.4} \approx -0.99$

   The 65 is the highest relative score.

11. Percentile of $455 = \dfrac{4256}{7210} \cdot 100 \approx 59$

    Shaylen's score was at the 59th percentile.

13. Kevin's percentile $= \dfrac{x}{9840} \cdot 100 = 65$

    $$x = \frac{65 \cdot 9840}{100} = 6396$$

    6396 students scored lower than Kevin.

15. a. The median is by definition the 50th percentile. Therefore, 50% of the four-person family incomes are more than $66,650.

    b. Because $178,500 is the 90th percentile, $100 - 90 = 10\%$ of the four-person family incomes are more than $178,500.

    c. $90 - 50 = 40\%$ of the four-person family incomes are between $66,650 and $178,500.

17. 1, 3, 4, 5, 5, 7, 8, 10, 10, 10, 12, 15, 18, 26, 28, 32, 41, 85

The median (or second quartile) is the average of the 9th and 10th data values. $\dfrac{10+10}{2}=10$

The first quartile is the median of the 1st through 9th data values. 5.

The third quartile is the median of the 10th through 18th data values. 26. $Q_1 = 5$, $Q_2 = 10$, $Q_3 = 26$

19. Northeast:
   minimum: 227,400, maximum: 346,000
   $Q_1 = 264,300$, $Q_2 = 315,800$, $Q_3 = 343,600$
   Midwest:
   minimum: 169,700, maximum: 216,900
   $Q_1 = 178,000$, $Q_2 = 197,600$, $Q_3 = 208,600$
   South:
   minimum: 148,000, maximum: 217,700
   $Q_1 = 163,400$, $Q_2 = 194,800$, $Q_3 = 203,700$
   West:
   minimum: 196,400, maximum: 337,700
   $Q_1 = 238,500$, $Q_2 = 263,700$, $Q_3 = 330,900$

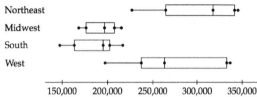

Answers will vary. Here are some possibilities. The region with the lowest median was the South; the region with the highest median was the Northeast. The range of prices was greatest for the West.

21. CBX-21:
   minimum: 44, maximum: 60
   $Q_1 = 46.5$, $Q_2 = 49.5$, $Q_3 = 52.5$
   PHT-34:
   minimum: 38, maximum: 50
   $Q_1 = 44$, $Q_2 = 46$, $Q_3 = 47$

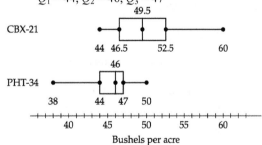

Bushels per acre

Answers will vary. Here is one possibility: The maximum number of bushels cultivated per acres for PHT-34 is approximately equal to the median number of bushels cultivated per acres for CBX-21.

23. a.   Using the formula:
   $$\dfrac{-1.101-0.944-0.157+0.629+1.573}{5}=0$$
   Mean $= 0 = \mu$
   $(x-\bar{x})_1 = -1.101$
   $(x-\bar{x})_2 = -0.944$
   $(x-\bar{x})_3 = -0.157$
   $(x-\bar{x})_4 = 0.629$
   $(x-\bar{x})_5 = 1.573$

   $(-1.101)^2 + (-0.944)^2 + (-0.157)^2$
   $+ (0.629)^2 + (1.573)^2 \approx 5$
   $\dfrac{5}{5} = 1$
   $\sqrt{1} = 1 = \sigma$
   The mean of the z-scores is 0, and their standard deviation is 1.

   b.   Using the formula:
   $$\dfrac{-1.459-0.973-0.243+0.365+0.973+1.377}{6}$$
   $=0$
   Mean $= 0 = \mu$
   $(x-\bar{x})_1 = -1.459$
   $(x-\bar{x})_2 = -0.973$
   $(x-\bar{x})_3 = -0.243$
   $(x-\bar{x})_4 = 0.365$
   $(x-\bar{x})_5 = 1.337$

   $(-1.459)^2 + (-0.973)^2 + (-0.243)^2$
   $+ (0.365)^2 + (0.973)^2 + (1.337)^2 \approx 6$
   $\dfrac{6}{6} = 1$
   $\sqrt{1} = 1 = \sigma$
   The mean of the z-scores is 0, and their standard deviation is 1.

   c.   The mean of any set of z-scores is always 0, and their standard deviation is always 1.

EXCURSION EXERCISES, SECTION 13.4

1. The z-score associated with A = 0.41 is 1.34.
   $$1.34 = \dfrac{x-\bar{x}}{s} = \dfrac{x-64}{10}$$
   $x = 13.4 + 64 = 77.4$
   A 77.4 would be the cut-off score for an A.

3. The z-score associated with A = 0.41 is 1.34.
   $$-1.34 = \dfrac{x-\bar{x}}{s} = \dfrac{x-7.2}{5.5}$$
   $x = -7.37 + 72 \approx 64.63$
   A runner needs a time of 64.63 seconds or less to be in the top 9% of runners.

EXERCISE SET 13.4

1. a. The percent of 19th-century boys who were at least 65 inches is 5.8%.

   b. The percent of data in all classes with a lower boundary of 55 inches and an upper boundary of 64 inches is the sum of the percents: 46% + 41% = 87%.
   The probability that a 19th-century boy selected at random will be at least 55 inches tall but less than 65 inches tall is 0.87.

3. a. 95% of the data lies within 2 standard deviations of the mean.

   b. 13.5 + 2.35 = 15.85%
   15.85% of the data lie more than 1 standard deviation above the mean.

   c. 34 + 34 + 13.5 = 81.5%
   81.5% of the data lies between 1 standard deviation below the mean and 2 standard deviations above the mean.

5. a. 12 ounces is 2 standard deviations below the mean, and 30 ounces is 1 standard deviation above the mean.
   34 + 34 + 13.5 = 81.5%.
   81.5% of the parcels weighed between 12 and 30 ounces.

   b. 42 ounces is 3 standard deviations above the mean. Therefore, 0.15% of the parcels weighed more than 42 ounces.

7. a. 68 miles per hour is 1 standard deviation above the mean. (0.16)(8000) = 1280.
   1280 vehicles had a speed of more than 68 miles per hour.

   b. 40 miles per hour is 3 standard deviation below the mean. (0.0015)(8000) = 12.
   12 vehicles had a speed of less than 40 miles per hour.

9. When z is 1.5, A is 0.433. When z is 0, A is 0.Therefore, the area of the distribution between $z = 0$ and $z = 1.5$ is 0.433 square unit.

11. The area of the distribution between $z = 0$ and $z = -1.85$ is equal to the area of the distribution between $z = 0$ and $z = 1.85$, due to symmetry. When z is 1.85, A is 0.468. When z is 0, A is 0.Therefore, the area of the distribution between $z = 0$ and $z = -1.85$ is 0.468 square unit.

13. When z is 1.9, A is 0.471. When z is 1, A is 0.341. 0.471 − 0.341 = 0.130.
   The area of the distribution between $z = 1$ and $z = 1.9$ is 0.130 square unit.

15. The area of the distribution between $z = 0$ and $z = -1.47$ is equal to the area of the distribution between $z = 0$ and $z = 1.47$, due to symmetry. When z is 1.47, A is 0.429. When z is 0, A is 0.When z is 1.64, A is 0.449. When z is 0, A is 0.0.429 + 0.449 = 0.878.
   The area of the distribution between $z = -1.47$ and $z = 1.64$ is 0.878 square unit.

17. The area of the distribution between $z = 0$ and $z = 1.3$ is 0.403 square unit. The area of the distribution to the right of $z = 0$ is 0.500 square unit.   0.500 − 0.403 = 0.097.
   The area of the distribution where $z > 1.3$ is 0.097 square unit.

19. The area of the distribution between $z = 0$ and $z = -2.22$ is 0.487 square unit. The area of the distribution to the left of $z = 0$ is 0.500 square unit. 0.500 − 0.487 = 0.013.
   The area of the distribution where $z < -2.22$ is 0.013 square unit.

21. The area of the distribution between $z = 0$ and $z = -1.45$ is 0.426 square unit. The area of the distribution to the right of $z = 0$ is 0.500 square unit. 0.426 + 0.500 = 0.926.
   The area of the distribution where $z > -1.45$ is 0.926 square unit.

23. The area of the distribution between $z = 0$ and $z = 2.71$ is 0.497 square unit. The area of the distribution to the left of $z = 0$ is 0.500 square unit. 0.497 + 0.500 = 0.997.
   The area of the distribution where $z < 2.71$ is 0.997 square unit.

25. If 0.200 square unit of the area of the distribution is to the right of z, then 0.300 square unit is between z and 0.
   At $A = 0.300$, $z = 0.84$.
   The z-score is 0.84.

27. If 0.184 square unit of the area of the distribution is to the left of z, then 0.316 square unit is between z and 0.
   At $A = 0.316$, $z = 0.90$. Because z is to the left of 0, the z-score is −0.90.

29. If 0.363 square unit of the area of the distribution is to the right of z, then 0.137 square unit is between z and 0.
   At $A = 0.137$, $z = 0.35$.
   The z-score is 0.35.

31. a. $z_{219} = \dfrac{219 - 185}{39} = 0.87$

    Table 13.10 indicates that 0.308 (30.8%) of the data in a normal distribution are between $z = 0$ and $z = 0.87$.
    Therefore, the percent of data to the right of $z = 0.87$ is 50% − 30.8% = 19.2%.
    Thus 19.2% of the women's cholesterol level is greater than 219.

    b. $z_{190} = \dfrac{190 - 185}{39} = 0.13$

    Table 13.10 indicates that 0.052 (5.2%) of the data in a normal distribution are between $z = 0$ and $z = 0.13$.

    $z_{225} = \dfrac{225 - 185}{39} = 1.03$

    Table 13.10 indicates that 0.348 (34.8%) of the data in a normal distribution are between $z = 0$ and $z = 1.03$.
    Thus the percent of women's cholesterol levels between 190 and 225 is
    34.8% − 5.2% = 29.6%.

33. a. $z_{950} = \dfrac{950 - 1025}{87} = -0.86$

    Table 13.10 indicates that 0.305 (30.5%) of the data in a normal distribution are between $z = -0.86$ and $z = 0$. Therefore, the percent of data to the right of $z = -0.86$ is 50% + 30.5% = 80.5%. Thus 80.5% of the light bulb's life span is at least 950 hours.

    b. $z_{800} = \dfrac{800 - 1025}{87} = -2.59$

    Table 13.10 indicates that 0.495 (49.5%) of the data in a normal distribution are between $z = -2.59$ and $z = 0$.

    $z_{900} = \dfrac{900 - 1025}{87} = -1.44$

    Table 13.10 indicates that 0.425 (42.5%) of the data in a normal distribution are between $z = -1.44$ and $z = 0$.
    Thus the percent of the light bulb's life span between 800 and 900 hours is
    49.5% − 42.5% = 7%.

35. a. $z_{14} = \dfrac{14 - 14.5}{0.4} = -1.25$

    0.394 square unit is between $z = -1.25$ and $z = 0$. Therefore, 0.106 square unit is to the left of $z = -1.25$.
    10.6% of the boxes will weigh less than 14 ounces.

    b. Calculating:

    $z_{13.5} = \dfrac{13.5 - 14.5}{0.4} = -2.5$

    $z_{15.5} = \dfrac{15.5 - 14.5}{0.4} = 2.5$

37. a. $z_{320} = \dfrac{320 - 350}{24} = -1.25$

    0.394 square unit is between $z = -1.25$ and $z = 0$. Therefore, 0.106 square unit is to the right of $z = -1.25$.
    The probability that a piece of this rope chosen at random will have a breaking point of less than 320 pounds is 0.106.

    b. Calculating:

    $z_{340} = \dfrac{340 - 350}{24} \approx -0.42$

    $z_{370} = \dfrac{370 - 350}{24} \approx 0.83$

    0.163 square unit is between $z = -0.42$ and $z = 0$. 0.297 square unit is between $z = 0$ and $z = 0.83$. 0.163 + 0.297 = 0.460.
    The probability of a piece of this rope chosen at random will have a breaking point of between 340 and 370 pounds is 0.460.

39. a. $z_3 = \dfrac{3 - 2.5}{0.75} \approx 0.67$

    0.249 square unit is between $z = 0$ and $z = 0.67$. 0.500 square unit is to the left of $z = 0$. 0.249 + 0.500 = 0.749.
    The probability that the time a customer spends waiting is at most 3 minutes is 0.749.

    b. $z_1 = \dfrac{1 - 2.5}{0.75} = -2$

    0.477 square unit is between $z = -2$ and $z = 0$. Therefore, 0.023 square unit is left of $z = -2$. The probability that the time a customer spends waiting is less than 1 minute is 0.023.

41. Answers will vary.

43. True.

45. True.

47. True.

49. False.

51. The first and third quartiles are at $A = 0.250$.
    The z-value for $A = 0.250$ is approximately 0.67.
    The two z-values are approximately −0.67 and 0.67.

EXCURSION EXERCISES, SECTION 13.5

1. Using a calculator, the exponential regression equation is $y \approx 10.14681746(0.8910371309)^x$, where x is the altitude above sea level in kilometers, and y is the atmospheric pressure in newtons per square centimeter.

Use the regression equation to find the pressure when the altitude is 11 km.

$$y \approx 10.14681746(0.8910371309)^x$$
$$y \approx 10.14681746(0.8910371309)^{11}$$
$$y \approx 2.9 \text{ newtons/cm}^2$$

EXERCISE SET 13.5

1. a. Graph b, as the points are clustered closer to the least-squares line.

   b. Graph c, as the points are clustered closer to the least-squares line.

3. a.

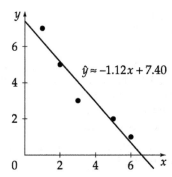

$$\hat{y} \approx -1.12x + 7.40$$

   b. $n = 5$
   $$\sum x = 1+2+3+5+6 = 17$$
   $$\sum y = 7+5+3+2+1 = 18$$
   $$\sum x^2 = 1+4+9+25+36 = 75$$
   $$\left(\sum x\right)^2 = 17^2 = 289$$
   $$\sum xy = 7+10+9+10+6 = 42$$

   c. Calculating:
   $$a = \frac{n\left(\sum xy\right)-\left(\sum x\right)\left(\sum y\right)}{n\left(\sum x^2\right)-\left(\sum x\right)^2}$$
   $$= \frac{5(42)-(17)(18)}{5(75)-289} = \frac{-48}{43} \approx -1.12$$
   $$b = \bar{y} - a\bar{x} \approx 7.40$$

   d. See the graph in part a.

   e. Yes.

   f. $\hat{y} = (-1.12)(3.4) + 7.40 \approx 3.6$

   g. Calculating:
   $$r = \frac{n\left(\sum xy\right)-\left(\sum x\right)\left(\sum y\right)}{\sqrt{n\left(\sum x^2\right)-\left(\sum x\right)^2}\cdot\sqrt{n\left(\sum y^2\right)-\left(\sum y\right)^2}}$$
   $$= \frac{5(42)-(17)(18)}{\sqrt{5(75)-289}\cdot\sqrt{5(88)-324}} \approx -0.96$$

5. $n = 5$
   $$\sum x = 2+3+4+6+8 = 23$$
   $$\sum y = 6+6+8+11+18 = 49$$
   $$\sum x^2 = 4+9+16+36+64 = 129$$
   $$\sum y^2 = 36+36+64+121+324 = 581$$
   $$\left(\sum x\right)^2 = 23^2 = 529, \ \left(\sum y\right)^2 = 49^2 = 2401$$
   $$\sum xy = 12+18+32+66+144 = 272$$
   $$a = \frac{n\left(\sum xy\right)-\left(\sum x\right)\left(\sum y\right)}{n\left(\sum x^2\right)-\left(\sum x\right)^2}$$
   $$= \frac{5(272)-(23)(49)}{5(129)-529} \approx -2.01$$
   $$b = \bar{y} - a\bar{x} \approx 0.56$$
   $$\hat{y} \approx 2.01x + 0.56$$
   $$r = \frac{n\left(\sum xy\right)-\left(\sum x\right)\left(\sum y\right)}{\sqrt{n\left(\sum x^2\right)-\left(\sum x\right)^2}\cdot\sqrt{n\left(\sum y^2\right)-\left(\sum y\right)^2}}$$
   $$= \frac{5(272)-(23)(49)}{\sqrt{5(129)-529}\cdot\sqrt{5(581)-2401}} \approx 0.96$$

7. $n = 5$
   $$\sum x = -3-1+0+2+5 = 3$$
   $$\sum y = 11.8+9.5+8.6+8.7+5.4 = 44$$
   $$\sum x^2 = 9+1+0+4+25 = 39$$
   $$\sum y^2 = 139.24+90.25+73.96+75.69+29.16$$
   $$= 408.3$$
   $$\left(\sum x\right)^2 = 3^2 = 9, \ \left(\sum y\right)^2 = 44^2 = 1936$$
   $$\sum xy = -35.4-9.5+0+17.4+27 = -0.5$$
   $$a = \frac{n\left(\sum xy\right)-\left(\sum x\right)\left(\sum y\right)}{n\left(\sum x^2\right)-\left(\sum x\right)^2}$$
   $$= \frac{5(-0.5)-(3)(44)}{5(39)-9} \approx -0.72$$
   $$b = \bar{y} - a\bar{x} \approx 9.23$$
   $$\hat{y} \approx -0.72x + 9.23$$
   $$r = \frac{n\left(\sum xy\right)-\left(\sum x\right)\left(\sum y\right)}{\sqrt{n\left(\sum x^2\right)-\left(\sum x\right)^2}\cdot\sqrt{n\left(\sum y^2\right)-\left(\sum y\right)^2}}$$
   $$= \frac{5(-0.5)-(3)(4)}{\sqrt{5(39)-9}\cdot\sqrt{5(408.3)-1936}} \approx -0.96$$

9. $n = 5$

$$\sum x = 1 + 2 + 4 + 6 + 8 = 21$$

$$\sum y = 4.1 + 6.0 + 8.2 + 11.5 + 16.2 = 46$$

$$\sum x^2 = 1 + 4 + 16 + 36 + 64 = 121$$

$$\sum y^2 = 16.81 + 36 + 67.24 + 132.25 + 262.44$$
$$= 514.74$$

$$\left(\sum x\right)^2 = 21^2 = 441$$

$$\left(\sum y\right)^2 = 46^2 = 2116$$

$$\sum xy = 4.1 + 12.0 + 32.8 + 69 + 129.6 = 247.5$$

$$a = \frac{n\left(\sum xy\right) - \left(\sum x\right)\left(\sum y\right)}{n\left(\sum x^2\right) - \left(\sum x\right)^2}$$

$$= \frac{5(137.9) - (-12)(-41.2)}{5(8) - 144} \approx 0.66$$

$$b = \bar{y} - a\bar{x} \approx -6.66$$

$$\hat{y} \approx 0.66x - 6.66$$

$$r = \frac{n\left(\sum xy\right) - \left(\sum x\right)\left(\sum y\right)}{\sqrt{n\left(\sum x^2\right) - \left(\sum x\right)^2} \cdot \sqrt{n\left(\sum y^2\right) - \left(\sum y\right)^2}}$$

$$= \frac{5(247.5) - (21)(46)}{\sqrt{5(121) - 441} \cdot \sqrt{5(514.74) - 2116}} \approx 0.99$$

11. a. Using a calculator, the slope (a) is –0.170 and the y-intercept is (b) is 54,545.585. Therefore the equation of the least-square line is $\hat{y} \approx -0.170x + 54,545.585$.

    b. $\hat{y} \approx -0.170(30,000) + 54,545.585$
        $= 49.445.59$
        Their car will be worth $49,446.

    c. Using a calculator, the linear correlation coefficient (r) is –0.999.

    d. As the number of miles the car is driven increases, the value of the car decreases.

13. a. Using a calculator, the slope (a) is 0.775 and the y-intercept is (b) is 20.707. Therefore the equation of the least-square line is $\hat{y} \approx 0.775x + 20.707$.

    b. $\hat{y} \approx 0.775(15) + 20.707 \approx 32.3$
        The predicted percentage of overweight males in 2015 is 32.3%.

15. a. Using a calculator, the linear correlation coefficient (r) is 0.99.

    b. Yes, at least for the years 2005 to 2010. The correlation coefficient is very close to 1, which indicates a strong linear correlation.

17. a. Using a calculator, the slope (a) is –0.91 and the y-intercept is (b) is 79.21. Therefore the equation of the least-square line is $\hat{y} \approx -0.91x + 79.21$.

    b. $\hat{y} \approx -0.91(25) + 79.21 \approx 56$
        The remaining lifetime of a woman of 25 years is about 56 years.

    c. interpolation

19. a. Using a calculator, the slope (a) is 1194.657143 and the y-intercept is (b) is 30,217.2. Therefore the equation of the least-square line for the tuition and fees at private 4-year colleges and universities is $\hat{y} \approx 1194.657x + 30,217.2$. Using a calculator, the linear correlation coefficient (r) is 0.999525765.

    b. Using a calculator, the slope (a) is 827.8857143 and the y-intercept is (b) is 14,233.4. Therefore the equation of the least-square line for the tuition and fees at public 4-year colleges and universities is $\hat{y} \approx 827.8857x + 14,233.4$. Using a calculator, the linear correlation coefficient (r) is 0.9993403551.

    c. Yes. The correlation coefficient is about 1, which indicates a strong linear correlation.

    d. The value of a indicates yearly increase in tuition and fees.
        For the private 4-year colleges and universities in part a, the number 1194.657143 means that tuition and fees increased by about $1195 each year.
        For the public 4-year colleges and universities in part b, the number 827.8857 means that tuition and fees increased by about $828 each year.

CHAPTER 13 REVIEW EXERCISES

1. $\dfrac{12+17+14+12+8+19+21}{7} \approx 14.7 = \mu$

   The mean of the population is about 14.7.
   The median is the fourth ranked number, or 14.
   The mode is the most common number, or 12.
   The range is $21 - 8 = 13$.

x	$x - \mu$	$(x-\mu)^2$
12	12 – 14.7	$(-2.7)^2 = 7.29$
17	17 – 14.7	$(2.3)^2 = 5.29$
14	14 – 14.7	$(-0.7)^2 = 0.49$
12	12 – 14.7	$(-2.7)^2 = 7.29$
8	8 – 14.7	$(-6.7)^2 = 44.89$
19	19 – 14.7	$(4.3)^2 = 18.49$
21	21 – 14.7	$(6.3)^2 = 39.69$
		123.43

   $\dfrac{\text{sum of the (deviations)}^2}{n} = \dfrac{123.43}{7}$
   ≈ 17.6

   The standard deviation of the population is
   $\sigma = \sqrt{17.6} \approx 4.2$.
   The variance of the population is
   $\sigma^2 = (\sqrt{17.6})^2 \approx 17.6$.

2. 14, as the mode is the most common number.

3. Answers will vary.

4. a. The median.

   b. The mode.

   c. The mean.

5. $(1235 + 1644 + 1576 + 1200 + 1200 + 1182 + 1212 + 1400) / 8 \approx 1331.125$
   The mean is about 1331.125 feet.
   The median is the average of the fourth and fifth ranked numbers, or 1223.5 feet.
   The mode is the most common number, or 1200 feet.
   The range is the difference between the highest and lowest values, or 462 feet.

6. $\dfrac{90}{2.5} = 36$ The average rate was 36 mph.

7. $[3(4) + 3(2.33) + 2(2.67) + 4(3) + 1(4)]/13$
   ≈ 3.10.
   The student has a grade point average of about 3.10.

8. a. $z_{82} = \dfrac{82 - 72}{8} = 1.25$

      Ann's z-score is 1.25

   b. Ann's percentile $= \dfrac{35}{40} \cdot 100 = 87.5$

      Ann was at the 88th percentile.

9. Mean $= \dfrac{12 + 18 + 20 + 14 + 16}{5} = 16$

   Sample variance $= \dfrac{\sum (x - \overline{x})^2}{n-1}$

   $= \dfrac{(12-16)^2 + (18-16)^2 + (20-16)^2 + (14-16)^2 + (16-16)^2}{4}$

   $= 10$
   $\sqrt{10} \approx 3.16$
   The sample variance is 10 minutes^2. The sample standard deviation is 3.16 minutes.

10. Mean
    $= \dfrac{6.55+6.88+7.18+7.50+7.89+7.93+7.96+8.15+8.17+8.12}{10}$

    $\approx 7.63 = \overline{x}$
    The sample mean is about $7.63.
    Median is $\dfrac{7.89 + 7.93}{2} = \$7.91$.

x	$x - \overline{x}$	$(x - \overline{x})^2$
6.55	6.55 – 7.63 = –1.08	1.1664
6.88	6.88 – 7.63 = –0.75	0.5625
7.18	7.18 – 7.63 = –0.45	0.2025
7.50	7.50 – 7.63 = –0.13	0.0169
7.89	7.89 – 7.63 = 0.26	0.0676
7.93	7.93 – 7.63 = 0.30	0.09
7.96	7.96 – 7.63 = 0.33	0.1089
8.15	8.15 – 7.63 = 0.52	0.2704
8.17	8.17 – 7.63 = 0.54	0.2916
8.12	8.12 – 7.63 = 0.49	0.2401
	Total =	3.0169

    Divide the sum by $n - 1$:
    $\dfrac{3.0169}{10-1} \approx 0.33521$
    Standard deviation for this sample is:
    $\sqrt{0.33521} \approx 0.58$

11. a. The second student's mean is 5 points higher than the first student's mean.

    b. The two standard deviations are the same.

12. a. $z_{72} = \dfrac{72 - 81}{5.2} \approx -1.73$

    b. $z_{84} = \dfrac{84 - 81}{5.2} \approx 0.58$

13. Minimum: 164, maximum: 310
$Q_1 = 172$, $Q_2 = 196.5$, $Q_3 = 221$
Drawing the plot:

14. a.   8 students scored at least an 84 on the test.

    b.   40 students took the test.

15. a. $28 + 8 + 18 + 4 + 8 + 4 + 6 + 2 = 78$.
         78% of the states paid an average teacher salary of at least $48,000.

    b. $18 + 4 + 8 + 4 = 34$. The probability that a state selected at random paid an average teacher salary of at least $56,000 but less than $72,000 is 0.34.

16. Using a calculator, the linear correlation coefficient (r) is –0.9004, which indicates a strong linear correlation.

17.
    a.   Using a calculator, the slope (a) is 7609.1 and the y-intercept is (b) is –72,019.1. Therefore the equation of the least-square line is $\hat{y} \approx 7609.1x - 72,019.1$.

    b. $\hat{y} \approx 7609.1(15) - 72,019.1 \approx 42,117.4$
         The predicted number of alternative fuel stations in the United States in 2015 is 42,117.

18. No. No information is given about how the scores are distributed below the mean.

19. a. $z_8 = \dfrac{8 - 6.5}{1} = 1.5$
         0.433 square unit lies between $z = 1.5$ and $z = 0$.
         0.500 square unit lies left of $z = 0$. $0.433 + 0.500 = 0.933$.
         The probability that a customer spends at most 8 minutes waiting is 0.933.

    b. $z_8 = \dfrac{6 - 6.5}{1} = -0.5$
         0.191 square unit lies between $z = -0.5$ and $z = 0$. Therefore, 0.309 square unit lies to the left of $z = -0.5$.
         The probability that a customer spends less than 6 minutes waiting is 0.309.

20. a. $z_{49.5} = \dfrac{49.5 - 50}{0.5} = -1$
         0.288 square units lie between $z = -1$ and $z = 0$. Therefore, 0.341 square units lies to the left of $z = -1$, and the percent is $50\% - 34.1\% = 15.9\%$.
         Thus 16% of the sacks will weigh less than 49.5 pounds.

    b.   95% of the sacks will weigh between 49 and 51 pounds, as 49 and 51 are two standard deviations below and above the mean, respectively.

21. a.   0.433 square unit lies between $z = 0$ and $z = 1.5$. Therefore, 0.067 square unit lies to the right of $z = 1.5$.
         6.7% of the manufacturer's telephones will last at least 7.25 years.

    b.   Calculating:
         $z_{5.8} = \dfrac{5.8 - 6.5}{0.5} = -1.4$
         $z_{6.8} = \dfrac{6.8 - 6.5}{0.5} = 0.6$
         0.419 square unit lies between $z = -1.4$ and $z = 0$.
         0.226 square unit lies between $z = 0$ and $z = 0.6$. $0.419 + 0.226 = 0.645$.
         64.5% of the manufacturer's telephones will last between 5.8 and 6.8 years.

    c. $z_{6.9} = \dfrac{6.9 - 6.5}{0.5} = 0.8$
         0.288 square unit lies between $z = 0$ and $z = 0.8$.
         0.500 square unit lies to the left of $z = 0$.
         $0.288 + 0.500 = 0.788$.
         78.8% of the company's phones will last less than 6.9 years.

22. $(91.4 + 92.6 + 94.5 + 94.6 + 94.3 + 91.5) / 6$
    $= 93.15$ million miles

23. a.

b. $n = 5$

$$\sum x = 10 + 12 + 14 + 15 + 16 = 67$$

$$\sum y = 8 + 7 + 5 + 4 + 1 = 25$$

$$\sum x^2 = 100 + 144 + 196 + 225 + 256 = 921$$

$$\left(\sum x\right)^2 = 67^2 = 4489$$

$$\sum xy = 80 + 84 + 70 + 60 + 16 = 310$$

c.   Calculating:

$$a = \frac{n\left(\sum xy\right) - \left(\sum x\right)\left(\sum y\right)}{n\left(\sum x^2\right) - \left(\sum x\right)^2}$$

$$= \frac{5(310) - (67)(25)}{5(921) - 4489} \approx -1.08$$

$$b = \overline{y} - a\overline{x} \approx 19.44$$

$$\hat{y} \approx -1.08x + 19.44$$

d.   See the graph in part a.

e.   Yes.

f. $\hat{y} \approx -1.08(8) + 19.44 \approx 10.8$

g. $\sum y^2 = 64 + 49 + 25 + 16 + 1 = 155$

$$\left(\sum y\right)^2 = 25^2 = 625$$

$$r = \frac{n\left(\sum xy\right) - \left(\sum x\right)\left(\sum y\right)}{\sqrt{n\left(\sum x^2\right) - \left(\sum x\right)^2} \cdot \sqrt{n\left(\sum y^2\right) - \left(\sum y\right)^2}}$$

$$= \frac{5(310) - (67)(25)}{\sqrt{5(921) - 4489} \cdot \sqrt{5(155) - 625}} \approx -0.95$$

24.   a.   Using a calculator, the linear correlation coefficient (r) is 0.999.

b.   Using a calculator, the slope (a) is 0.07 and the y-intercept is (b) is 0.29. Therefore the equation of the least-square line is

$$\hat{y} \approx 0.07x + 0.29.$$

c. $\hat{y} \approx 0.07(195) + 0.29 = 13.94$

A weight of 195 pounds should stretch the spring 13.94 inches.

25.   a.   Using a calculator, the slope (a) is 0.018 and the y-intercept is (b) is 0.0005. Therefore the equation of the least-square line is $\hat{y} \approx 0.018x + 0.0005.$

b.   Using a calculator, the linear correlation coefficient (r) is 0.999.
Yes, a linear model of the data is a reasonable model, as r is very close to 1.

c. $\hat{y} \approx 0.018(100) + 0.0005 \approx 1.80$

The expected download time is 1.80 seconds.

26.   a.   Using a calculator, the slope (a) is 58.89921372 and the y-intercept is (b) is 0.1924231594. Therefore the equation of the least-square line is

$$\hat{y} \approx 58.89921372x + 0.1924231594.$$

Using a calculator, the linear correlation coefficient (r) is 0.9911051145.

b. $\hat{y} \approx 58.89921372(0.06) + 0.1924231594$
≈ 3.7
It would take about 3.7 h.

CHAPTER 13 TEST

1. $\dfrac{2 + 7 + 11 + 12 + 7 + 9 + 15}{7} \approx 9.1$

The mean is 9.1. The median is the fourth rank number, or 9. The mode is the most common number, or 7.

2. $GPA = \dfrac{(4 \times 3) + (3 \times 3) + (2 \times 4) + (2 \times 3) + (4 \times 2)}{3 + 3 + 4 + 3 + 2}$

$= \dfrac{43}{15} \approx 2.87$

Justin's GPA for the fall semester is 2.87.

3.   The range is the difference between the largest and smallest numbers, or 24.

$$\text{Mean} = \frac{7 + 11 + 12 + 15 + 22 + 31}{6} \approx 16.3$$

$$[(7 - 16.3)^2 + (11 - 16.3)^2 + (12 - 16.3)^2$$
$$+ (15 - 16.3)^2 + (22 - 16.3)^2$$
$$+ (31 - 16.3)^2]/5$$
$$\approx 76.7$$

The sample variance is 76.7.
The standard deviation is 8.76.

4.   a. $z_{77} = \dfrac{77 - 65}{10.2} \approx 1.18$

b. $z_{60} = \dfrac{60 - 65}{10.2} \approx -0.49$

5.   Minimum: 422, maximum: 729
$Q_1 = 481.5$, $Q_2 = 531.5$, $Q_3 = 612$
The plot is shown below:

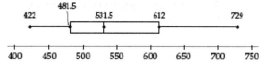

6. a. $15 + 11 + 11 = 37$
        37% of the movie attendees are at least 40 years old.

    b. $12 + 23 = 35$
        35% of the movie attendees are at least 21 but less than 40 years old.

7. a. 54 ounces is 2 standard deviations away from 34 ounces. Therefore, 47.5% of the parcels weighed between 34 and 54 ounces.

    b. 24 ounces is 1 standard deviation away from 34 ounces. Therefore, $13.5 + 2.35 + 0.15 = 16\%$ of the parcels weighed less than 24 ounces.

8. a. $z_{17} = \dfrac{17 - 18}{0.8} = -1.25$
        0.394 square unit lies between $z = -1.25$ and $z = 0$. Therefore, 0.106 square unit lies to the left of $z = -1.25$.
        10.6% of the boxes will weigh less than 17 ounces.

    b. Calculating:
        $$z_{18.4} = \frac{18.4 - 18}{0.8} = 0.5$$
        $$z_{19} = \frac{19 - 18}{0.8} = 1.25$$
        0.191 square unit lies between $z = 0$ and $z = 0.5$.
        0.394 square unit lies between $z = 0$ and $z = 1.25$. Therefore, 0.203 square unit lies between $z = 0.5$ and $z = 1.25$. 20.3% of the boxes weigh between 18.4 and 19 ounces.

9. a. Using a calculator, the linear correlation coefficient (r) is -0.9308039961.

    b. Yes. The absolute value of the correlation coefficient, $|-0.93080309961| > 0.9$, which indicates a strong linear relationship.

10. a. Using a calculator, the slope (a) is -7.98 and the y-intercept is (b) is 767.12. Therefore the equation of the least-square line is $\hat{y} \approx 7.98x + 767.12$.

    b. $\hat{y} \approx 7.98(89) + 767.12 \approx 57$
        57 calories are expected in a soup that is 89% water.